Student Solutions Manual
for Yandl's

Applied Calculus

Mary B. Ehlers
Seattle University

Brooks/Cole Publishing Company
Pacific Grove, California

Brooks/Cole Publishing Company
A Division of Wadsworth, Inc.

Printed in the United States of America

10   9   8   7   6   5   4   3   2   1

ISBN 0-534-14204-4

Sponsoring Editor: *Faith B. Stoddard*
Editorial Assistant: *Nancy Champlin*
Production Coordinator: *Dorothy Bell*
Cover Design: *Katherine Minerva*
Cover Photo: *Lee Hocker*
Art Coordinator: *Lisa Torri*
Printing and Binding: *Malloy Lithographing, Inc.*

PREFACE

This manual contains the detailed solutions to the more than **1200** odd-numbered problems in the text **Applied Calculus** by Andre L. Yandl.

In order to develop the ability to solve problems, the student should first attempt the problem on his/her own before comparing the solution with the solution given in this manual. The solution process is more important than the answer, and understanding is more important than memorization. Many problems can be solved by more than one method and sometimes the answer may be expressed in different forms. Thus some solutions may differ from the one in the manual and still be correct.

I wish to thank the members of the team that helped in the preparation and production for this project. Mike Allen, Bob Heighton, and Jim Riester assisted with the typing of the various versions and revisions of the manuscript. Professor Yandl checked many of the solutions for correctness. At Brooks/Cole, Dorothy Bell served as Production Coordinator and Lisa Torri as Art Coordinator, and Faith Stoddard, Assistant Mathematics Editor, provided editorial supervision and encouragement. I especially thank Frank Arena, who also assisted with the typing of the manuscript, assembled it for production, and remained with the project from beginning to end.

The manual was prepared using EXP: The Scientific Word Processor for use on MS-DOS computers. EXP is available from Brooks/Cole Publishing Company, Pacific Grove, California 93950.

Mary B. Ehlers

Seattle University

Seattle, Washington 98122

CONTENTS

GETTING STARTED

1. $x^7 - y^7 = (x - y)(x^6 + x^5y + x^4y^2 + x^3y^3 + x^2y^4 + xy^5 + y^6)$

3. $8x^3 - 125y^3 = (2x)^3 - (5y^3) = (2x - 5y)(4x^2 + 10xy + 25y^2)$

5. $32a^5 - 243c^5 = (2a)^5 - (3c)^5 = (2a - 3c)[(2a)^4 + (2a)^3(3c) + (2a)^2(3c)^2$
$+ (2a)(3c)^3 + (3c)^4] = (2a - 3c)(16a^4 + 24a^3c + 36a^2c^2 + 54ac^3 + 81c^4)$

7. $x^3 + y^3 = \left(x^3 - (-y)^3\right) = (x + y)(x^2 - xy + y^2)$

9. $\frac{x^3}{8} + 27y^3 = \left(\frac{x}{2}\right)^3 - \left(-3y\right)^3 = \left(\frac{x}{2} + 3y\right)\left(\frac{x^2}{4} - \frac{3}{2}xy + 9y^2\right)$

11.

```
                    1
                 1     1
              1     2     1
           1     3     3     1
        1     4     6     4     1
     1     5    10    10     5     1
  1     6    15    20    15     6     1
```

13. $(x + y)^6 = x^6 + 6x^5y + 15x^4y^2 + 20x^3y^3 + 15x^2y^4 + 6xy^5 + y^6$

15. $(x + y)^7 = x^7 + 7x^6y + 21x^5y^2 + 35x^4y^3 + 35x^3y^4 + 21x^2y^5 + 7xy^6 + y^7$

17. $(2x + 3y)^3 = (2x)^3 + 3(2x)^2(3y) + 3(2x)(3y)^2 + (3y)^3$
$= 8x^3 + 36x^2y + 54xy^2 + 27y^3$

19. $\quad \bullet \quad (3a - 2b)^4 = (3a)^4 - 4(3a)^3(2b) + 6(3a)^2(2b)^2 - 4(3a)(2b)^3 + (2b)^4$
$= 81a^4 - 216a^3b + 216a^2b^2 - 96ab^3 + 16b^4$

21.
$$\left(2x+\frac{y}{2}\right)^8 = (2x)^8 + 8(2x)^7\left(\frac{y}{2}\right) + 28(2x)^6\left(\frac{y}{2}\right)^2 + 56(2x)^5\left(\frac{y}{2}\right)^3 + 70(2x)^4\left(\frac{y}{2}\right)^4$$
$$+ 56(2x)^3\left(\frac{y}{2}\right)^5 + 28(2x)^2\left(\frac{y}{2}\right)^6 + 8(2x)\left(\frac{y}{2}\right)^7 + \left(\frac{y}{2}\right)^8 = 256x^8 + 512x^7y + 448x^6y^2$$
$$+ 224x^5y^3 + 70x^4y^4 + 14x^3y^5 + \frac{7}{4}x^2y^6 + \frac{xy^7}{8} + \frac{y^8}{256}$$

23. (a) $n = 8 + 4 = 12$

(b) $\frac{495}{5}(8)a^7b^5 = 792a^7b^5$

(c) The preceeding term is ka^9b^3 where $\frac{9k}{3+1} = 495$ and $k = 220$. The preceeding term is $220a^9b^3$.

25. (a) $n = 4 + 10 = 14$

(b) $\frac{1001(4)}{11}a^3b^{11} = 364a^3b^{11}$

(c) $\frac{1001(10)}{5}a^5b^9 = 2002a^5b^9$

27. (a) $n = 5 + 12 = 17$

(b) $6188\,a^4b^{13}\frac{(5)}{13} = 2380a^4b^{13}$

(c) $6188\left(\frac{12}{6}\right)a^6b^{11} = 12376a^6b^{11}$

29. (a) $n = 16 + 4 = 20$

(b) $\frac{4845(16)}{5}a^{15}b^5 = 15504a^{15}b^5$

(c) $\frac{4845(4)}{17}a^{17}b^3 = 1140a^{17}b^3$

31. (a) $n = 7 + 8 = 15$

(b) $\frac{6435(7)}{9}a^6b^9 = 5005a^6b^9$

(c) $\frac{6435(8)}{8}a^8b^7 = 6435a^8b^7$

33. $\frac{(3+101)}{2}(50) = \frac{104}{2}(50) = 52(50) = 2600$

35. $1 + 2 + 3 + \cdots + 50 = \frac{(1+50)}{2}(50) = (51)(25) = 1275$

37. (a) $1^3 + 2^3 = 1 + 8 = 9, \quad 1 + 2 = 3$

(b) $1^3 + 2^3 + 3^3 = 36, \quad 1 + 2 + 3 = 6$

(c) $1^3 + 2^3 + 3^3 + 4^3 = 100, \quad 1 + 2 + 3 + 4 = 10$

(d) $1^3 + 2^3 + 3^3 + 4^3 + 5^3 = 225, \quad 1 + 2 + 3 + 4 + 5 = 15$

(e) $1^3 + 2^3 + 3^3 + 4^3 + 5^3 + 6^3 = 441, \quad 1 + 2 + 3 + 4 + 5 + 6 = 21$

(f) $1^3 + 2^3 + 3^3 + 4^3 + 5^3 + 6^3 + 7^3 = 784, \quad 1 + 2 + 3 + 4 + 5 + 6 + 7 = 28$

Noting $3^2 = 9$, $6^2 = 36$, $10^2 = 100$, $15^2 = 225$, etc., we guess

$$1^3 + 2^3 + \cdots + n^3 = (1 + 2 + \cdots + n)^2 = \left(\frac{n(n+1)}{2}\right)^2 = \frac{n^2(n+1)^2}{4}$$

39. To find $53(47) + 36(72) = 5083$, press the following calculator keys in order: (53 × 47) + (36 × 72) = 5083

41. Press the following calculator keys in order: (72 × 23.5) + (64 × 90.2) − (13.3 × 14.7) = 7269.29

43. ((23.5 × 45.3) − (17.6 × 34.8)) ÷ 17.5 = 25.83257143

45. Press the following calculator keys in order: ((59.5 × 23.6) + (27.5 × 41.7)) ÷ (19.8 + 64.2) = 30.36845238

47. Calculator keys in order are (45 × (13.2 + 17.8)) − (16 × (14.9 + 13.1)) = 947

49. Calculator keys in order are 32 x^2 + 53 x^2 = 3833

51. By calculator, 45 x^2 − 78 x^2 = −4059

53. By calculator, 5.2 y^x 5 − 7.4 y^x 4 = 803.3827

55. By calculator, (31.2 x^2 + 14.7 y^x 3 =) × (6.2 y^x 3) = 989052.3819

57. By calculator, (8.3 x^2 + 63.2 y^x 3 =) × (51.3 x^2 + 63.2 y^x 3 − 17.4 y^x 4) = 41260314020

59. By calculator, (1.04 y^x 40 − 1.03 y^x 38 =) × (1.02 y^x 12 − 23 y^x 3) = −21000.93812

61. By calculator, the operations (5.1 y^x 5 + 6.2 y^x 4 =) ÷ (3.2 y^x 2 + 4.1 y^x 3 =) give 62.25143833

63. $\dfrac{32^3 + .43^4}{5.2^2 + 4.5^4} = (32^3 + .43^4)/(5.2^2 + 4.5^4)$. The operations (32 y^x 3 + .43 y^x 4 =) ÷ (5.2 y^x 2 + 4.5 y^x 4 =) give 74.96647626

65. $\dfrac{4^{12}+1.02^{42}}{1.04^{25}-13.2}=(4^{12}+1.02^{42})/(1.04^{25}-13.2)$. The operations

$(\ 4\ y^x\ 12\ +\ 1.02\ y^x\ 42\ =\ )\div(\ 1.04\ y^x\ 25\ -\ 13.2\ =\ )$ give -1592648.342

67. The operations $(\ 1.6\ y^x\ 20\ +\ 1.05\ y^x\ 30\ =\ )\div(\ 1.015\ y^x\ 20$
$+\ 1.025\ y^x\ 10\ -\ 2313.2\ =\ )$ give -5.234017633

69. (a) To find $3500(1.08)^{15}$ by calculator the operations $1.08\ y^x\ 15\ =\ \times\ 3500\ =$ give $\$11,102.59$

  (b) To find $3500(1.04)^{30}$ by calculator the operations $1.04\ y^x\ 30\ =\ \times\ 3500\ =$ give $\$11,351.89$

  (c) To find $3500(1.02)^{60}$ by calculator the operations $1.02\ y^x\ 60\ =\ \times\ 3500\ =$ give $\$11,483.61$

  (d) To find $3500\left(1+\dfrac{.08}{12}\right)^{15(12)}$ by calculator the operations $.08\ \div\ 12\ =\ +\ 1$
  $=\ y^x\ 15\ =\ y^x\ 12\ =\ \times\ 3500\ =$ give $\$11,574.23$

71. With Bank A, $500\left(1+\dfrac{.0695}{12}\right)^{96}=\$870.44$ (By calculator $.0695\ \div\ 12$
$=\ +\ 1\ =\ y^x\ 96\ =\ \times\ 500\ =$ give 870.44). With Bank B, $500\left(1+\dfrac{.07}{4}\right)^{32}$
$=\$871.11$ (By calculator $.07\ \div\ 4\ =\ +\ 1\ =\ y^x\ 32\ =\ \times\ 500\ =$ give 871.11)

Choose Bank B to obtain $\$871.11$

73. $\dfrac{350\left[(1+.045)^{40}-1\right]}{.045}=\$37460.61$ will be the balance.

(By calculator, $1\ +\ .045\ =\ y^x\ 40\ =\ -\ 1\ =\ \times\ 350\ =\ \div\ .045\ =$ give 37460.61)

75. $77,313.47=\dfrac{R\left[\left(1+\dfrac{.09}{12}\right)^{144}-1\right]}{.09/12}\Rightarrow 77,313.47\left(\dfrac{.09}{12}\right)\cdot\dfrac{1}{\left[\left(1+\dfrac{.09}{12}\right)^{144}-1\right]}$

$=R=300.00.$ $\$300.00$ was deposited monthly. (By calculator, $.09\ \div\ 12$
$=\ +\ 1\ =\ y^x\ 144\ =\ -1\ =\ 1\ /\ x\ \times\ 77313.47\ =\ \times\ .09\ =\ \div\ 12$
$=$ give 300.00)

BASIC ALGEBRAIC CONCEPTS

EXERCISE SET 0.1 THE REAL NUMBERS

1. $5(7-9) = 5(-2) = -10$

3. $-[3-(4-8)] = -[3-(-4)] = -[3+4] = -7$

5. $-\dfrac{5}{9} + \dfrac{7}{9} = \dfrac{-5+7}{9} = \dfrac{2}{9}$

7. $\dfrac{-15}{19} - \dfrac{-17}{23} = \dfrac{-15}{19} + \dfrac{17}{23} = \dfrac{-15(23)+(17)(19)}{(19)(23)} = \dfrac{-345+323}{19(23)} = \dfrac{-22}{437}$

9. $\dfrac{-7}{65} - \dfrac{-8}{169} = \dfrac{-7}{13(5)} + \dfrac{8}{13(13)} = \dfrac{-7(13)+8(5)}{(13)(13)(5)} = \dfrac{-91+40}{(13)(13)(5)} = \dfrac{-51}{(13)(13)(5)} = \dfrac{-51}{845}$

11. $\dfrac{-21/55}{7/25} = \dfrac{-21}{55} \cdot \dfrac{25}{7} = \dfrac{-3 \cdot 5}{11} = \dfrac{-15}{11}$

13. $\dfrac{18}{\frac{15}{7}} = 18 \cdot \dfrac{7}{15} = \dfrac{6 \cdot 3 \cdot 7}{3 \cdot 5} = \dfrac{42}{5}$

15.

$$6 \overline{)\,11.000} \quad \text{(quotient } 1.83\ldots\text{)}$$

answer: $1.8\overline{3}$

```
      1.83 ...
  6 ) 11.000
      6
      50
      48
      20
      18
      20
```

17.

```
        1.625
8 ) 13.000        answer:  1.625
    8
    50
    48
     20
     16
      40
      40
       0
```

19.

```
        1.54 ...
11 ) 17.000        answer:  1.5̄4̄
     11
     60
     55
      50
      44
       60
```

21.

```
           16.538461538461 ...
13 ) 215.0        answer:  16.5̄38461̄
     13
     85
     78
      70
      65
       50
       39
       110
       104
         60
         52
          80
          78
           20
           13
            70
```

23.

$$12.8823529411764705\ldots$$

$17\overline{)219.0}$ answer: $12.\overline{8823529411764705}$

$\underline{17}$

49
$\underline{34}$

150
$\underline{136}$

140
$\underline{136}$

40
$\underline{34}$

60
$\underline{51}$

90
$\underline{85}$

50
$\underline{34}$

160
$\underline{153}$

70
$\underline{68}$

20
$\underline{17}$

30
$\underline{17}$

130
$\underline{119}$

110
$\underline{102}$

80
$\underline{68}$

120
$\underline{119}$

100
$\underline{85}$

150

25. $-x^2(9x^2) = -9x^4$

27. $\dfrac{2^{15}a^{15}b^{10}}{2^8 a^{12}b^{12}} = \dfrac{2^7 a^3}{b^2} = \dfrac{128a^3}{b^2}$

29.
$$\frac{x^{12}y^8}{-8x^9y^9} = \frac{-x^3}{8y}$$

31.
$$\frac{-7^5x^{20}y^{15}}{7^6x^9y^{12}} = \frac{-x^{11}y^3}{7}$$

33.
$$x = \frac{2 \pm \sqrt{4-8}}{4} = \frac{2 \pm \sqrt{-4}}{4} = \frac{2 \pm 2i}{4} = \frac{1 \pm i}{2}.$$ These are 2 imaginary solutions.

35.
$$x = \frac{-5 \pm \sqrt{25+144}}{-6} = \frac{-5 \pm \sqrt{169}}{-6} = \frac{-5 \pm 13}{-6} \qquad x = \frac{-18}{-6} = 3, \text{ or } x = \frac{8}{-6} = \frac{-4}{3}$$

These are 2 real solutions.

37.   Suppose p and q are rational numbers. $p = \frac{a}{b}$ and $q = \frac{c}{d}$ where a, b, c and d are integers and $b \neq 0$ and $d \neq 0$, $p + q = \frac{a}{b} + \frac{c}{d} = \frac{ad + bc}{bd}$. Since a, b, c and d are integers, $ad + bc$ is an integer, bd is an integer and $bd \neq 0$ since $b \neq 0$ and $d \neq 0$. Hence $\frac{ad + bc}{bd}$ is a rational number.

39.   Suppose p and q are rational numbers. Then $p = \frac{a}{b}$ and $q = \frac{c}{d}$ where a, b, c and d are integers and $b \neq 0$ and $d \neq 0$. $pq = \frac{a}{b} \cdot \frac{c}{d} = \frac{ac}{bd}$. Since a, b, c and d are integers, ac and bd are integers, and $bd \neq 0$ since $b \neq 0$ and $d \neq 0$. Hence pq is a rational number.

41.   Suppose p is a rational number and q is irrational and $p + q = r$. Then the real number $q = r - p$ is irrational and r must be irrational since if r were rational $r - p$ would be rational by exercise 38.

43. $5 - 3 = 2 \neq 3 - 5 = -2$

45.   No. $1 + 3 = 4$ and 4 is not an odd integer.

47.   Assume a and b are real numbers and $ab = 0$. If $a = 0$, then $ab = 0 \cdot b = 0$ and we are done. If $a \neq 0$, $\frac{1}{a}(ab) = \frac{1}{a} \cdot 0$ and $\frac{1}{a}(a) \cdot b = b = \frac{1}{a} \cdot 0 = 0$ by problem 46. Hence if $ab = 0$, $a = 0$ or $b = 0$.

EXERCISE SET 0.2 ELEMENTARY INTRODUCTION TO SETS

1. $\{1,2,3,4,5,6\}$

3. $\{1\}$

5. {Wisconsin, Washington, West Virginia, Wyoming}

7. $\{c,d,e\}$

9. {Johnson, Nixon, Ford, Carter, Reagan, Bush}

11. $2x - x = 10 - 3,\quad x = 7.$ The solution set is $\{7\}$.

13. $x = 2x - 6,\quad 6 = x.$ The solution set is $\{6\}$.

15. $10 - 6 = 5x - 3x,\quad 4 = 2x,\quad x = 2.$ The solution set is $\{2\}$.

17. $1 + 4 = 5x - 4x,\quad 5 = x$ but 5 is not in the replacement set. The solution set is $\emptyset$.

19. $8 = 8x,\quad x = 1.$ The solution set is $\{1\}$.

21. Lincoln, Jefferson, Truman

23. 3 and 4 are the only members of the set.

25. $5, 9, 233$

27. $1, 2, 4$

29. Harry Reasoner, Barbara Walters, Phil Donahue

31. False,   since $2 < 3$

33. True, $\pi \doteq 3.14 > 2$

35. False, m is not a vowel in the English alphabet.

37. $2^2 = 4;$ The subsets are: $\emptyset, \{1,2\}, \{1\}, \{2\}$

39. $2^5 = 32;$ The subsets are: $\emptyset, \{1\}, \{2\}, \{3\}, \{4\}, \{5\}, \{1,2\}, \{1,3\}, \{1,4\}, \{1,5\}, \{2,3\},$
$\{2,4\}, \{2,5\}, \{3,4\}, \{3,5\}, \{4,5\}, \{3,4,5\}, \{2,4,5\}, \{2,3,5\}, \{2,3,4\}, \{1,4,5\}, \{1,3,5\},$
$\{1,3,4\}, \{1,2,5\}, \{1,2,4\}, \{1,2,3\}, \{1,2,3,4\}, \{1,2,3,5\}, \{1,2,4,5\}, \{1,3,4,5\},$
$\{2,3,4,5\}, \{1,2,3,4,5\}$

41. $2^2 = 4,\quad 4 - 1 = 3.$ There are 3 proper subsets.

43. $2^4 = 16,\quad 16 - 1 = 15.$ There are 15 proper subsets.

45. $2^n - 1$

47. (a) 1, (b) 6, (c) 15, (d) 20, (e) 15, (f) 6, (g) 1

$$
\begin{array}{ccccccccccccc}
 & & & & & & 1 & & & & & & \\
 & & & & & 1 & & 1 & & & & & \\
 & & & & 1 & & 2 & & 1 & & & & \\
 & & & 1 & & 3 & & 3 & & 1 & & & \\
 & & 1 & & 4 & & 6 & & 4 & & 1 & & \\
 & 1 & & 5 & & 10 & & 10 & & 5 & & 1 & \\
\end{array}
$$

| 1 | 6 | 15 | 20 | 15 | 6 | 1 |

49. $\{0, 7, 8, 9\}$

51. $\{0, 1, 2, 5, 6, 7, 9\}$

53. $A \bigcup B = \{1, 2, 3, 4, 5, 6\}$
$A \bigcap B = \{2, 5\}$

55. $B \bigcup C = \{2, 3, 4, 5, 8\}$
$B \bigcap C = \emptyset$

57. $(A \bigcap B)' = \{2, 5\}' = \{0, 1, 3, 4, 6, 7, 8, 9\}$
$A' \bigcup B' = \{0, 7, 8, 9\} \bigcup \{0, 1, 3, 4, 6, 7, 8, 9\} = \{0, 1, 3, 4, 6, 7, 8, 9\}$

59. $(A \bigcap C)' = \{3, 4\}' = \{0, 1, 2, 5, 6, 7, 8, 9\}$
$A' \bigcup C' = \{0, 7, 8, 9\} \bigcup \{0, 1, 2, 5, 6, 7, 9\} = \{0, 1, 2, 5, 6, 7, 8, 9\}$

EXERCISE SET 0.3 VENN DIAGRAMS

1.

$$A \cap (B \cup C)$$

3.

$$A \cup (B' \cap C)$$

Shade $B' \cap C$ and then shade the region not already shaded which lies within A.

5.

$$(A \cap B)' \cup C'$$

Shade the region outside $A \cap B$, and then also shade all points outside of C.

7.

$$A \cap B$$

$$(A \cap B)'$$

$$A'$$

$$B'$$

$$A' \cup B'$$

The second and fifth Venn diagrams are the same.

9.

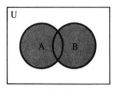

A is within $A \bigcup B$

$$A \subseteq (A \bigcup B)$$

11.

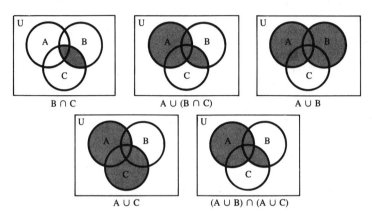

B ∩ C   A ∪ (B ∩ C)   A ∪ B

A ∪ C   (A ∪ B) ∩ (A ∪ C)

The second and fifth Venn diagrams are the same.

13.

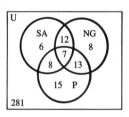

Let SA, NG and P denote the sets of students who read Scientific American, National Geographic, and Playboy respectively.

(a) 6   (b) 8   (c) $350 - 69 = 281$

15.

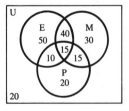

Let E, M and P denote the sets of students who took English, Mathematics and Philosophy respectively.

(a) 15
(b) $40 + 15 = 55$
(c) $50 + 40 + 10 + 15 = 115$
(d) $40 + 30 + 15 + 15 = 100$
(e) $10 + 15 + 15 + 20 = 60$

17.

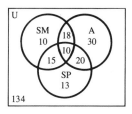

Let SM, A and SP denote the sets of students who smoked, drank alcohol, and played sports respectively.

(a)13
(b)20
(c)$250 - 116 = 134$

19.

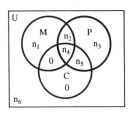 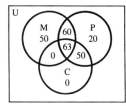

Let M be the set of male customers, P be the set of customers who made a purchase, and C be the set of customers who use a credit card. <u>Note</u>: One can't use a credit card unless they make a purchase.

$n_1 + n_2 + n_3 + n_4 + n_5 + n_6 = 230$

$n_1 + n_2 + n_4 = 173, \quad n_2 + n_3 + n_4 + n_5 = 193$

$n_4 + n_5 = 113$

$n_2 + n_4 = 123$

$n_4 = 63$

a)50
b)60
c)50
d)20

Although $50 + 60 + 63 + 50 + 20 = 243$, only 230 customers were surveyed.

EXERCISE SET 0.4 POLYNOMIALS

1.
$$f(x) + g(x) = (2x^3 + 5x^2 - 10) + (-3x^4 + 4x^3 + 10x - 7)$$
$$= -3x^4 + (2x^3 + 4x^3) + 5x^2 + 10x + (-10 - 7)$$
$$= -3x^4 + 6x^3 + 5x^2 + 10x - 17$$

$$f(x) - g(x) = (2x^3 + 5x^2 - 10) - (-3x^4 + 4x^3 + 10x - 7)$$
$$= 2x^3 + 5x^2 - 10 + 3x^4 - 4x^3 - 10x + 7$$
$$= 3x^4 + (2x^3 - 4x^3) + 5x^2 - 10x + (7 - 10)$$
$$= 3x^4 - 2x^3 + 5x^2 - 10x - 3$$

3.

$$f(x) + g(x) = (7x^6 + 5x - 8) + (5x^3 + 5x - 13)$$
$$= 7x^6 + 5x^3 + (5x + 5x) + (-8 - 13)$$
$$= 7x^6 + 5x^3 + 10x - 21$$

$$f(x) - g(x) = (7x^6 + 5x - 8) - (5x^3 + 5x - 13)$$
$$= 7x^6 - 5x - 8 - 5x^3 - 5x + 13$$
$$= 7x^6 - 5x^3 + (5x - 5x) + (-8 + 13)$$
$$= 7x^6 - 5x^3 + 5$$

5.

$$f(x) + g(x) = (4x^5 + 6x^2 + 2x - 5) + (-3x^4 + x^3 + 5x^2 + x + 5)$$
$$= 4x^5 - 3x^4 + x^3 + (6x^2 + 5x^2) + (2x + x) + (-5 + 5)$$
$$= 4x^5 - 3x^4 + x^3 + 11x^2 + 3x$$

$$f(x) - g(x) = (4x^5 + 6x^2 + 2x - 5) - (-3x^4 + x^3 + 5x^2 + x + 5)$$
$$= 4x^5 + 6x^2 + 2x - 5 + 3x^4 - x^3 - 5x^2 - x - 5$$
$$= 4x^5 + 3x^4 - x^3 + (6x^2 - 5x^2) + (2x - x) + (-5 - 5)$$
$$= 4x^5 + 3x^4 - x^3 + x^2 + x - 10$$

7.

$$
\begin{array}{rrrrrrr}
f(x) = & & x^4 & & +3x & - & 6 \\
g(x) = & 2x^6 + & x^4 & +7x^2 & -6x & + & 7 \\
\hline
f(x) + g(x) = & 2x^6 + & 2x^4 & +7x^2 & -3x & + & 1
\end{array}
$$

$$
\begin{array}{rrrrrrr}
f(x) = & & x^4 & & +3x & - & 6 \\
-g(x) = & -2x^6 - & x^4 & -7x^2 & +6x & - & 7 \\
\hline
f(x) - g(x) = & -2x^6 & & -7x^2 & +9x & - & 13
\end{array}
$$

9.

$$
\begin{array}{rrrrr}
f(x) = & \frac{2}{3}x^3 + & \frac{5}{6}x^2 - & \frac{7}{8}x + & \frac{12}{13} \\
g(x) = & \frac{3}{4}x^3 - & \frac{5}{7}x^2 + & \frac{8}{7}x - & \frac{5}{4} \\
\hline
f(x) + g(x) = & \frac{17}{12}x^3 + & \frac{5}{42}x^2 + & \frac{15}{56}x - & \frac{17}{52}
\end{array}
$$

$$
\begin{array}{rrrrr}
f(x) = & \frac{2}{3}x^3 + & \frac{5}{6}x^2 - & \frac{7}{8}x + & \frac{12}{13} \\
-g(x) = & -\frac{3}{4}x^3 + & \frac{5}{7}x^2 - & \frac{8}{7}x + & \frac{5}{4} \\
\hline
f(x) - g(x) = & -\frac{1}{12}x^3 + & \frac{65}{42}x^2 - & \frac{113}{56}x + & \frac{103}{52}
\end{array}
$$

11. $(3x-4)(4x+5) = 3x(4x+5) - 4(4x+5) = 12x^2 + 15x - 16x - 20$

$= 12x^2 - x - 20$

13. $(x-2)(x+2) = x(x+2) - 2(x+2) = x^2 + 2x - 2x - 4 = x^2 - 4$

15. $(x-3)(x^2+6x-4) = x(x^2+6x-4) - 3(x^2+6x-4)$

$= x^3 + 6x^2 - 4x - 3x^2 - 18x + 12 = x^3 + 3x^2 - 22x + 12$

17. $(2x+3)(3x^2-5x+7) = 2x(3x^2-5x+7) + 3(3x^2-5x+7)$

$= 6x^3 - 10x^2 + 14x + 9x^2 - 15x + 21 = 6x^3 - x^2 - x + 21$

19.

$$
\begin{array}{r}
-3x^4 + 4x^3 + 10x - 7 \\
\times \quad 2x^3 + 5x^2 - 10 \\
\hline
30x^4 - 40x^3 \qquad - 100x + 70 \\
-15x^6 + 20x^5 \qquad + 50x^3 - 35x^2 \\
-6x^7 + 8x^6 \qquad + 20x^4 - 14x^3 \\
\hline
\end{array}
$$

$f(x)g(x) = -6x^7 - 7x^6 + 20x^5 + 50x^4 - 4x^3 - 35x^2 - 100x + 70$

21.

$$
\begin{array}{r}
f(x) = 7x^6 + 5x - 8 \\
g(x) = 5x^3 + 5x - 13 \\
\hline
-91x^6 \qquad - 65x + 104 \\
35x^7 \qquad + 25x^2 - 40x \\
35x^9 \qquad + 25x^4 - 40x^3 \\
\hline
\end{array}
$$

$f(x)g(x) = 35x^9 + 35x^7 - 91x^6 + 25x^4 - 40x^3 + 25x^2 - 105x + 104$

23.

$$
\begin{array}{r}
g(x) = -3x^4 + x^3 + 5x^2 + x + 5 \\
f(x) = 4x^5 + 6x^2 + 2x - 5 \\
\hline
15x^4 - 5x^3 - 25x^2 - 5x - 25 \\
-6x^5 + 2x^4 + 10x^3 + 2x^2 + 10x \\
-18x^6 + 6x^5 + 30x^4 + 6x^3 + 30x^2 \\
-12x^9 + 4x^8 + 20x^7 + 4x^6 + 20x^5 \\
\hline
-12x^9 + 4x^8 + 20x^7 - 14x^6 + 20x^5 + 47x^4 + 17x^3 + 7x^2 + 5x - 25
\end{array}
$$

25.

$$g(x) = 2x^6 + x^4 + 7x^2 - 6x + 7$$

$$f(x) = \underline{\qquad\qquad x^4 + 3x - 6}$$

$$-12x^6 \qquad -6x^4 \qquad -42x^2 + 36x - 42$$

$$+6x^7 \qquad +3x^5 \qquad +21x^3 - 18x^2 + 21x$$

$$2x^{10} + x^8 \qquad +7x^6 - 6x^5 + 7x^4 \qquad\qquad\qquad\qquad$$

$$f(x)g(x) = 2x^{10} + x^8 + 6x^7 - 5x^6 - 3x^5 + x^4 + 21x^3 - 60x^2 + 57x - 42$$

27.

$$\begin{array}{r} 3x + 14 \\ x - 3 \overline{)\,3x^2 + 5x - 6} \\ \underline{3x^2 - 9x} \\ 14x - 6 \\ \underline{14x - 42} \\ 36 \end{array}$$

$$q(x) = 3x + 14$$
$$r(x) = 36$$

29.

$$\frac{x^2 - 4}{x - 2} = \frac{(x-2)(x+2)}{x-2} \qquad q(x) = x + 2 \qquad r(x) = 0$$

31.

$$\begin{array}{r} 5x^3 + 22x^2 + 112x + 555 \\ x - 5 \overline{)\,5x^4 - 3x^3 + 2x^2 - 5x + 11} \\ \underline{5x^4 - 25x^3} \\ 22x^3 + 2x^2 \\ \underline{22x^3 - 110x^2} \\ 112x^2 - 5x \\ \underline{112x^2 - 560x} \\ 555x + 11 \\ \underline{555x - 2775} \\ 2786 \end{array}$$

$$q(x) = 5x^3 + 22x^2 + 112x + 555$$
$$r(x) = 2786$$

33.

$$\begin{array}{r} x^2 - 3x + 8 \\ x^2 + 3x - 2 \overline{)\,x^4 + 0 - 3x^2 + 0 + 6} \\ \underline{x^4 + 3x^2 - 2x^2} \\ -3x^3 - x^2 + 0 \\ \underline{-3x^3 - 9x^2 + 6x} \\ 8x^2 - 6x + 6 \\ \underline{8x^2 + 24x - 16} \\ -30x + 22 \end{array}$$

$$q(x) = x^2 - 3x + 8$$
$$r(x) = -30x + 22$$

35.

$$\require{enclose}\begin{array}{r}
3x^4 + 12x^3 + 46x^2 + 160x + 543 \\
x^2 - 4x + 2 \enclose{longdiv}{3x^6 + 0 + 4x^4 + 0 - 5x^2 + 6x - 7}
\end{array}$$

$$
\begin{aligned}
&\underline{3x^6 - 12x^5 + 6x^4} \\
&\quad 12x^5 - 2x^4 + 0 \\
&\quad \underline{12x^5 - 48x^4 + 24x^3} \\
&\qquad 46x^4 - 24x^3 - 5x^2 \\
&\qquad \underline{46x^4 - 184x^3 + 92x^2} \\
&\qquad\quad 160x^3 - 97x^2 + 6x \\
&\qquad\quad \underline{160x^3 - 640x^2 + 320x} \\
&\qquad\qquad 543x^2 - 314x - 7 \\
&\qquad\qquad \underline{543x^2 - 2172x + 1086} \\
&\qquad\qquad\qquad + 1858x - 1093
\end{aligned}
$$

$q(x) = 3x^4 + 12x^3 + 46x^2 + 160x + 543$

$r(x) = 1858x - 1093$

37. $p(5) = 5^5 - 4(125) + 3(25) - 35 + 13 = 3678$

$$\require{enclose}\begin{array}{r}
x^4 + 5x^3 + 29x^2 + 148x + 733 \\
x - 5 \enclose{longdiv}{x^5 + 0 - 4x^3 + 3x^2 - 7x + 13}
\end{array}$$

$$
\begin{aligned}
&\underline{x^5 - 5x^4} \\
&\quad 5x^4 - 4x^3 \\
&\quad \underline{5x^4 - 25x^3} \\
&\qquad 29x^3 + 3x^2 \\
&\qquad \underline{29x^3 - 145x^2} \\
&\qquad\quad 148x^2 - 7x \\
&\qquad\quad \underline{148x^2 - 740x} \\
&\qquad\qquad 733x + 13 \\
&\qquad\qquad \underline{733x - 3665} \\
&\qquad\qquad\qquad 3678 \qquad R(x) = 3678
\end{aligned}
$$

39. $p(-3) = 3(-3)^7 + 5(3)^6 + 4(3^5) - 4(27) - 6(9) - 15 - 7 = -2128$

$$
\begin{array}{r}
3x^6 - 4x^5 + 8x^4 - 24x^3 + 76x^2 - 234x + 707 \\
x+3 \overline{\smash{\big)}\ 3x^7 + 5x^6 - 4x^5 + 0 + 4x^3 - 6x^2 + 5x - 7}
\end{array}
$$

$\underline{3x^7 + 9x^6}$

$-4x^6 - 4x^5$

$\underline{-4x^6 - 12x^5}$

$8x^5 + 0$

$\underline{8x^5 + 24x^4}$

$-24x^4 + 4x^3$

$\underline{-24x^4 - 72x^3}$

$76x^3 - 6x^2$

$\underline{76x^3 + 228x^2}$

$-234x^2 + 5x$

$\underline{-234x^2 - 702x}$

$707x - 7$

$\underline{707x + 2121}$

$-2128 \qquad r = -2128$

EXERCISE SET 0.5 SYNTHETIC DIVISION

1.

$$
3\ \overline{\big|\ \begin{array}{ccc} 5 & 5 & -7 \\ & 15 & 60 \\ \hline 5 & 20 & 53 \end{array}}
$$

$q(x) = 5x + 20$

$r(x) = 53$

3.

$$
2\ \overline{\big|\ \begin{array}{cccc} 4 & 5 & -6 & 7 \\ & 8 & 26 & 40 \\ \hline 4 & 13 & 20 & 47 \end{array}}
$$

$q(x) = 4x^2 + 13x + 20$

$r(x) = 47$

5.

$$
-2\ \overline{\big|\ \begin{array}{cccc} 3 & 7 & -4 & 8 \\ & -6 & -2 & 12 \\ \hline 3 & 1 & -6 & 20 \end{array}}
$$

$q(x) = 3x^2 + x - 6$

$r(x) = 20$

7.
$$
\begin{array}{r|rrrrr}
6 & 4 & 0 & 5 & -5 & 3 & -2 \\
  &   & 24 & 144 & 894 & 5334 & 32022 \\
\hline
  & 4 & 24 & 149 & 889 & 5337 & 32020
\end{array}
$$

$q(x) = 4x^4 + 24x^3 + 149x^2 + 889x + 5337$

$r(x) = 32020$

9.
$$
\begin{array}{r|rrrrrrrr}
4 & 6 & 5 & -3 & 6 & -8 & 5 & -6 & 1 \\
  &   & 24 & 116 & 452 & 1832 & 7296 & 29204 & 116792 \\
\hline
  & 6 & 29 & 113 & 458 & 1824 & 7301 & 29198 & 116793
\end{array}
$$

$q(x) = 6x^6 + 29x^5 + 113x^4 + 458x^3 + 1824x^2 + 7301x + 29198$

$r(x) = 116{,}793$

11.
$$
\begin{array}{r|rrrrrrrrrr}
-5 & 6 & 0 & -8 & 0 & 4 & 0 & -5 & 0 & 6 & -5 \\
   &   & -30 & 150 & -710 & 3550 & -17770 & 88850 & -444225 & 2221125 & -11105655 \\
\hline
   & 6 & -30 & 142 & -710 & 3554 & -17770 & 88845 & -444225 & 2221131 & -11105660
\end{array}
$$

$q(x) = 6x^8 - 30x^7 + 142x^6 - 710x^5 + 3554x^4 - 17770x^3 + 88845x^2 - 444225x + 2221131$

$r(x) = -11105660$

13. Using synthetic division the procedure is long. However with perseverence one can show

$r(x) = -29{,}245{,}885$ and

$q(x) = 7x^{10} - 28x^9 + 112x^8 - 448x^7 + 1785x^6 - 7140x^5 + 28560x^4 - 114240x^3 + 456966x^2 - 1827869x + 7311472$

15.
$$
\begin{array}{r|rrrr}
3 & 5 & -4 & -8 & -10 \\
  &   & 15 & 33 & 75 \\
\hline
  & 5 & 11 & 25 & 65
\end{array}
\qquad p(3) = 65
$$

17.
$$
\begin{array}{r|rrrr}
2 & -2 & +4 & -5 & +14 \\
  &   & -4 & 0 & -10 \\
\hline
  & -2 & 0 & -5 & 4
\end{array}
\qquad p(2) = 4
$$

19.
$$
\begin{array}{r|rrrrrr}
-3 & 1 & 4 & 0 & -6 & 2 & -6 \\
   &   & -3 & -3 & 9 & -9 & 21 \\
\hline
   & 1 & 1 & -3 & 3 & -7 & 15
\end{array}
\qquad p(-3) = 15
$$

21.

-2	5	0	0	0	-6	0	5	10	-17	$p(-2) = 1167$
		-10	20	-40	80	-148	$+296$	-602	$+1184$	
	5	-10	20	-40	74	-148	301	-592	1167	

23.

4	1	0	0	0	4	0	0	0	4	-4	10
		4	16	64	256	1040	4160	16640	66560	266256	1065008
	1	4	16	64	260	1040	4160	16640	66564	266252	1065018

$$p(4) = 1065018$$

25.

1	5	3	7	-15
		5	8	15
	5	8	15	0

Dividing $5x^3 + 3x^2 + 7x - 15$ by $x - 1$, the remainder is 0; hence $x - 1$ is a factor of $5x^3 + 3x^2 + 7x - 15$.

27.

-1	3	0	5	-6	-14
		-3	3	-8	14
	3	-3	8	-14	0

Dividing $3x^4 + 5x^2 - 6x - 14$ by $x + 1$, the remainder is 0; hence $x + 1$ is a factor of $3x^4 + 5x^2 - 6x - 14$.

EXERCISE SET 0.6 FACTORING

1. (a) Yes, the last digit is even

   (b) Yes, $5 + 4 = 9 = 3 \cdot 3$

   (c) Yes, the last digit is zero

   (d) Yes, $5 + 4 = 9 = 9 \cdot 1$

   (e) No, $5 + 0 - 4 = 1$ which is not divisible by 11.

   $540 = 2^2 \cdot 3^3 \cdot 5$

3. (a) Yes, the last digit is even

   (b) Yes, $1 + 7 + 8 + 2 = 18 = 6 \cdot 3$

   (c) Yes, the last digit is zero

   (d) Yes, $1 + 7 + 8 + 2 = 18 = 9 \cdot 2$

   (e) Yes, $1 + 8 - (7 + 2) = 0 = 11(0)$

   $178{,}200 = 2^3 \cdot 3^4 \cdot 5^2 \cdot 11$

5. (a) Yes, the last digit is even

(b) Yes, $3 + 3 + 7 + 5 = 18 = 6 \cdot 3$

(c) Yes, the last digit is zero

(d) Yes, $3 + 3 + 7 + 5 = 18 = 9 \cdot 2$

(e) No, $3 + 3 + 5 - 7 = 4$ which is not divisible by 11.
$303,750 = 2 \cdot 3^5 \cdot 5^4$

7. $(x - 7)^2$

9. $(x - 3)(x + 2)$

11. $(x - 7)(x + 5)$

13. $(x - 2)(x + 2)$

15. $(2x + 3)(x - 4)$

17. $(3x + 5)(x + 3)$

19. $(x + 1)(x^2 + 3x - 10) = (x + 1)(x + 5)(x - 2)$

-1	1	4	-7	-10
		-1	-3	10
	1	3	-10	0

21. $(x + 2)(x^2 - 4x - 21) = (x + 2)(x - 7)(x + 3)$

-2	1	-2	-29	-42
		-2	8	42
	1	-4	-21	0

23. $(x + 1)(3x^2 + 10x + 3) = (x + 1)(3x + 1)(x + 3)$

-1	3	13	13	3
		-3	-10	-3
	3	10	3	0

25. $(x+1)(x^3 - 6x^2 - x + 30) = (x+1)(x+2)(x^2 - 8x + 15)$

$$= (x+1)(x+2)\cdot(x-3)\cdot(x-5)$$

$$
\begin{array}{r|rrrrr}
-1 & 1 & -5 & -7 & 29 & 30 \\
   &   & -1 & 6 & 1 & -30 \\
\hline
   & 1 & -6 & -1 & 30 & 0
\end{array}
\qquad
\begin{array}{r|rrrr}
-2 & 1 & -6 & -1 & 30 \\
   &   & -2 & 16 & -30 \\
\hline
   & 1 & -8 & 15 & 0
\end{array}
$$

27. $\left(x - \sqrt{3}\right)\left(x + \sqrt{3}\right)$

29. $x^2 - 4x + 1 = 0$ if $x = \dfrac{4 \pm \sqrt{16 - 4}}{2} = 2 \pm \sqrt{3}$

$x^2 - 4x + 1 = \left(x - 2 - \sqrt{3}\right)\left(x - 2 + \sqrt{3}\right)$

31. $x^4 - 9 = (x^2 - 3)(x^2 + 3) = \left(x - \sqrt{3}\right)\left(x + \sqrt{3}\right)(x^2 + 3)$

33. $(x+1)(x^2 - 4x - 1) = (x+1)(x^2 - 4x + 4 - 5) = (x+1)\left[(x-2)^2 - 5\right]$

$$= (x+1)\left(x - 2 - \sqrt{5}\right)\left(x - 2 + \sqrt{5}\right)$$

$$
\begin{array}{r|rrrr}
-1 & 1 & -3 & -5 & -1 \\
   &   & -1 & 4 & 1 \\
\hline
   & 1 & -4 & -1 & 0
\end{array}
$$

35. $(x-5)(x^2 - 2x - 4) = (x-5)(x^2 - 2x + 1 - 5) = (x-5)\left[(x-1)^2 - 5\right]$

$$= (x-5)\left(x - 1 - \sqrt{5}\right)\left(x - 1 + \sqrt{5}\right)$$

$$
\begin{array}{r|rrrr}
5 & 1 & -7 & 6 & 20 \\
  &   & 5 & -10 & -20 \\
\hline
  & 1 & -2 & -4 & 0
\end{array}
$$

37. Not factorable over the real numbers since $x^4 + 4x^2 + 16 > 0$ for all real numbers x. $x^4 + 4x^2 + 16 = 0$ has no real solution.

39. $xz + 6wy + 3wx + 2yz = xz + 2yz + 3wx + 6wy = z(x + 2y) + 3w(x + 2y)$
$= (z + 3w)(x + 2y)$

41. $2xz - 2wy + yz - 4wx = 2xz + yz - 2wy - 4wx = z(2x + y) - 2w(2x + y)$
$= (2x + y)(z - 2w)$

43. $6wx - 6yz - 9wz + 4xy = 6wx + 4xy - 6yz - 9wz = 2x(3w + 2y) - 3z(2y + 3w)$
$= (2y + 3w)(2x - 3z)$

45. $x^3 + x^2 - y^3 - y^2 = x^3 - y^3 + x^2 - y^2 = (x - y)(x^2 + xy + y^2) + (x - y)(x + y)$
$= (x - y)(x^2 + xy + y^2 + x + y)$

EXERCISE SET 0.7 EQUATIONS IN ONE VARIABLE

1. All real numbers except $-2, 3$ and -5.

3. All real numbers except $-1, 1$ and -6.

5. All real numbers greater than or equal to 4.

7. $5 - 3y = 2y + 20$
$-15 = 5y, \quad y = -3.$ The solution set is $\{-3\}$.

9. $2x - 3 - 5x = 4x - 17$
$-3x - 3 = 4x - 17$
$14 = 7x, \quad x = 2.$ The solution set is $\{2\}$.

11. $3y - 2 + 4y = 2y + 6 - 18$
$7y - 2 = 2y - 12, \quad 5y = -10, \quad y = -2.$ The solution set is $\{-2\}$.

13.

$$30\left(\frac{x+3}{5} - \frac{x-8}{3}\right) = \left(\frac{x+4}{2}\right) \cdot 30$$

$$6x + 18 - 10x + 80 = 15x + 60, \quad -4x + 98 = 15x + 60,$$

$$38 = 19x, \quad x = 2. \text{ The solution set is } \{2\}.$$

15.

$$x^2 - 6x + 5 = 0$$

$$(x-5)(x-1) = 0, \quad x = 5 \text{ or } x = 1. \text{ The solution set is } \{1,5\}.$$

17.

$$z^2 + 7z - 18 = 0$$

$$(z+9)(z-2) = 0, \quad z = -9 \text{ or } z = 2. \text{ The solution set is } \{2, -9\}.$$

19.

$$y^2 + 10y - 75 = 0$$

$$(y+15)(y-5) = 0, \quad y = -15 \text{ or } y = 5. \text{ The solution set is } \{5, -15\}.$$

21.

$$x^2 + x - 2 = x^2 - 9 + 13$$

$$(x-2) = 4, \quad x = 6. \text{ The solution set is } \{6\}.$$

23.

$$2z^2 + 3z - z^2 + 1 = 11$$

$$z^2 + 3z - 10 = 0, \quad (z+5)(z-2) = 0$$

$$z = -5 \text{ or } z = 2. \text{ The solution set is } \{-5, 2\}.$$

25.

$$5x^2 - 7x - 196 = 0$$

$$(5x + 28)(x - 7) = 0, \quad x = \frac{-28}{5} \text{ or } x = 7. \text{ The solution set is } \{\frac{-28}{5}, 7\}.$$

27.

$$3z^2 + 11z - 60 = 0$$

$$(3z + 20)(z - 3) = 0, \quad z = \frac{-20}{3} \text{ or } z = 3. \text{ The solution set is } \{\frac{-20}{3}, 3\}.$$

29. $6(x+3)+5(x+1)=3(x+1)(x+3), \quad 6x+18+5x+5=3x^2+12x+9,$

$11x+23=3x^2+12x+9, \quad 0=3x^2+x-14, \quad 0=(3x+7)(x-2),$

$x=\dfrac{-7}{3}$ or $x=2$ both of which are in the the replacement set.

The solution set is $\{2, \dfrac{-7}{3}\}$.

31. $\sqrt{x+1}=x-5.$ Squaring both sides of the equation,

$x+1=x^2-10x+25, \quad 0=x^2-11x+24, \quad 0=(x-8)(x-3),$

$x=8$ or 3 but only 8 satisfies the original equation. The solution set is $\{8\}$.

33.   Squaring both sides of the equation we obtain $x+6+6\sqrt{x+6}+9=27+x,$

$6\sqrt{x+6}=12, \quad \sqrt{x+6}=2, \quad x+6=4, \quad x=-2$ which satisfies the

original equation.   The solution set is $\{-2\}$.

35. $5-x=\sqrt{5x-1}.$ Squaring both sides of the equation,

$25-10x+x^2=5x-1, \quad x^2-15x+26=0, \quad (x-2)(x-13)=0,$

$x=2$ or $x=13$ but only 2 satisfies the original equation. The solution set is $\{2\}$.

37.   Let x be the amount invested at 12%. Then $40{,}000-x$ is the amount

invested at 8%. $40000(.09)=x(.12)+.08(40000-x),$

$400(9)=.12x+400(8)-.08x, \quad 3600-3200=.04x,$

$\dfrac{400}{.04}=\dfrac{40000}{4}=\$10{,}000.$ Invest $\$10{,}000$ at 12% and $\$30{,}000$ at 8%.

39. $1600-\dfrac{1}{4}p^2=\dfrac{37}{11}p+\dfrac{1}{2}p^2, \quad 0=\dfrac{3}{4}p^2+\dfrac{37}{11}p-1600,$

$$p=\dfrac{-\dfrac{37}{11}\pm\sqrt{\left(\dfrac{37}{11}\right)^2+3(1600)}}{\dfrac{3}{2}}=\$44 \ (p \text{ must be positive so the other}$$

solution to the equation is not considered).

41.

number of trees	yield per tree
30	475
$30 + x$	$475 - 7x$

for $x \leq 20$. x is the number of additional trees planted.

$(30 + x)(475 - 7x) = 16318,$ $14250 + 265x - 7x^2 = 16318,$

$0 = 7x^2 - 265x + 2068,$ $x = \dfrac{265 \pm \sqrt{265^2 - 4(7)(2068)}}{14} = \dfrac{265 \pm 111}{14}$

$x = 26.86$ or 11. Only 11 is a feasible solution and 41 trees should be planted per acre.

43.   If $\$p$ is the price per hour and x is the number of hours per month, $p = 120 - 1.5x$ where $p \leq 75$.

profit $=$ income $-$ expense, $1264 = px - 100 - 5x,$

$1264 = (120 - 1.5x)x - 100 - 5x,$ $1264 = 115x - 1.5x^2 - 100,$

$1.5x^2 - 115x + 1364 = 0,$ $15x^2 - 1150x + 13640 = 0,$

$3x^2 - 230x + 2728 = 0,$ $x = \dfrac{230 \pm \sqrt{230^2 - 4(3)(2728)}}{6}.$

$x = 62$ or $x = \dfrac{44}{3},$ If $x = 62$, $p = \$27$. If $x = \dfrac{44}{3}$, $p = \$98$ which is not a feasible

solution. Therefore the price per hour was $\$27$ and 62 hours were taught per month.

45.   Let x the number of pounds of the first kind. Then $120 - x$ is the number of pounds of the second kind.

Value of mixture $= x(5.10) + (120 - x)4.50 = 4.90(120),$

$5.10x + 540 - 4.5x = 588,$ $6x = 48,$ $x = \dfrac{480}{6} = 80.$ Therefore, use 80 pounds

of the first kind and 40 pounds of the second.

47.   Let x be the number of quarts withdrawn. Then $30 - x$ quarts remain leaving $.1(30 - x) = 3 - .1x$ quarts of antifreeze. Since x quarts of antifreeze are added, the new amount of antifreeze is $x + 3 - .1x$.

We need $.2(30) = 6$ quarts of antifreeze. $x + 3 - .1x = 6,$ $.9x = 3,$

$x = \dfrac{3}{.9} = \dfrac{30}{9} = \dfrac{10}{3} = 3\tfrac{1}{3}$qt. Drain $3\tfrac{1}{3}$ quarts.

49.   Let x be the number of spark plugs sold. They cost $66\tfrac{2}{3}$¢ each.

profit $=$ income $-$ expense,   profit $= \tfrac{x}{2}(80) + \tfrac{x}{2}(60) - \dfrac{200}{3}x = 1000,$

$40x + 30x - \dfrac{200}{3}x = 1000,$ $210x - 200x = 3000,$ $10x = 3000,$ $x = 300.$

He sold 300 spark plugs.

51.   Let x be the number of gallons of ginger ale. Then $18 - x$ is the number of gallons of cranberry juice. The cost of the punch is $x(4) + 6(18 - x) = 84,$

$4x + 108 - 6x = 84,$ $2x = 24,$ $x = 12.$

12 gallons of ginger ale were in the 18 gallons of punch. Hence, 2/3 gallon of ginger ale was in one gallon of punch.

53. Let x be the length as shown

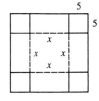

$$5x^2 = 18000$$
$$x^2 = 3600$$
$$x = 60$$

The dimensions of the sheet were 70 cm by 70 cm.

55. Let x and y be the lengths as labeled.

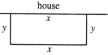

$$x \leq 35$$
$$xy = 500$$

The cost is $(x + 2y)120 + x(40) = 9200,\quad 120x + 240y + 40x = 9200,$

$160x + 240\left(\dfrac{500}{x}\right) = 9200,\quad 160x^2 + 120{,}000 = 9200x,\quad 160x^2 - 9200x + 120{,}000 = 0,$

$2x^2 - 115x + 1500 = 0,\quad x = \dfrac{115 \pm \sqrt{115^2 - 8(1500)}}{4}.\quad x = 37.5 \text{ or } 20.$

Since $x \leq 35$, the dimensions are 20 ft (along existing wall) by 25 ft.

EXERCISE SET 0.8 INEQUALITIES

1. $2x + 13 < 5x + 28,\quad 13 - 28 < 5x - 2x,\quad -15 < 3x,\quad -5 < x$

The solution set is $(-5, +\infty)$.

3. $6x + 5 \leq 2x + 33,\quad 6x - 2x \leq 33 - 5,\quad 4x \leq 28,\quad x \leq 7$

The solution set is $(-\infty, 7]$.

5. $\dfrac{x}{2} + \dfrac{1}{3} < \dfrac{3x}{5} - \dfrac{2}{7},\quad \dfrac{1}{3} + \dfrac{2}{7} < \dfrac{3x}{5} - \dfrac{x}{2},\quad \dfrac{7+6}{21} < \dfrac{6x - 5x}{10},\quad \dfrac{13}{21} < \dfrac{x}{10},\quad \dfrac{130}{21} < x$

The solution set is $\left(\dfrac{130}{21}, +\infty\right)$.

7. $3x + 1 < 6x + 4 < 4x + 10,\quad 3x + 1 < 6x + 4$ and $6x + 4 < 4x + 10$

$-3 < 3x$ and $2x < 6,\quad -1 < x$ and $x < 3,\quad -1 < x < 3$

The solution set is $(-1, 3)$.

9. $15 + 4x < 8x + 3 < 6x + 5,\quad 12 < 4x$ and $2x < 2,\quad 3 < x$ and $x < 2$

which is impossible. No solutions

11. $x^2 + 8x - 20 \leq 0,$
$(x+10)(x-2) \leq 0,$
$-10 \leq x \leq 2$

x		-10			2	
$x+10$	$-$	0	$+$	$+$	$+$	$+$
$x-2$	$-$	$-$	$-$	$-$	0	$+$
$(x+10)(x-2)$	$-$	0	$-$	$-$	0	$+$

The solution set is $[-10, 2]$.

13. $x^2 + 3x - 10 \leq 0,$
$(x-5)(x+2) \leq 0,$
$-2 \leq x \leq 5$

x		-2			5	
$x-5$	$-$	$-$	$-$	0	$+$	
$x+2$	$-$	0	$+$	$+$	$+$	
$(x-5)(x+2)$	$+$	0	$-$	0	$+$	

The solution set is $[-2, 5]$.

15. $8x^3 < 27, \quad x^3 < \dfrac{27}{8} = \left(\dfrac{3}{2}\right)^3, \quad x < \dfrac{3}{2}$

The solution set is $(-\infty, 1.5)$.

17. $16x^4 - 81 \leq 0, \quad 16x^4 \leq 81, \quad x^4 \leq \dfrac{81}{16} = \left(\dfrac{3}{2}\right)^4, \quad \dfrac{-3}{2} \leq x \leq \dfrac{3}{2}$

The solution set is $[-1.5, 1.5]$.

19. $3x^2 + 24x - 9 > 0, \quad x^2 + 8x - 3 > 0, \quad x^2 + 8x + 16 > 3 + 16,$
$(x+4)^2 > 19, \quad x+4 > \sqrt{19} \text{ or } x+4 < -\sqrt{19}, \quad x > \sqrt{19} - 4 \text{ or } x < -\sqrt{19} - 4$
The solution set is $(-\infty, -\sqrt{19} - 4) \bigcup (\sqrt{19} - 4, +\infty)$.

21. $\dfrac{x^2 + 3x - 4}{x+5} = \dfrac{(x+4)(x-1)}{x+5} > 0$

x		-5		-4		1	
$x+4$	$-$	$-$	$-$	0	$+$	$+$	$+$
$x-1$	$-$	$-$	$-$	$-$	$-$	0	$+$
$x+5$	$-$	0	$+$	$+$	$+$	$+$	$+$
$\dfrac{(x+4)(x-1)}{x+5}$	$-$	U	$+$	0	$-$	0	$+$

$-5 < x < -4 \text{ or } x > 1.$ The solution set is $(-5, -4) \bigcup (1, +\infty)$.

23.

$$\frac{(x-3)(x-2)}{5-x} > 0$$

x				2		3		5	
$x-3$	$-$		$-$		$-$	0	$+$	$+$	$+$
$x-2$	$-$		0		$+$	$+$	$+$	$+$	$+$
$5-x$	$+$		$+$		$+$	$+$	$+$	0	$-$
$\dfrac{(x-3)(x-2)}{5-x}$	$+$		0		$-$	0	$+$	U	$-$

The solution set is $(-\infty, 2) \cup (3, 5)$.

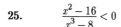

25.

$$\frac{x^2-16}{x^3-8} < 0$$

x		-4		2			4	
x^2-16	$+$	0	$-$	$-$		$-$	0	$+$
x^3-8	$-$		$-$		0	$+$	$+$	$+$
$\dfrac{x^2-16}{x^3-8}$	$-$	0	$+$		U	$-$	0	$+$

The solution set is $(-4, 2) \cup (4, +\infty)$

27.

$$\frac{x^2+3x+8}{x+5} < 0$$

$x^2 + 3x + 8 = 0$

if $x = \dfrac{-3 \pm \sqrt{9-32}}{2}$

x		-5	
x^2+3x+8	$+$	$+$	$+$
$x+5$	$-$	0	$+$
$\dfrac{x^3+3x+8}{x+5}$	$-$	U	$+$

$x^2 + 3x + 8$ has no real solutions. The solution set is $(-\infty, -5)$

29.

$$\frac{2x^2+4x-4}{x+1} < 0 \qquad \frac{x^2+2x-2}{x+1} < 0 \qquad x^2+2x-2 = 0$$

if $x^2 + 2x + 1 = 3 \qquad (x+1)^2 = 3 \qquad x+1 = \sqrt{3} \qquad x = \sqrt{3}-1$

or $\quad x+1 = -\sqrt{3} \qquad x = -\sqrt{3}-1$

x		$-\sqrt{3}-1$		-1		$\sqrt{3}-1$	
$x-(\sqrt{3}-1)$	$-$	$-$	$-$	$-$	$-$	0	$+$
$x-(-\sqrt{3}-1)$	$-$	0	$+$	$+$	$+$	$+$	$+$
$x+1$	$-$	$-$	$-$	0	$+$	$+$	$+$
$\dfrac{2x^2+4x-4}{x+1}$	$-$	0	$+$	U	$-$	0	$+$

Solution set is $(-\infty, -\sqrt{3}-1) \cup (-1, \sqrt{3}-1)$.

31. $\dfrac{x^2+3x-6}{x+3} > 0$

$x^2 + 3x - 6 = 0$ if $x = \dfrac{-3 \pm \sqrt{9+24}}{2} = \dfrac{-3 \pm \sqrt{33}}{2}$

$\dfrac{\left(x - \left(\frac{-3 + \sqrt{33}}{2}\right)\right)\left(x - \left(\frac{-3 - \sqrt{33}}{2}\right)\right)}{x+3} > 0$

x		$\frac{-3-\sqrt{33}}{2}$		-3		$\frac{-3+\sqrt{33}}{2}$	
$x - \left(\dfrac{-3+\sqrt{33}}{2}\right)$	$-$	$-$	$-$	$-$	$-$	0	$+$
$x - \left(\dfrac{-3-\sqrt{33}}{2}\right)$	$-$	0	$+$	$+$	$+$	$+$	$+$
$x + 3$	$-$	$-$	$-$	0	$+$	$+$	$+$
$\dfrac{x^2+3x-6}{x+3}$	$-$	0	$+$	U	$-$	0	$+$

The solution set is $\left(\dfrac{-3-\sqrt{33}}{2}, -3\right) \cup \left(\dfrac{-3+\sqrt{33}}{2}, +\infty\right)$

33. $\dfrac{3x^2 - 6x + 2}{x^2 - 5} \le 0$

$3x^2 - 6x + 2 = 0$ if $x = \dfrac{6 \pm \sqrt{36-24}}{6} = \dfrac{6 \pm \sqrt{12}}{6} = \dfrac{3 \pm \sqrt{3}}{3}$

$x^2 - 5 = 0$ if $x = \pm\sqrt{5}$

x		$-\sqrt{5}$		$\frac{3-\sqrt{5}}{3}$		$\frac{3+\sqrt{5}}{3}$		$\sqrt{5}$	
$3x^2 - 6x + 2$	$+$	$+$	$+$	0	$-$	0	$+$	$+$	$+$
$x^2 - 5$	$+$	0	$-$	$-$	$-$	$-$	$-$	0	$+$
$\dfrac{3x^2 - 6x + 2}{x^2 - 5}$	$+$	U	$-$	0	$+$	0	$-$	U	$+$

The solution set is $\left(-\sqrt{5}, \dfrac{3-\sqrt{3}}{3}\right] \cup \left[\dfrac{3+\sqrt{3}}{3}, \sqrt{5}\right)$

35. $\dfrac{x^2 + 5x + 2}{x+5} - (2x-3) > 0$

$\dfrac{x^2 + 5x + 2 - (2x-3)(x+5)}{x+5} = \dfrac{x^2 + 5x + 2 - 2x^2 - 7x + 15}{x+5} > 0$

$\dfrac{-x^2 - 2x + 17}{x+5} > 0, \ \dfrac{x^2 + 2x - 17}{x+5} < 0.$

$x^2 + 2x - 17 = 0$ if $x = -1 \pm 3\sqrt{2}$

x		$-1-3\sqrt{2}$		-5		$-1+3\sqrt{2}$	
$x+5$	$-$	$-$	$-$	0	$+$	$+$	$+$
$x^2+2x-17$	$+$	0	$-$	$-$	$-$	0	$+$
$\dfrac{x^2+2x-17}{x+5}$	$-$	0	$+$	U	$-$	0	$+$

The solution set is $(-\infty,-1-3\sqrt{2})\cup(-5,-1+3\sqrt{2})$

37. $\dfrac{3x+2}{x-3}-\dfrac{6x+4}{x-2}=\dfrac{(3x+2)(x-2)-(x-3)(6x+4)}{(x-3)(x-2)}>0$

$\dfrac{3x^2-10x-8}{(x-3)(x-2)}=\dfrac{(3x+2)(x-4)}{(x-3)(x-2)}<0$

x		$-\frac{2}{3}$		2		3		4	
$3x+2$	$-$	0	$+$	$+$	$+$	$+$	$+$	$+$	$+$
$x-4$	$-$	$-$	$-$	$-$	$-$	$-$	$-$	0	$+$
$x-2$	$-$	$-$	$-$	0	$+$	$+$	$+$	$+$	$+$
$x-3$	$-$	$-$	$-$	$-$	$-$	0	$+$	$+$	$+$
$\dfrac{(3x+2)(x-4)}{(x-3)(x-2)}$	$+$	0	$-$	U	$+$	U	$-$	0	$+$

The solution set is $(-2/3,2)\cup(3,4)$

39. $-x^2+3x-9=0$ if $x=\dfrac{-3\pm\sqrt{9-36}}{-2}$. There are no real solutions. Thus,

$-x^2+3x-9$ is always positive or always negative and if $x=0$, $-x^2+3x-9$ is negative.

x		-2		-1	
$-x^2+3x-9$	$-$	$-$	$-$	$-$	$-$
$x+2$	$-$	0	$+$	$+$	$+$
$x+1$	$-$	$-$	$-$	0	$+$
$\dfrac{-x^2+3x-9}{(x+2)(x+1)}$	$-$	U	$+$	U	$-$

The solution set is $(-\infty,-2)\cup(-1,+\infty)$

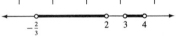

41. Assume a, b and c are real numbers, $a<b$ and $c<0$.
$b-a$ is positive, c is negative. Hence, $(b-a)(c)$ is negative.
$(b-a)c<0$
$bc-ac<0$

$$bc - ac+ + ac < 0 + ac$$
$$bc < ac$$

43. The solution shown is not correct as $x = -5$ is not a solution which as may be easily verified. In the last step, both sides of the inequality are multiplied by $\frac{1}{x+2}$. This step is invalid since we don't know if $\frac{1}{x+2}$ is positive, negative, or worse yet, undefined.

45. Let x be the hours Betty must work.

$$(x + 10)6 + 7x \geq 385$$

$$13x \geq 385 - 60 = 325$$

$$x \geq \frac{325}{13} = 25 \text{ hours}$$

47. Profit = revenue − cost > 0

$$p(252 - 7p) - 1092 - 4q > 0$$

$$p(252 - 7p) - 4(252 - 7p) - 1092 > 0$$

$$252p - 7p^2 - 1008 + 28p > 1092$$

$$0 > 7p^2 - 280p + 2100 = 7(p^2 - 40p + 300)$$

$$0 > (p - 30)(p - 10)$$

$$\$10 < p < \$30$$

p		10		30	
$p - 10$	$-$	0	$+$	$+$	$+$
$p - 30$	$-$	$-$	$-$	0	$+$
$(p - 10)(p - 30)$	$+$	0	$+$	0	$+$

49. We want the revenue to exceed the cost.

$$p(935 - 17p) > 5100 + 15(935 - 17p)$$

$$935p - 17p^2 > 5100 + 14025 - 255p$$

$$0 > 17p^2 - 1190p + 19125$$

$$0 > 17(p^2 - 70p + 1125)$$

$$0 > (p - 45)(p - 25)$$

$$\$25 < p < \$45$$

51. Revenue−cost ≥ 400,000

$q(1000)(30 - .25q) - (200 + 5q)(1000) \geq 400{,}000$ where q is the number of units sold in thousands.

$q(30 - .25q) - 200 - 5q \geq 400$

$30q - .25q^2 - 200 - 5q \geq 400$

$0 \geq .25q^2 - 25q + 600$

$0 \geq q^2 - 100q + 2400$

$0 \geq (q - 40)(q - 60)$

q		40		60	
$q - 40$	$-$	0	$+$	$+$	$+$
$q - 60$	$-$	$-$	$-$	0	$+$
$(q - 40)(q - 60)$	$+$	0	$-$	0	$+$

$40 \leq q \leq 60$

At least 40,000 units but no more than 60,000 units must be produced and sold.

53. Let x be the number of quarts to be drained.

$.35(15) \leq$ quarts of antifreeze $\leq .45(15)$

$5.25 \leq (15 - x)(.25) + x \leq 6.75$

$5.25 \leq \frac{15}{4} - \frac{x}{4} + x \leq 6.75$

$5.25 \leq \frac{15}{4} + \frac{3}{4}x \leq 6.75$

$21 \leq 15 + 3x \leq 27$
$6 \leq 3x \leq 12$
2 quarts $\leq x \leq$ 4 quarts

55. $x^2 + 2x + 6 = (x+1)^2 + 5$ and $x^2 + 8x + 20 = (x+4)^2 + 4$ are always positive. Thus, the solution is found by examining the signs of the other factors.

x			−5				−3				5	
$(x+3)^3$	$-$	$-$	$-$	$-$	$-$	$-$	0	$+$	$+$	$+$	$+$	$+$
$(x-5)^7$	$-$	$-$	$-$	$-$	$-$	$-$	$-$	$-$	$-$	$-$	0	$+$
$(x+5)^3$	$-$	$-$	0	$+$	$+$	$+$	$+$	$+$	$+$	$+$	$+$	$+$
fraction	$-$	$-$	U	$+$	$+$	$+$	0	$-$	$-$	$-$	0	$+$

The solution set is $(-\infty, -5) \cup (-3, 5)$

EXERCISE SET 0.9   ABSOLUTE VALUE

1. $|x-3| < 4$

   $-4 < x - 3 < 4$

   $-1 < x < 7$

   The solution set is $(-1,7)$

3. $x+7 > 3$ or $x+7 < -3$

   $x > -4$ or $x < -10$

   The solution set is $(-\infty,-10) \cup (-4,+\infty)$

5. $|3x - 2| < 4$

   $-4 < 3x - 2 < 4$

   $-2 < 3x < 6$

   $-2/3 < x < 2$

   The solution set is $(-2/3,2)$

7. $2+4x \geq 14$ or $2+4x \leq -14$

   $4x \geq 12$ or $4x \leq -16$

   $x \geq 3$ or $x \leq -4$

   The solution set is $(-\infty,-4] \cup [3,+\infty)$

9. No solutions, $|2 - 3x| \geq 0$ for all x.   The solution set is ϕ.

11. $|2 - 5x| \leq 4x + 2$

    If $4x + 2 < 0$, $4x < -2$, and $x < -1/2$.   There are no solutions where $x < -1/2$.

    If $4x + 2 \geq 0$ or $x \geq -1/2$

    $-4x - 2 \leq 2-5x \leq 4x + 2$

    $-4x - 2 \leq 2 - 5x$ and $2 - 5x \leq 4x + 2$

    $x \leq 4$ and $0 \leq 9x$

    $x \leq 4$ and $0 \leq x$

    The solution set is $[0,4]$

13. If $5x + 5 \leq 0$, $5x \leq -5$ and $x \leq -1$.   There are no solutions when $x \leq -1$.

If $5x + 5 > 0$, $-5x - 5 < x + 13 < 5x + 5$

$-5x - 5 < x + 13$ and $x + 13 < 5x + 5$

$-18 < 6x$ and $8 < 4x$

$-3 < x$ and $2 < x$ (and $x \geq -1$)

The solution set is $(2, +\infty)$.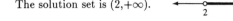

15. $|1 - 4x| > |3x + 1|$ if and only if

$(1 - 4x)^2 > (3x + 1)^2$

$1 - 8x + 16x^2 > 9x^2 + 6x + 1$

$7x^2 - 14x > 0$

$7x(x - 2) > 0$

$x > 2$ or $x < 0$

The solution set is $(-\infty, 0) \cup (2, +\infty)$

17. $|x + 2| + |x - 3| > 0$ is true for all real numbers since $|x + 2| \geq 0$ and $|x - 3| \geq 0$. Furthermore $|x - 3|$ and $|x + 2|$ cannot both be zero simultaneously. The solution set is $(-\infty, \infty)$.

19. $x + 2 = 0$ if $x = -2$; $x - 8 = 0$ if $x = 8$.

Case 1. $x \leq -2$.   Then $x < 8$

$$= |x + 2| + |x - 8| = -x - 2 - x + 8 = 16$$
$$-2x = 10$$
$$x = -5$$

Case 2. $-2 < x < 8$
$x > -2$ and $x < 8$
$x + 2 > 0$ and $x - 8 < 0$

$x + 2 - x + 8 = 16$ leading to
$10 = 16$.
There are no solutions in this case.

Case 3. $x \geq 8$
$x \geq -2$ and $x \geq 8$
$x + 2 \geq 0$ and $x - 8 \geq 0$

$x + 2 + x + 8 = 16$, $2x = 22$, $x = 11$.

The solution set is $\{-5, 11\}$

21. $|x+3| + |x-5| = 8$

$x+3 = 0$ if $x = -3$; $x-5 = 0$ if $x = 5$

Case 1. $x \leq -3$
$$-x-3-x+5 = 8; \quad -2x = 6, \quad x = -3.$$

Case 2. $-3 < x < 5$
$$x+3-x+5 = 0$$
$$8 = 8$$

All real numbers in the interval $(-3,5)$ are solutions.

Case 3. $x \geq 5$
$$x+3+x-5 = 8$$
$$2x = 10$$
$$x = 5$$

The solution set is $\{-3,5\}$

23. $3x+2 = 0$ if $x = -2/3$; $5x-7 = 0$ if $x = 7/5$

Case 1. $x \leq -2/3$
$$-3x-2-5x+7 = 19$$
$$-8x = 14$$
$$x = -14/8 = -7/4$$

Case 2. $-2/3 < x < 7/5$
$$3x+2-5x+7 = 19$$
$$-2x = 10$$
$$x = -5 \quad \text{which is not included within this case.}$$

Case 3. $3x+2+5x-7 = 19$
$$8x = 24$$
$$x = 3$$

The solution set is $\{-7/4,3\}$

25.   (a) $7-(-6) = 13$ (b) $-6-7 = -13$ (c) $|13| = 13$

27.   (a) $1/3-(-1/2) = 5/6$ (b) $-1/2-1/3 = -5/6$ (c) $|5/6| = 5/6$

29.   If $|a| = 0$ and $a \geq 0$, $|a| = a = 0$
If $|a| = 0$ and $a \leq 0$, $|a| = -a = 0 \Rightarrow a = 0$
Hence, in either case, $a = 0$.
Conversely, if $a = 0$, $|a| = 0$ by definition.

31. $\left|\frac{a}{b}\right| = \left|a \cdot \frac{1}{b}\right| = |a| \cdot \left|\frac{1}{b}\right|$ by problem 30.

If b is positive, $\frac{1}{b}$ is positive and $\left|\frac{1}{b}\right| = \frac{1}{b} = \frac{1}{|b|}$

If b is negative, $\frac{1}{b}$ is negative and $\left|\frac{1}{b}\right| = \frac{-1}{b} = \frac{1}{-b} = \frac{1}{|b|}$

Hence, $\left|\frac{a}{b}\right| = |a| \cdot \frac{1}{|b|} = \frac{|a|}{|b|}$.

33. If $c \geq 0$, $|c|^2 = (c)^2$. If $c < 0$, $|c|^2 = (-c)^2 = c^2$. Thus, for all real numbers c, $c^2 = |c|^2$.

   $0 \leq |a| \leq |b|$ implies $|a|^2 \leq |b|^2$, or $a^2 \leq b^2$. Conversely, if $a^2 \leq b^2$, $|a|^2 \leq |b|^2$ and $|a| \leq |b|$.

EXERCISE SET 0.10   CHAPTER REVIEW

1. (a) -32

   (b) $-3[2+1] = -9$

   (c) $\frac{34}{5}$

   (d) $\frac{12+35}{21} = \frac{47}{21}$

   (e) $\frac{-297}{43}$

   (f) $\frac{63}{56} + \frac{14}{56} = \frac{77}{56} = \frac{11}{8}$

   (g) $\frac{-3/5}{12} = \frac{-3}{5} \cdot \frac{1}{12} = -\frac{1}{20}$

   (h) $\frac{-5}{7} \cdot \frac{28}{15} = \frac{-4}{3}$

3. (a) 3

   (b) $5x + 15 = 4x + 4 + 20$
   $$x = 9$$

   (c) $1/3$

5. (a) $\sqrt{x+4} = 2x - 7$
   $x + 4 = (2x-7)^2 = 4x^2 - 28x + 49$
   $0 = 4x^2 - 29x + 45 = (x-5)(4x-9)$
   $x = 5$ or $x = 9/4$

   $\sqrt{5+4} = 10 - 7$ so 5 checks

   $\sqrt{\frac{9}{4}+4} = \frac{5}{2} \neq 2(\frac{9}{4}) - 7 = \frac{9}{2} - 7$

   5 is the only solution.

   (b) $\sqrt{7x+22} = x + 4$
   $7x + 22 = x^2 + 8x + 16$
   $0 = x^2 + x - 6 = (x+3)(x-2)$
   $x = -3$, $x = 2$ and both check

7. (a)

2	2	5	−6	−5	−35	−6
		4	18	24	38	6
	2	9	12	19	3	0

Since this remainder is zero, $x=2$ is a solution.

(b)

−3	5	0	−4	0	3	6	−4	7
		−15	45	−123	−369	−1116	3330	−9978
	5	−15	41	−123	372	−1110	3326	−9971

$p(-3) = -9971$

9. The remainder is 7 by the Remainder Theorem.

11. (a) $x^3 - x^2 - 3x + 3 = (x-1)(x^2-3) = (x-1)(x-\sqrt{3})(x+\sqrt{3})$

(b) $x^3 - 3x^2 + 5x - 15 = x^2(x-3) + 5(x-3) = (x^2+5)(x-3)$

(c) $x^3 - x^2 - 3x + 2 = (x-2)(x^2+x-1) = (x-2)(x+\frac{1}{2}-\frac{\sqrt{5}}{2})(x+\frac{1}{2}+\frac{\sqrt{5}}{2})$

$x^2 + x - 1 = 0$ if $x = \dfrac{-1 \pm \sqrt{1+4}}{2} = \dfrac{-1 \pm \sqrt{5}}{2}$

13. (a)
$$3x - 2 < 5x - 10$$
$$8 < 2x$$
$$4 < x$$

The solution set is $(4,\infty)$

(b) $\dfrac{x+1}{3} + 1 \geq \dfrac{2x-3}{7} + 2$

$7x + 7 + 21 \geq 6x - 9 + 42$
$7x + 28 \geq 6x + 33$
$x \geq 5$

The solution set is $[5,\infty)$

(c) $|x - 2| < 5$
$-5 < x - 2 < 5$
$-3 < x < 7$

The solution set is $(-3,7)$

(d) $|x + 3| \leq 2$
$-2 \leq x + 3 \leq 2$
$-5 \leq x \leq -1$

The solution set is $[-5,-1]$

(e) $|2 - x| \geq 5$
$2 - x \geq 5$ or $2 - x \leq -5$
$-3 \geq x$ or $7 \leq x$

The solution set is $(-\infty, -3] \cup [7, \infty)$

(f) $|3 + x| + |5 - 2x| < 0$ has no solution as $|3 + x|$ and $|5 - 2x|$ are nonnegative for all real numbers x.

(g) $|2x - 3| < 5/4$
$-5/4 < 2x - 3 < 5/4$
$7/4 < 2x < /17/4$
$7/8 < x < 17/8$

The solution set is $(7/8, 17/8)$

(h) $x^2 + 13x - 14 < 0$
$(x + 14)(x - 1) < 0$

x			-14			1	
$x + 14$	$-$	$-$	0	$+$	$+$	$+$	$+$
$x - 1$	$-$	$-$	$-$	$-$	$-$	0	$+$
$x^2 + 13x - 14$	$+$	$+$	0	$-$	$-$	0	$+$

The solution set is $(-14, 1)$

(i) $\dfrac{(x - 2)(x + 2)}{x + 3} \geq 0$

x		-3		-2		2	
$x - 2$	$-$	$-$	$-$	$-$	$-$	0	$+$
$x + 2$	$-$	$-$	$-$	0	$+$	$+$	$+$
$x + 3$	$-$	0	$+$	$+$	$+$	$+$	$+$
$\dfrac{x^2 - 4}{x + 3}$	$-$	U	$+$	0	$-$	0	$+$

The solution set is $(-3, -2] \cup [2, +\infty)$

(j) $\dfrac{(x + 2)(x - 1)}{(3 + x)(5 - x)} < 0$

x		-3		-2		1		5	
$x+2$	$-$	$-$	$-$	0	$+$	$+$	$+$	$+$	$+$
$x-1$	$-$	$-$	$-$	$-$	$-$	0	$+$	$+$	$+$
$3+x$	$-$	0	$+$	$+$	$+$	$+$	$+$	$+$	$+$
$5-x$	$+$	$+$	$+$	$+$	$+$	$+$	$+$	0	$-$
$\dfrac{x^2+x-2}{15+2x-x^2}$	$-$	U	$+$	0	$-$	0	$+$	U	$-$

The solution is $(-\infty,-3)\cup(-2,1)\cup(5,+\infty)$

(k) $\dfrac{2x+3-9(x-2)}{x-2}<0$

$\dfrac{2x+3-9x+18}{x-2}=\dfrac{-7x+21}{x-2}=\dfrac{7(3-x)}{x-2}<0$

x		2		3	
$3-x$	$+$	$+$	$+$	0	$-$
$x-2$	$-$	0	$+$	$+$	$+$
$\dfrac{21-7x}{x-2}$	$-$	U	$+$	0	$-$

The solution set is $(-\infty,2)\cup(3,+\infty)$

(l) $|2-3x|<5x-6$

if $5x-6\geq 0$, $5x\geq 6$ and $x\geq 6/5$ and $|2-3x|<5x-6$ is equivalent to
$-5x+6<2-3x<5x-6$.   Thus,
$-5x+6<2-3x$ and $2-3x<5x-6$.
$4<2x$ and $8<8x$
$2<x$ and $1<x$.   Hence, if $x>2$, x is a solution.
If $5x-6<0$, there are no solutions.

The solution set is $(2,+\infty)$

15.
(a) False   (f) True
(b) True   (g) True
(c) False   (h) False
(d) False   (i) False
(e) True

17. $2^5-1=31$

19.

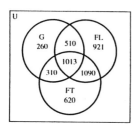

$$260 + 510 + 310 + 1013 + 921 + 1090 + 620 = 4724$$

The pollster polled only 4724 voters and should not be paid.

CHAPTER 1

FUNCTIONS

EXERCISE SET 1.1 FUNCTION

1. (a) The domain is $\{2, 4, 6, 8\}$ (b) $f(2) = y$, $f(6) = z$.

3. (a) The domain is $\{0, 1, 2, 3, 4\}$ (b) $f(1) = 2$, $f(3) = 6$, $f(4) = 8$.

5. (a) The domain is $\{0, 1, 2, 3, 4\}$ (b) $f(0) = 0$, $f(2) = 8$, $f(4) = 64$.

7. (a) The domain is S.   (b) $f(-2) = 4 + 1 = 5$, $f(2) = 4 + 1 = 5$,
$f(6) = 36 + 1 = 37$.

9. (a) The domain is S.   (b) $f(-2.3) = -3$, $f(\pi) = 3$, $f(4.2) = 4$,
$f(9) = 9$.

11.   No. a is paired with both 1 and 4.

13. $\{x \mid x \neq 4\}$

15. $\{x \mid x \neq 4 \text{ or } -4\}$

17. $x^3 - 8 \geq 0 \rightarrow x^3 \geq 8 \rightarrow x \geq 2$.   The domain is $[2, +\infty)$

19. $f(5) = \dfrac{2}{5-4} = 2$, $f(0) = \dfrac{2}{0-4} = -.5$, $f(-3) = \dfrac{2}{-3-4} = -\dfrac{2}{7}$.

21. $h(-2) = \dfrac{-2-2}{4-16} = \dfrac{-4}{-12} = \dfrac{1}{3}$, $h(1) = \dfrac{1-2}{1-16} = \dfrac{-1}{-15} = \dfrac{1}{15}$, $h(5) = \dfrac{5-2}{25-16} = \dfrac{3}{9} = \dfrac{1}{3}$.

23. $g(2) = \sqrt{8-8} = 0$, $g(3) = \sqrt{27-8} = \sqrt{19}$, $g(5) = \sqrt{125-8} = \sqrt{117} = 3\sqrt{13}$.

25. $S(x) = 2x + 3 + x^2 + 3x - 4 = x^2 + 5x - 1$.   The domain is $(-\infty, +\infty)$.

$D(x) = 2x + 3 - x^2 - 3x + 4 = -x^2 - x + 7$.   The domain is $(-\infty, +\infty)$.

$P(x) = (2x+3)(x^2 + 3x - 4) = 2x^3 + 3x^2 + 6x^2 + 9x - 8x - 12$

$= 2x^3 + 9x^2 + x - 12$.   The domain is $(-\infty, +\infty)$.

$Q(x) = \dfrac{2x+3}{x^2 + 3x - 4} = \dfrac{2x+3}{(x+4)(x-1)}$.   The domain is

$(-\infty, -4) \bigcup (-4, 1) \bigcup (1, +\infty)$.

27.

$$S(x) = \frac{2}{x-3} + \frac{3}{x+2} = \frac{2(x+2)+3(x-3)}{(x-3)(x+2)} = \frac{5x-5}{(x-3)(x+2)}.$$

The domains of $S(x)$, $D(x)$ and $P(x)$ are $(-\infty, -2) \cup (-2, 3) \cup (3, +\infty)$.

$$D(x) = \frac{2}{x-3} - \frac{3}{x+2} = \frac{2(x+2)-3(x-3)}{(x-3)(x+2)} = \frac{-x+13}{(x-3)(x+2)}.$$

$$P(x) = \frac{2}{x-3} \cdot \frac{3}{x+2} = \frac{6}{x^2-x-6} \qquad Q(x) = \frac{2/(x-3)}{3/(x+2)} = \frac{2(x+2)}{3(x-3)} = \frac{2x+4}{3x-9}.$$

The domain is the same as for $S(x)$ as $\dfrac{3}{x+2}$ is never zero.

29.

$$S(x) = \frac{x-2}{x+5} + \frac{x+5}{(x-2)(x+2)} = \frac{(x-2)(x^2-4)+(x+5)^2}{(x+5)(x-2)(x+2)}$$

$$= \frac{x^3 - 2x^2 - 4x + 8 + x^2 + 10x + 25}{(x+5)(x-2)(x+2)} = \frac{x^3 - x^2 + 6x + 33}{(x+5)(x^2-4)}.$$

The domains of $S(x)$, $D(x)$, $P(x)$ and $Q(x)$ are

$$(-\infty, -5) \cup (-5, -2) \cup (-2, 2) \cup (2, +\infty).$$

$$D(x) = \frac{(x-2)(x^2-4)-(x+5)^2}{(x+5)(x-2)(x+2)}$$

$$= \frac{x^3 - 2x^2 - 4x + 8 - x^2 - 10x - 25}{(x+5)(x^2-4)} = \frac{x^3 - 3x^2 - 14x - 17}{(x+5)(x^2-4)}.$$

$$P(x) = \frac{(x-2)(x+5)}{(x+5)(x-2)(x+2)} = \frac{1}{x+2}.$$

$$Q(x) = \frac{x-2}{x+5} \cdot \frac{(x-2)(x+2)}{(x+5)} = \frac{(x-2)^2(x+2)}{(x+5)^2}.$$

31.

$$S(x) = \frac{x^2-25}{x+7} + \frac{x-3}{x-5} = \frac{(x^2-25)(x-5)+(x+7)(x-3)}{(x+7)(x-5)}$$

$$= \frac{x^3 - 5x^2 - 25x + 125 - x^2 + 4x - 21}{(x+7)(x-5)} = \frac{x^3 - 4x^2 - 21x + 104}{(x+7)(x-5)}.$$

The domains of $S(x)$, $D(x)$ and $P(x)$ exclude -7 and 5.
The domain of $Q(x)$ excludes -7, 5 and 3.

$$D(x) = \frac{x^3 - 5x^2 - 25x + 125 - x^2 - 4x + 21}{(x+7)(x-5)} = \frac{x^3 - 6x^2 - 29x + 146}{(x+7)(x-5)}.$$

$$P(x) = \frac{(x-5)(x+5)}{(x+7)} \cdot \frac{(x-3)}{x-5} = \frac{(x+5)(x-3)}{x+7} = \frac{x^2 + 2x - 15}{x+7}.$$

$$Q(x) = \frac{(x-5)(x+5)}{x+7} \cdot \frac{(x-5)}{x-3} = \frac{(x-5)^2(x+5)}{(x+7)(x-3)}.$$

33.

x	1	6	8
$f(x)$	2	-1	4
$g(x)$	3	0	5
$S(x)$	5	-1	9
$D(x)$	-1	-1	-1
$P(x)$	6	0	20
$Q(x)$	$\frac{2}{3}$		$\frac{4}{5}$

The domains of $S(x)$, $D(x)$ and $P(x)$ are $\{1, 6, 8\}$.
The domain of $Q(x)$ is $\{1, 8\}$.

35.

x	1	3	5
$f(x)$	4	1	2
$g(x)$	-2	5	0
$S(x)$	2	6	2
$D(x)$	6	-4	2
$P(x)$	-8	5	0
$Q(x)$	-2	.2	

The domains of $S(x)$, $D(x)$ and $P(x)$ are $\{1, 3, 5\}$.
The domain of $Q(x)$ is $\{1, 3\}$.

37.

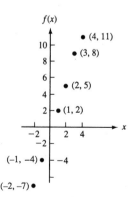

39.

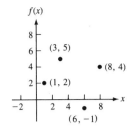

41.

$f(x)$ graph with points $(1, 4)$, $(2, 3)$, $(3, 1)$, $(5, 2)$, $(4, 0)$, $(6, -1)$

43.

$(f \circ g)(x) = f(x^2 + 1) = 2(x^2 + 1) - 3 = 2x^2 - 1$

$(g \circ f)(x) = g(2x - 3) = (2x - 3)^2 + 1 = 4x^2 - 12x + 10$

$(f \circ f)(x) = f(2x - 3) = 2(2x - 3) - 3 = 4x - 9$

$(g \circ g)(x) = g(x^2 + 1) = (x^2 + 1)^2 + 1 = x^4 + 2x^2 + 2$

45.

$(f \circ g)(x) = f(-x^2 + 3x - 2) = (-x^2 + 3x - 2)^2 + 2(-x^2 + 3x - 2) - 3$

$\quad = x^4 - 6x^3 + 13x^2 - 12x + 4 - 2x^2 + 6x - 4 - 3 = x^4 - 6x^3 + 11x^2 - 6x - 3$

$(g \circ f)(x) = g(x^2 + 2x - 3) = -(x^2 + 2x - 3)^2 + 3(x^2 + 2x - 3) - 2$

$\quad = -x^4 - 4x^3 + 2x^2 + 12x - 9 + 3x^2 + 6x - 9 - 2 = -x^4 - 4x^3 + 5x^2 + 18x - 20$

$(f \circ f)(x) = f(x^2 + 2x - 3) = (x^2 + 2x - 3)^2 + 2(x^2 + 2x - 3) - 3$

$\quad = x^4 + 4x^3 - 2x^2 - 12x + 9 + 2x^2 + 4x - 6 - 3 = x^4 + 4x^3 - 8x$

$(g \circ g)(x) = g(-x^2 + 3x - 2) = -(-x^2 + 3x - 2)^2 + 3(-x^2 + 3x - 2) - 2$

$\quad = -x^4 + 6x^3 - 13x^2 + 12x - 4 - 3x^2 + 9x - 6 - 2 = -x^4 + 6x^3 - 16x^2 + 21x - 12$

47.

$(f \circ g)(x) = f\left(|x + 2|\right) = \left||x + 2| - 3\right|$

$(g \circ f)(x) = g\left(|x - 3|\right) = \left||x - 3| + 2\right|$

$(f \circ f)(x) = f\left(|x - 3|\right) = \left||x - 3| - 3\right|$

$(g \circ g)(x) = g\left(|x + 2|\right) = \left||x + 2| + 2\right|$

49.

$h(x) = 5x + 2$ and $g(x) = \sqrt{x}, \quad x \geq -.4$ Then $g \circ h(x) = g(5x + 2) = \sqrt{5x + 2}$

51. $h(x) = 3x + 2$ and $g(x) = x^9$,   Then $g \circ h(x) = g(3x + 2) = (3x + 2)^9$

53. Yes. For each value of x there is a unique second value.

55. Yes. For each value of x there is a unique second value.

EXERCISE SET 1.2 INVERSE OF A FUNCTION

1. If $f(x) = x^4$, $f(2) = f(-2) = 16$,   Hence f is not one-to-one.

3. $h(4) = 7^2 = 49$, $h(-10) = (-7)^2 = 49$.   Hence h is not one-to-one.

5. $g(c) = g(k) = 5$.   Hence g is not one-to-one.

7. (a) Since no two values of x are associated with the same value by the function f, f is a one-to-one function.

(b)

x	-7	1	2	3	5
$f^{-1}(x)$	e	d	a	c	b

9. (a) Since no two values of x are associated with the same value by the function f, f is a one-to-one function.

(b)

x	s	t	u	v	w	y	z
$f^{-1}(x)$	1	2	3	4	5	6	7

11. (a) $f(-3) = 9$, $f(-1) = 1$, $f(0) = 0$, $f(2) = 4$, $f(4) = 16$, $f(5) = 25$, all of which differ. Hence f is one-to-one.

(b)

x	0	1	4	9	16	25
$f^{-1}(x)$	0	-1	2	-3	4	5

13. (a) Suppose $f(x_1) = f(x_2)$. Then $\sqrt{x_1 - 3} = \sqrt{x_2 - 3}$ and $\left(\sqrt{x_1 - 3}\right)^2 = \left(\sqrt{x_2 - 3}\right)^2$.

This implies $x_1 - 3 = x_2 - 3$ or $x_1 = x_2$ since $x_1 \geq 3$ and $x_2 \geq 3$.
Hence f is one-to-one.

(b) Let $y = \sqrt{x - 3}$; interchanging x and y we have $x = \sqrt{y - 3}$, $x^2 = y - 3$, $y = x^2 + 3$.
Thus $f^{-1}(x) = 3 + x^2$ for x in the interval $[0, +\infty)$ which is the range of f.

15. (a) If $f(x_1) = f(x_2)$, $-2x_1 + 7 = -2x_2 + 7$, $-2x_1 = -2x_2$, and $x_1 = x_2$.
Hence f is one-to-one.

(b) Let $y = -2x + 7$. Interchanging x and y we obtain $x = -2y + 7$.
$2y = 7 - x$ $y = \dfrac{7 - x}{2}$. Hence $f^{-1}(x) = \dfrac{7 - x}{2}$.

17. $y = \dfrac{x + 5}{x + 3}$. Interchanging x and y, $x = \dfrac{y + 5}{y + 3}$ $x(y + 3) = y + 5$ $xy + 3x = y + 5$

$xy - y = 5 - 3x$ $y(x - 1) = 5 - 3x$ $y = f^{-1}(x) = \dfrac{5 - 3x}{x - 1}$ if $x \neq 1$.

19. Interchanging x and y, $x = \dfrac{3y + 2}{1 - y}$ $x(1 - y) = 3y + 2$ $x - xy = 3y + 2$

$x - 2 = 3y + xy = y(3 + x)$ $y = h^{-1}(x) = \dfrac{x - 2}{3 + x}$.

21. Interchanging x and y $x = \dfrac{y}{y - 1}$. $x(y - 1) = y$ $xy - x = y$

$xy - y = x$ $(x - 1)y = x$ $y = g^{-1}(x) = \dfrac{x}{x - 1}$
which is the same as the formula for $g(x)$.

23. (a) $B = \{-1, 3, 7, 8, 15\}$

(b)

x	a	b	c	d	e
$f^{-1}(x)$	3	7	-1	8	15

25. (a) $f(x) = x^2 - 4x + 5 = x^2 - 4x + 4 + 1 = (x - 2)^2 + 1$. $f(2 + c) = f(2 - c) = (\pm c)^2 + 1$
$= c^2 + 1$. Hence let $B = \{2 + c \mid c \geq 0\} = [2, +\infty)$

(b) Interchanging x and y, we obtain $x = (y - 2)^2 + 1$ $x - 1 = (y - 2)^2$
$\sqrt{x - 1} = y - 2$ $f^{-1}(x) = y = 2 + \sqrt{x - 1}$, $x \geq 1$.

27. (a) $y = h(x) = x^2 - 10x + 27 = x^2 - 10x + 25 + 2 = (x-5)^2 + 2$ $h(5+c) = h(5-c) = c^2 + 2$
Hence let $B = \{5 + c \mid c \geq 0\} = [5, +\infty)$.

(b) Interchanging x and y, $x = (y-5)^2 + 2$ $x - 2 = (y-5)^2$, $\sqrt{x-2} = y - 5$,
$y = h^{-1}(x) = \sqrt{x-2} + 5$, $x \geq 2$.

29. (a) $y = g(x) = -(x^2 - 4x) - 3 = -(x^2 - 4x + 4) + 1 = -(x-2)^2 + 1$
Since $g(2-c) = g(2+c)$, let $B = \{2 + c \mid c \geq 0\} = [2, \infty)$

(b) Interchanging x and y, $x = -(y-2)^2 + 1$ $(y-2)^2 = -x + 1$ $y - 2 = \sqrt{1-x}$
$y = g^{-1}(x) = 2 + \sqrt{1-x}$

31. (a) $y = f(x) = (x-3)^4 + 2$ $f(3+c) = f(3-c) = c^4 + 2$
Hence let $B = \{3 + c \mid c \geq 0\} = [3, +\infty)$

(b) Interchanging x and y, $x = (y-3)^4 + 2$ $(y-3)^4 = x - 2$ $y - 3 = \sqrt[4]{x-2}$
$f^{-1}(x) = y = 3 + \sqrt[4]{x-2}$

33.

x	-1	0	1	2	3	4	5
$g(x)$	16	9	4	1	0	1	4

By the horizontal line test the function is not one-to-one.

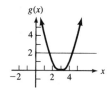

35.

x	-8	-7	-6	-5	-4	-3	-2
$f(x)$	64	49	36	25	16	9	4

x	4	9	16	25	36	49	64
$f^{-1}(x)$	-2	-3	-4	-5	-6	-7	-8

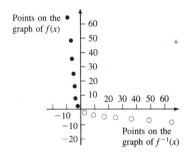

37. $f(x) = \sqrt{x}$ for x in $[4, \infty)$. $f^{-1}(x) = x^2$ for x in $[2, \infty)$

x	4	9	16
$f(x)$	2	3	4

x	2	3	4
$f^{-1}(x)$	4	9	16

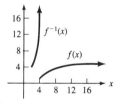

39. $f(x) = 3x - 4$ $f^{-1}(x) = \dfrac{x+4}{3}$

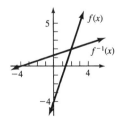

x	0	3	2
$f(x)$	-4	5	2

x	2	-1	5
$f^{-1}(x)$	2	1	3

41. $f(x) = x^3$ $f^{-1}(x) = \sqrt[3]{x}$

x	-2	-1	0	1	2
$f(x)$	-8	-1	0	1	8

x	-8	-1	0	1	8
$f^{-1}(x)$	-2	-1	0	1	2

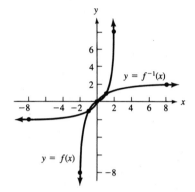

43.   By Exercise 25, restrict the domain of

$f(x) = x^2 - 4x + 5$ to $[2, +\infty)$.   Then $f^{-1}(x) = 2 + \sqrt{x-1}$

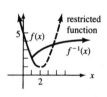

x	0	1	2	3	4
$f(x)$	5	2	1	2	5

x	1	2	5
$f^{-1}(x)$	2	3	4

45. $h(x) = x^2 - 10x + 27.$

By Exercise 27, the restricted domain is

$[5, +\infty)$ and $h^{-1}(x) = 5 + \sqrt{x-2}.$

x	4	5	6	7
$h(x)$	3	2	3	6

x	2	3	6
$h^{-1}(x)$	5	6	7

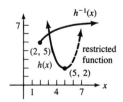

47. $g(x) = -x^2 + 4x - 3.$

By Exercise 29, the restricted domain is

$[2, \infty)$ and $g^{-1}(x) = 2 + \sqrt{1-x}.$

x	0	1	2	3	4
$g(x)$	-3	0	1	0	-3

x	-3	0	1
$g^{-1}(x)$	4	3	2

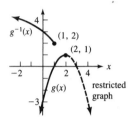

EXERCISE SET 1.3 SOME SPECIAL FUNCTIONS

1.   The slope of the line $m = \dfrac{5-(-3)}{3-(-1)} = \dfrac{8}{4} = 2$

The equation of the line is $y-(-3) = 2\big(x-(-1)\big)$ $y+3 = 2(x+1)$

$y + 3 = 2x + 2$ $y = 2x - 1.$

3.   The slope of the line $m = \dfrac{6-(-2)}{-3-(-5)} = \dfrac{6+2}{-3+5} = \dfrac{8}{2} = 4$

The equation of the line is $y-(-2) = 4\big(x-(-5)\big)$ $y+2 = 4(x+5)$

$y + 2 = 4x + 20$ $y = 4x + 18$

5.   Since the two points have the same second coordinate of 2 the line
is horizontal and has the equation $y = 2.$

7.   By Exercise 2, the slope of the line is -1. The equation of the line
is $y - 8 = -(x-3)$ $y - 8 = -x + 3$ $y = -x + 11.$

9. By Exercise 4, the line that is perpendicular is vertical.

Hence this line is horizontal and has the equation $y = 9$.

11. The perpendicular line has equation $12x - 6y = 36$ Hence $6y = 12x - 36$

$y = 2x - 6$ and its slope is 2. Hence our line has slope $-\frac{1}{2}$

and since $(-7, 0)$ lies on the line its equation is

$$y - 0 = -\frac{1}{2}\left(x - (-7)\right) \qquad 2y = -x - 7 \qquad x + 2y + 7 = 0$$

13. $f(x) = x^2 + 10x - 4$ gives a parabola which is concave up. The vertex

has first coordinate $x = -\frac{10}{2} = -5$. The axis of symmetry is $x = -5$

Some points on the graph are

x	-7	-6	-5	-4
$f(x)$	-25	-28	-29	-28

15. $h(x) = 3x^2 - 18x + 5$ gives a parabola which is concave up. The first

coordinate of the vertex is $x = -\frac{(-18)}{6} = 3$. Some points on the graph are

x	1	2	3	4
$h(x)$	-10	-19	-22	-19

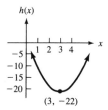

17. $g(x) = -4x^2 + 24x - 9$. The parabola is concave down. The vertex

has first coordinate $x = \frac{-24}{2(-4)} = 3$. Some points on the graph are

x	1	2	3	4
$g(x)$	11	23	27	23

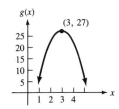

19. (a) Since $f(x) = x^3 + 3x^2 - x - 3$ is a polynomial of degree 3, the maximum number of critical points is 2.

(b) Some points on the graph are

x	-3	-2	-1	0	1	2
$f(x)$	0	3	0	-3	0	15

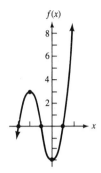

21. (a) Since $f(x)$ is a polynomial of degree 3, the maximum number of critical points is 2.

(b) Some points on the graph are

x	-4	-3	-2	-1	0	1	2
$f(x)$	0	4	0	-6	-8	0	24

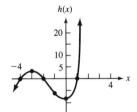

23. (a) Since f is a polynomial of degree 3 it has at most 2 critical points

(b) Some points on the graph are

x	-3	-2	-1	0	1	2	3
$f(x)$	-27	-8	-1	0	1	8	27

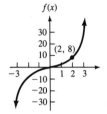

25. (a) Since h is a polynomial of degree 4, it has at most 3 critical points.

(b) Some points on the graph are

x	-2	-1	0	1	2
$h(x)$	16	1	0	1	16

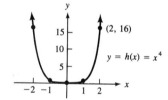

27. (a) $f(x) = \dfrac{x^2 - 3x + 4}{(x+3)(x-1)}$.   The domain excludes 1 and -3.

(b) $f(2) = \dfrac{2^2 - 3 \cdot 2 + 4}{2^2 + 2 \cdot 2 - 3} = \dfrac{4 - 6 + 4}{4 + 4 - 3} = \dfrac{2}{5}$

$f(4) = \dfrac{4^2 - 3 \cdot 4 + 4}{4^2 + 2 \cdot 4 - 3} = \dfrac{16 - 12 + 4}{16 + 8 - 3} = \dfrac{8}{21}$

$f(6) = \dfrac{6^2 - 3 \cdot 6 + 4}{6^2 + 2 \cdot 6 - 3} = \dfrac{36 - 18 + 4}{36 + 12 - 3} = \dfrac{22}{45}$

29. (a) $f(x) = \dfrac{x^2 - 5x + 3}{(x-1)^2(x+1)}$.   The domain excludes 1 and -1.

(b) $f(-2) = \dfrac{(-2)^2 - 5(-2) + 3}{(-2)^3 - (-2)^2 - (-2) + 1} = \dfrac{4 + 10 + 3}{-8 - 4 + 2 + 1} = \dfrac{-17}{9}$

$f(0) = \dfrac{0 - 0 + 3}{0 - 0 - 0 + 1} = \dfrac{3}{1} = 3$ $f(2) = \dfrac{4 - 10 + 3}{8 - 4 - 2 + 1} = \dfrac{-3}{3} = -1$

31. (a) $(f \circ g)(x) = f\big(g(x)\big) = f\big(|x|\big) = [\![\,|x|\,]\!]$

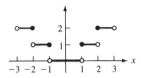

(b) $(g \circ f)(x) = g\big(f(x)\big) = g\big([\![x]\!]\big) = |[\![x]\!]|$

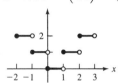

(c) $(f \circ h)(x) = f\big(h(x)\big) = f(x-3) = [\![x-3]\!]$ The function shifts the graph of the greatest integer function to the right 3 units.

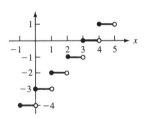

(d) $(h \circ f)(x) = h\big(f(x)\big) = h\big([\![x]\!]\big) = [\![x]\!] - 3$ The function lowers the graph of the greatest integer function by 3 units.

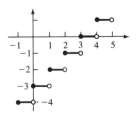

33. $f(x) = (-1)^{\left[\frac{x}{3}\right]}$, $\quad f(0) = (-1)^0 = 1$, $\quad f(1) = (-1)^0 = 1$, $\quad f(2) = (-1)^0 = 1$,

$f(3) = (-1)^1 = -1$, $\quad f(4) = (-1)^1 = -1$, $\quad f(5) = (-1)^1 = -1$,

$f(6) = (-1)^2 = 1$,

$f(7) = (-1)^2 = 1$, $\quad f(8) = (-1)^2 = 1$, $\quad f(9) = (-1)^3 = -1$, $\quad f(10) = (-1)^3 = -1$,

$f(11) = -1$, $\quad f(12) = 1$

35. $h(x) = |x - 2|$. Some points on the graph are

x	-1	0	1	2	3	4
$h(x)$	3	2	1	0	1	2

The function shifts the graph of the absolute value function 2 units to the right.

37.

Hours worked x	Earnings $I(x)$.
$0 \le x \le 40$	$14x$
$40 \le x \le 48$	$40(14) + (x - 40)21 = 21x - 280$
$168 \ge x \ge 48$	$40(14) + 8(21) + (x - 48)28$
	$= 728 + 28x - 1344 = 28x - 616$

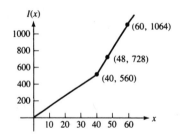

39. (a) $p = mx + b$ $\quad$ $(90, 45)$ and $(135, 30)$ are on the graph.

Hence $m = \dfrac{45 - 30}{90 - 135} = \dfrac{15}{-45} = -\dfrac{1}{3}$

$p - p_1 = m(x - x_1)$

$p - 30 = -\dfrac{1}{3}(x - 135)$

$p - 30 = -\dfrac{1}{3}x + 45$

$p = -\dfrac{1}{3}x + 75$

(b) revenue = no sold · price = $x\left(-\frac{1}{3}x + 75\right)$

$$= -\frac{1}{3}x^2 + 75x$$

(c) profit = revenue − expense = $-\frac{1}{3}x^2 + 75x - (200 + 21x)$

$P(x) = -\frac{1}{3}x^2 + 54x - 200$. The graph is a parabola which is concave

down. Hence $P(x)$ has a maximum when $x = \dfrac{-54}{-\frac{2}{3}} = 27(3) = 81$ razors.

The maximum monthly profit is $-\frac{1}{3}(81)^2 + 54(81) - 200 = \$1,987$

41.

Let x be the length of the side costing

$13 per foot. Cost of fencing is

$$x(13) + (x + 2y)8 = 840 \qquad 21x + 16y = 840$$

$$y = \frac{840 - 21x}{16} = \frac{840}{16} - \frac{21}{16}x$$

$$\text{Area} = xy = x\left(\frac{840}{16} - \frac{21}{16}x\right) = -\frac{21}{16}x^2 + \frac{105}{2}x$$

The graph of this function is a parabola which is concave downward

and has a maximum when $x = \dfrac{-\frac{105}{2}}{-\frac{42}{16}} = \dfrac{105(16)}{42(2)} = 20$. The dimensions of the

rectangle that encloses the largest area are as shown.

43.

tuition/credit hour	credit hours taken
$160	225000
$160 + x$	$225000 - 1250x$

revenue $= (160 + x)(225000 - 1250x) = 36000000 + 25000x - 1250x^2$

The graph is a parabola which is concave downward. Hence the maximum revenue

occurs when $x = \dfrac{-25000}{-2(1250)} = \$10$. Increase tuition by \$10 per credit hour.

45. (a)

Price per paddle (p)	Number sold (x)
$ 15	10
$ 5	30

$p = mx + b$ $(10, 15)$ and $(30, 5)$ are on the graph. Hence $m = \dfrac{15 - 5}{10 - 30}$

$= \dfrac{10}{-20} = -\frac{1}{2}$ $p - p_1 = -\frac{1}{2}(x - x_1)$ $p - 5 = -\frac{1}{2}(x - 30)$

$p = -\frac{1}{2}x + 15 + 5$ $p = -\frac{1}{2}x + 20$

(b) Revenue $= x\left(-\frac{1}{2}x + 20\right) = -\frac{1}{2}x^2 + 20x$

(c) Profit $=$ revenue $-$ expense $= -\frac{1}{2}x^2 + 20x - 25 - 4x = -\frac{1}{2}x^2 + 16x - 25$.

The maximum occurs when $x = \dfrac{-16}{-1} = 16$. Selling 16 paddles will maximize the profit.

EXERCISE SET 1.4 THE EXPONENTIAL FUNCTION

1. $5\left(8^{\frac{2}{3}}\right)\left(9^{\frac{3}{2}}\right) = 5 \cdot \left(\sqrt[3]{8}\right)^2 \cdot \left(\sqrt{9}\right)^3 = 5 \cdot 2^2 \cdot 3^3 = 5 \cdot 4 \cdot 27 = 540.$

3. $7\left(\sqrt[3]{27}\right)^4 / \left(\sqrt{49}\right)^3 = 7(3^4)/7^3 = 3^4/7^2 = 81/49$

5. $\left(\sqrt[4]{16}\right)^{-5}\left(\sqrt[3]{8}\right)^7 / \left(\sqrt[3]{27}\right)^{-5} = \dfrac{2^{-5} \cdot 2^7}{3^{-5}} = 2^2 \cdot 3^5 = 4 \cdot 243 = 972$

7. $\sqrt[5]{59049} \cdot \left(\sqrt[3]{4096}\right)^2 = 9(16)^2 = 2304$

9. $\sqrt[3]{81} \cdot \sqrt[5]{729} / \left(\sqrt[5]{2187}\right)^6 = \left(3^4\right)^{\frac{1}{3}} \cdot \left(3^6\right)^{\frac{1}{5}} / \left(3^7\right)^{\frac{6}{5}} = \dfrac{3^{\frac{4}{3}+\frac{6}{5}}}{3^{\frac{42}{5}}}$

$= 3^{\frac{4}{3}+\frac{6}{5}-\frac{42}{5}} = 3^{\frac{20+18-126}{15}} = 3^{-\frac{88}{15}}$

11. $f(x) = 3^x$. Some of the points on the graph are

x	-2	-1	0	1	2
$f(x)$	$\frac{1}{9}$	$\frac{1}{3}$	1	3	9

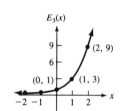

13. $f(x) = \left(\frac{1}{2}\right)^x$. Some points on the graph are

x	-3	-2	-1	0	1	2
$f(x)$	8	4	2	1	$\frac{1}{2}$	$\frac{1}{4}$

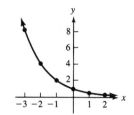

15. $f(x) = \left(\frac{1}{4}\right)^x$. Some points on the graph are

x	-2	-1	0	1	2
$f(x)$	16	4	1	$\frac{1}{4}$	$\frac{1}{16}$

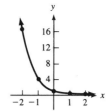

17. (a) $\left(h \circ g\right)(x) = h(2x - 3) = 3^{2x-3}$. Some points on the graph are

x	-1	0	1	2	3
$\left(h \circ g\right)(x)$	$\frac{1}{243}$	$\frac{1}{27}$	$\frac{1}{3}$	3	27

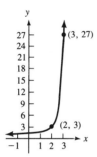

(b) $\left(g \circ h\right)(x) = g\left(3^x\right) = 2\left(3^x\right) - 3$. Some points on the graph are

x	-1	0	1	2
$g \circ h(x)$	$-\frac{7}{3}$	-1	3	15

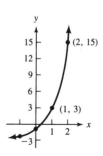

19. $A = 5000(1 + .08/4)^{9 \cdot 4} = 5000(1.02)^{36} = \$10,199.44$. The interest is earned is $\$10,199.44 - \$5000 = \$5199.44$.

21. $A = 8000\left(1 + \frac{.09}{12}\right)^{28} = \$9861.69$. Compound interest earned is $\$9861.69 - \$8000 = \$1,861.69$.

23. The amounts of the deposit at the three banks after 3 months are:

Bank A: $A = 3000\left(1 + \frac{.0795}{365}\right)^{90} = \$3060.21$

Bank B: $A = 3000\left(1 + \frac{.08}{12}\right)^{3} = \$3060.40$

Bank C: $A = 3000\left(1 + \frac{.081}{4}\right) = \$3060.75$

Hence choose Bank C for greatest interest.

25. Effective interest rate $= 1\left(1 + \frac{.08}{4}\right)^{4} - 1 = (1.02)^4 - 1 = .0824$. Hence \$1 grows to \$1.0824 in one year and the effective interest rate is 8.24%.

27. If \$1 is invested for one year, it grows to $1\left(1 + \frac{.07}{365}\right)^{365} = \$1.0725$. Hence the effective interest rate is 7.25%.

29. $A = 4000 e^{(.09)(15)} = 4000(3.8574) = \$15,429.70$

31. $P = 75000(1 + .025)^{10} = 75000(1.28008) = 96006$

33. $P = 250000(1.035)^{18} = 250000(1.8575) = 464,372$

35. Since growth is exponential and the initial population is 150,000, the population after t hours is given by $P(t) = 150,000 b^t$ where b is a constant. Since the population after 23 hours is 250,000, $250,000 = 150,000 b^{23}$ $\frac{25}{15} = \frac{5}{3} = b^{23}$, $b = \sqrt[23]{\frac{5}{3}}$ and $P(t) = 150,000\left(\frac{5}{3}\right)^{\frac{t}{23}}$ $P(86) = 150,000\left(\frac{5}{3}\right)^{\frac{74}{23}} = 1,013,007$

37. $P(t) = P_0 b^t$. We let $P_0 = 5$ so $P(t) = 5b^t$. But $P(5730) = 2.5$ and $2.5 = 5\left(b^{5730}\right)$ Hence $(.5)^{\frac{1}{5730}} = b$ $P(t) = 5(.5)^{\frac{t}{5730}}$ $P(17190) = 5(.5)^{\frac{17190}{5730}} = 5(.5)^3 = .625\,\text{gm}.$

39. $P(t) = P_0 b^t$. Let $t = 0$ at the beginning of 1975. Then $P_0 = 3$ and $P(t) = 3b^t$. We know that $P(12.3) = \frac{3}{2} = 1.5$. Hence $1.5 = 3b^{12.3}$, $\frac{1}{2} = b^{12.3}$, $(.5)^{\frac{1}{12.3}} = b$. $P(t) = 3\left((.5)^{\frac{1}{12.3}}\right)^t$

or $P(t) = 3(.5)^{\frac{t}{12.3}}$ If observation begins in 1975 at the end

of year 2000 there will be $P(t) = 3(.5)^{\frac{26}{12.3}} = .6931$ gm.

41. (a) $W(t) = 110\left(1 - e^{-.2(3)}\right) = 49.63$ words/minute

     (b) $W(t) = 110\left(1 - e^{-.2(6)}\right) = 76.87$ words/minute

     (c) $W(t) = 110\left(1 - e^{-.2(12)}\right) = 100.02$ words/minute

43.   If the value is declining at the rate of 15% per year, the value
          after t years is $V(t) = 30{,}000(1 - .15)^t = 30{,}000(.85)^t$

     (a) $30000(1 - .15)^2 = 30000(.85)^2 = \$21{,}675$

     (b) $V(4) = 30000(.85)^4 = \$15{,}660$

     (c) $V(8) = 30000(.85)^8 = \$8{,}175$

45. (a) $15{,}000(1 - .18)^4 = 15{,}000(.82)^4 = \$6{,}782$

     (b) $15{,}000(.82)^6 = \$4{,}560$

     (c) $15{,}000(.82)^8 = \$3{,}066$

47.

m	2	5	15	50	150	500	1500
$\left(1 + \frac{1}{m}\right)^m$	2.25	2.48832	2.63288	2.69159	2.70928	2.71557	2.71738

EXERCISE SET 1.5 THE LOGARITHMIC FUNCTION

1. (a) $\log 64 = \log 2^6 = 6 \log 2 = 6(.3010) = 1.8060$

    (b) $\log 36 = \log 2^2 \cdot 3^2 = \log 2^2 + \log 3^2 = 2 \log 2 + 2 \log 3$

         $= 2(.3010) + 2(.4771) = 1.5562$

    (c) $\log 54 = \log 27(2) = \log 3^3 + \log 2 = 3 \log 3 + \log 2 = 3(.4771) + .3010 = 1.7323$

3. (a) $\log 4^{.2} = .2 \log 4 = .2 \log 2^2 = .4 \log 2 = .4(.3010) = .1204$

(b) $\log \sqrt[4]{27} = \frac{1}{4} \log 3^3 = \frac{3}{4} \log 3 = \frac{3}{4}(.4771) = .357825$

(c) $\log \sqrt[7]{.75} = \frac{1}{7} \log\left(\frac{3}{4}\right) = \frac{1}{7}(\log 3 - \log 2^2) = \frac{1}{7}(.4771 - 2(.3010)) = -.0178428$

5. (a) $\log_3 2 = \dfrac{\log 2}{\log 3} = \dfrac{.3010}{.4771} = .6309$

(b) $\log_4 27 = \dfrac{\log 27}{\log 4} = \dfrac{\log 3^3}{\log 2^2} = \dfrac{3 \log 3}{2 \log 2} = \dfrac{3(.4771)}{2(.3010)} = 2.3776$

(c) $\log_9 64 = \dfrac{\log 64}{\log 9} = \dfrac{\log 2^6}{\log 3^2} = \dfrac{6 \log 2}{2 \log 3} = \dfrac{6(.3010)}{2(.4771)} = 1.8927$

7. (a) $\ln(1.5) = \ln\left(\frac{3}{2}\right) = \ln 3 - \ln 2 = 1.0986 - .6931 = .4055$

(b) $\ln .75 = \ln\left(\frac{3}{4}\right) = \ln 3 - \ln 2^2 = \ln 3 - 2 \ln 2 = 1.0986 - 2(.6931) = -.2876$

(c) $\ln\left(\frac{27}{16}\right) = \ln 27 - \ln 16 = \ln 3^3 - \ln 2^4$

$= 3 \ln 3 - 4 \ln 2 = 3(1.0986) - 4(.6931) = .5234$

9. (a) $\ln 5e^3 = \ln 5 + \ln e^3 = \ln 5 + 3 \ln e = \ln 5 + 3 \cdot 1 = 3 + \ln 5$

(b) $\ln \sqrt[3]{e^2} = \ln e^{\frac{2}{3}} = \left(\frac{2}{3}\right)\ln e = \frac{2}{3}(1) = \frac{2}{3}$

(c) $\ln\left(\frac{27}{e}\right) = \ln 27 - \ln e = 3 \ln 3 - 1 = 3(1.0986) - 1 = 2.2958$

11. $\log_3(2x+3) = 4.$ $2x + 3 > 0$ if $2x > -3$ or $x > -1.5$.

$3^{\log_3(2x+3)} = 3^4,$ $2x + 3 = 81$ $2x = 78$

$x = 39$ which is larger than -1.5 The solution set is $\left\{39\right\}$.

13. $\log_5(5 - 3x) = 2.$ $5 - 3x > 0$ if $5 > 3x$ or $x < \frac{5}{3}.$ $5^{\log_5(5-3x)} = 5^2$

$5 - 3x = 25;$ $-20 = 3x$ $x = -\frac{20}{3}$ which is less than $\frac{5}{3}$. The solution set

is $\left\{-\frac{20}{3}\right\}$.

15. $\log_2(5 - 6x) = 3 + \log_2(3x - 7)$ $5 - 6x > 0$ if $5 > 6x$ or $x < \frac{5}{6}$

$3x - 7 > 0$ if $3x > 7$ or $x > \frac{7}{3}$ Since no number is greater than $\frac{7}{3}$

and less than $\frac{5}{6}$, there are no solutions to the equation.

17. $\log_{12}(x-2) = 1 - \log_{12}(x-3)$ $x-2 > 0$ if $x > 2$ $x-3 > 0$ if $x > 3$.

Hence any solution must be larger than 3. $\log_{12}(x-2)(x-3) = 1$

$(x-2)(x-3) = 12^1$, $(x-2)(x-3) = 12$ $x^2 - 5x - 6 = 0$

$(x-6)(x+1) = 0$ if $x = 6$ or $x = -1$. -1 is not a solution and

the solution set is $\left\{6\right\}$.

19. $\log_2(x+7) = 2 - \log_2(x+4)$ $x+7 > 0$ if $x > -7$ $x+4 > 0$ if $x > -4$

Hence any solution must be larger than -4. $\log_2(x+7) + \log_2(x+4) = 2$

$\log_2(x+7)(x+4) = 2$ $(x+7)(x+4) = 2^2$

$(x+7)(x+4) = 4$ $x^2 + 11x + 28 = 4$ $x^2 + 11x + 24 = 0$

$(x+3)(x+8) = 0$ $x = -3$ or $x = -8$.   Hence -3 is the only solution

since $-8 < -4$.

21. $\log|x+4| + \log|x+1| = 1$, $x \neq -4$ and $x \neq -1$.

$\log|x+4||x+1| = 1$ $|x+4||x+1| = 10$ $|(x+4)(x+1)| = 10$

$|x^2 + 5x + 4| = 10$ $x^2 + 5x + 4 = 10$ or $x^2 + 5x + 4 = -10$

$x^2 + 5x - 6 = 0$ or $x^2 + 5x + 14 = 0$ $(x+6)(x-1) = 0$ or

$x = \dfrac{-5 \pm \sqrt{25 - 56}}{2}$ which are not real. $x = -6$ or $x = 1$ are the solutions.

23. $x > 0$, $x \neq 1$, and $x > -12$. $x + 12 = x^2$

$x^2 - x - 12 = 0$ $(x-4)(x+3) = 0$ $x = 4$ or -3 4 is the only

solution as -3 can not be a base for a logarithmic function.

25. $x \neq 1$ and $3 + x - 3x^2 > 0$,   and $x > 0$. $3 + x - 3x^2 = x^3$

$x^3 + 3x^2 - x - 3 = x^2(x+3) - (x+3) = (x^2 - 1)(x+3)$

$= (x-1)(x+1)(x+3) = 0$ $x = 1$, $x = -3$ or $x = -1$ However none

of these can serve as bases for logarithms. So there are no solutions.

27. $3^{2x-1} = 2e^{3x+5}$ $\ln 3^{2x-1} = \ln\left(2e^{3x+5}\right)$ $(2x-1)\ln 3 = \ln 2 + (3x+5)\ln e$

$2x \ln 3 - \ln 3 = \ln 2 + 3x + 5$ $2x \ln 3 - 3x = \ln 2 + \ln 3 + 5$

$x(2 \ln 3 - 3) = 5 + \ln 6$ $x = \dfrac{5 + \ln 6}{2 \ln 3 - 3} = \dfrac{5 + \ln 6}{-3 + \ln 9} = -8.4603$

29. $A = P(1 + \frac{.08}{4})^t$ where t the number of quarters since the deposit
was made. $3P_0 = P_0(1.02)^t$ $3 = (1.02)^t$ $\ln 3 = \ln(1.02)^t = t \ln(1.02)$
$t = \dfrac{\ln 3}{\ln(1.02)} = 55.48$ quarters. Thus 56 quarters are needed.

31. $3P_0 = P_0(1.005)^t$ where t is the number of months since the deposit
was made. $3 = (1.005)^t$ $\ln 3 = t \ln(1.005)$ $t = \dfrac{\ln 3}{\ln(1.005)} = 220.27$ months.
221 months are needed.

33. $3P_0 = P_0\, e^{.08t}$ where t is the number of years since the deposit was
made. $3 = e^{.08t}$ $\ln 3 = .08t \ln e,$ $t = \dfrac{\ln 3}{.08} = 13.73$ years.

35. $P = P_0(1 + .30)^t$ $60000 = 20000(1.3)^t$ $3 = (1.3)^t$ $\dfrac{\ln 3}{\ln (1.3)} = t = 4.187.$
Thus the population will reach 60000 in March of 1991.

37. $W(t) = 110\left(1 - e^{-.2t}\right).$ We want t such that $W(t) = 80 = 110\left(1 - e^{-.2t}\right)$
$\frac{8}{11} = 1 - e^{-.2t}$ $e^{-.2t} = \frac{3}{11}$ $\ln\left(e^{-.2t}\right) = \ln\left(\frac{3}{11}\right)$ $-.2t \ln e = \ln\left(\frac{3}{11}\right)$
$t = \dfrac{\ln\left(\frac{3}{11}\right)}{-.2} = 6.496$ weeks. The student should attend school $6\frac{1}{2}$ weeks.

39. $P(t) = P_0\, b^t.$ Since the population grew to 195,000 in 15 years,
$195 = 150b^{15}.$ $\dfrac{195}{150} = b^{15}$ and $\left(\dfrac{195}{150}\right)^{\frac{1}{15}} = b.$ $P(t) = 150{,}000\left(\dfrac{195}{150}\right)^{\frac{t}{15}}.$
We want to find t such that $P(t) = 225000 = 150000\left(\dfrac{195}{150}\right)^{\frac{t}{15}}.$
$\dfrac{225}{150} = \left(\dfrac{195}{150}\right)^{\frac{t}{15}}.$ $\ln\left(\dfrac{225}{150}\right) = \ln\left(\dfrac{195}{150}\right)^{\frac{t}{15}}.$ $\ln\left(\dfrac{225}{150}\right) = \left(\dfrac{t}{15}\right)\ln\left(\dfrac{195}{150}\right).$
$t = \dfrac{15\ln\left(\frac{225}{150}\right)}{\ln\left(\frac{195}{150}\right)} = \dfrac{6.082}{.262} = 23.18$ years or in the year 1993.

41. $P = P_0\, b^t$ Since the half life is 1690 years, $.5 = 1 \cdot b^{1690}$ $b = (.5)^{\frac{1}{1690}}.$
Since we want $\frac{3}{4}$ of the mass to remain, $.75 = 1(.5)^{\frac{t}{1690}}$
$\ln(.75) = \ln\left(.5^{\frac{t}{1690}}\right) = \left(\dfrac{t}{1690}\right)\ln(.5)$ $t = \dfrac{1690 \ln(.75)}{\ln(.5)} = 701.4$ years

43. Let $z = y \log_b x$, $x > 0$. $b^z = b^{y \log_b x} = \left(b^{\log_b x}\right)^y$ [by property 9 of

exponential functions] $= x^y$ [by property 3 of logarithmic functions]. Hence

$\log_b x^y = \log_b b^z = z$ [by property 9 of logarithmic functions] and $z = y \log_b x = \log_b x^y$.

45. Let $x = \log_a b$, $y = \log_b c$, $z = \log_c d$ and $w = \log_a d$. Then

$a^x = b$, $b^y = c$, $c^z = d$ and $a^w = d$. Since $a^x = b$ and $b^y = c$,

$\left(a^x\right)^y = a^{xy} = c$. $c^z = \left(a^{xy}\right)^z = a^{xyz} = d = a^w$. Equating exponents

$w = xyz$ or $\log_a d = \log_a b \log_b c \log_c d$, for $a,b,c,d > 0$ and a,b and $c \neq 1$.

A generalization is $\log_a f = (\log_a b)(\log_b c)(\log_c d)(\log_d f)$

for $a,b,c,d,f > 0$ and a,b,c and $d \neq 1$.

47. $x = \dfrac{72}{5.5} = 13.1$ years

49. $x = \dfrac{72}{7.3} = 9.9$ years

EXERCISE SET 1.6 MORE ON GRAPHS

1. 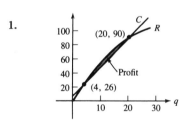 The graphs of C and R intersect at the points $(4, 26)$ and $(20, 90)$. $R > C$ when the graph of R lies above the graph of C. This occurs when $4 < q < 20$, so the company operates at a profit.

3. C_1 is the cost of option 1 and C_2 is the cost of option 2.

$C_1 = 30 + .4m$ and $C_2 = 40 + .35m$ where m is the number of miles driven.

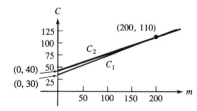 The graphs intersect at point $(200, 110)$.
(a) Choose option 1 if less than 200 miles are driven, (b) Choose option 2 if more than 200 miles are driven.

5. $C = 4q + 10 + 12 = 4q + 22$ is the new cost function.

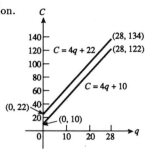

7. The previous revenue function was $R = -1.2q^2 + 60\,q$.

The new revenue function is $R_1 = -1.2(q+1)^2 + 60(q+1)$.

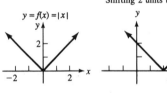

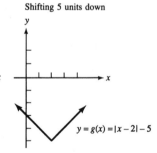

9.

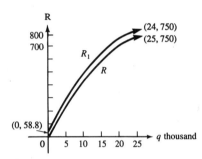

Shifting 2 units to the right

Shifting 5 units down

$y = g(x) = |x - 2| - 5$

11.

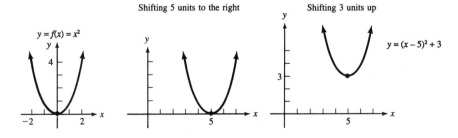

Shifting 5 units to the right

Shifting 3 units up

$y = f(x) = x^2$

$y = (x - 5)^2 + 3$

13.

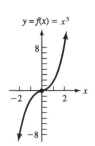

$y = f(x) = x^3$

Shifting 2 units left

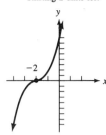

y

Shifting 3 units down

$y = g(x) = (x+2)^3 - 3$

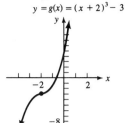

15.

$y = f(x) = [x]$

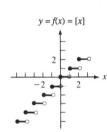

Shifting 5 units to the right

y

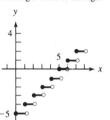

Shifting 2 units up

$y = g(x) = [x-5] + 2$

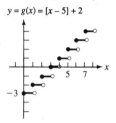

17. $g(x) = |x+1| - 3$ as the graph is shifted 1 unit to the left and then down 3 units.

19. $g(x) = (x+2)^2 + 5$ as the graph is shifted 2 units to the left and up 5 units.

21. $g(x) = x^2 - 4x + 5 = x^2 - 4x + 4 + 1$

$g(x) = (x-2)^2 + 1$

Shifting the graph of g two units to the right
and then up one unit we obtain

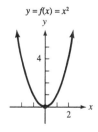

$y = f(x) = x^2$

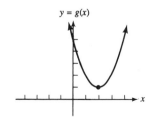

$y = g(x)$

23. $g(x) = x^3 + 3x^2 + 3x + 6$

$= x^3 + 3x^2 + 3x + 1 + 5 = (x+1)^3 + 5$

Shifting the graph one unit to the left
and up five units we obtain

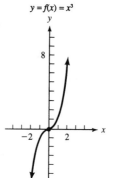

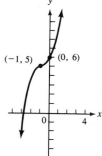

25. $g(x) = x^4 + 8x^3 + 24x^2 + 32x + 19$

$= (x+2)^4 + 3$

Shifting the graph two units to the left
and up three units we obtain

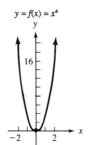

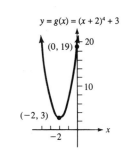

27.   Unemployment was increasing in February through May, November and December,
and decreasing in other months.

29.

The graphs of $f(x)$ and $-f(x)$ are mirror
images of each other across the x-axis.

31.

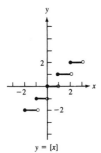

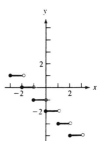

$y = [x]$ $f(x) = [x - 3] + 5$ $-f(x) = -[x - 3] - 5$

EXERCISE SET 1.7 THE SIGMA NOTATION

1. $\displaystyle\sum_{x \varepsilon A} f(x) = f(1) + f(3) + f(7) + f(10)$

$\qquad = 4 + 10 + 22 + 31 = 67$

3. $\displaystyle\sum_{x \varepsilon A} f(x) = f(-2) + f(-1) + f(0) + f(3) + f(6)$

$\qquad = 5 + 2 + 1 + 10 + 37 = 55$

5. $\displaystyle\sum_{x \varepsilon A} f(x) = f(-4) + f(0) + f(2) + f(4) + f(5) + f(9)$

$\qquad = -64 + 0 + 8 + 64 + 125 + 729 = 862$

7. $\displaystyle\sum_{x \varepsilon A} f(x) = \sum_{i=-4}^{4} i^3 = -64 - 27 - 8 - 1 + 0 + 1 + 8 + 27 + 64 = 0$

9. $\displaystyle\sum_{i=2}^{6} (3i + 5) = 11 + 14 + 17 + 20 + 23 = 85$

11. $\displaystyle\sum_{i=5}^{8} (i^2 + 3i - 2) = (25 + 15 - 2) + (36 + 18 - 2) + (49 + 21 - 2) + (64 + 24 - 2)$

$\qquad = 38 + 52 + 68 + 86 = 244$

13. $\displaystyle\sum_{i=-2}^{2} (i^2 + 1) = 5 + 2 + 1 + 2 + 5 = 15$

15. $\displaystyle\sum_{i=1}^{150} i^2 = \frac{150(150 + 1)(2(150) + 1)}{6} = \frac{150(151)(301)}{6} = 1,136,275$

17. $\displaystyle\sum_{i=41}^{160} i = \sum_{i=1}^{160} i - \sum_{i=1}^{40} i = \frac{(160)(161)}{2} - \frac{40(41)}{2} = 12880 - 820 = 12060$

19. $\displaystyle\sum_{i=61}^{170} i^3 = \sum_{i=1}^{170} i^3 - \sum_{i=1}^{60} i^3 = \frac{(170)^2(171)^2}{4} - \frac{60^2(61)^2}{4} = 207,917,325$

21. $\displaystyle\sum_{i=1}^{120} (2i^3 - 5i^2 + 6i - 7) = 2\sum_{i=1}^{120} i^3 - 5\sum_{i=1}^{120} i^2 + 6\sum_{i=1}^{120} i - \sum_{i=1}^{120} 7$

$\qquad = \frac{2(120)^2(121)^2}{4} - \frac{5(120)(121)(241)}{6} + \frac{6(120)(121)}{2} - 7(120) =$

$\qquad 105415200 - 2916100 + 43560 - 840 = 102,541,820$

23. $\displaystyle\sum_{i=0}^{185} -9 = -9(186) = -1674$

25. Terms of an arithmetic progression are being added with a first term of 1 and common difference of 2. Hence the i th term is

$$1 + 2(i-1) = 2i - 1 \text{ and the sum is } \sum_{i=1}^{9} (2i-1).$$

27. Terms of an arithmetic progression are being added with a first term of 5 and common difference of 3. Hence the i th term is

$$5 + (i-1)3 = 2 + 3i \text{ and the sum is } \sum_{i=1}^{7} (2+3i).$$

29. $\displaystyle 1^2 + 2^2 + 3^2 + 4^2 + 5^2 + 6^2 + 7^2 + 8^2 = \sum_{i=1}^{8} i^2$

31. $\displaystyle 2^3 + 3^3 + 4^3 + 5^3 + 6^3 = \sum_{i=2}^{6} i^3$

33. $(-1)^2 + (-1)^3 + (-1)^4 + (-1)^5 + (-1)^6 + (-1)^7 + (-1)^8$

$$+ (-1)^9 + (-1)^{10} + (-1)^{11} = \sum_{i=2}^{11} (-1)^i$$

35. $\displaystyle 1^2 - 2^2 + 3^2 - 4^2 + 5^2 - 6^2 + 7^2 - 8^2 + 9^2 = \sum_{i=1}^{9} i^2 (-1)^{i+1}$

EXERCISE SET 1.8 CHAPTER REVIEW

1. (a) $\{a, b, c, d\}$ (b) $\{3, 9, 1\}$ (c) f is not one-to-one as $f(a) = 3 = f(d)$

3. (a) The domain is the set of all real numbers.

  (b) The range is the set of all real numbers.

  (c) Graphing $y = f(x) = x^3 - 2$, some points on the graph are

x	-2	-1	0	1	2
$f(x)$	-10	-3	-2	-1	6

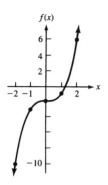

The function is one-to-one by the horizontal line test.

To find the inverse interchange x and y and solve for y.

$x = y^3 - 2$ $y^3 = x + 2$, $y = \sqrt[3]{x+2}$ and $f^{-1}(x) = \sqrt[3]{x+2}$

5. (a) The domain is the set of all humans.

(b) The range is the set of all human females who have had a child.

(c) The function is not one-to-one as some females have had more than one child.

7. $x - 2 \geq 0$ if $x \geq 2$. The domain is $[2, +\infty)$

9. $x^2 + 7x - 18 = (x+9)(x-2) > 0$ if $x > 2$ or $x < -9$.

The domain is $(-\infty, -9) \cup (2, +\infty)$

11. $x - 3 \neq 0$ if $x \neq 3$. The domain is $(-\infty, 3) \cup (3, +\infty)$

13. $S(x) = x^2 + x - 2 + \dfrac{x+3}{x-1} = \dfrac{(x^2 + x - 2)(x - 1) + (x + 3)}{x - 1}$

$= \dfrac{x^3 + x^2 - 2x - x^2 - x + 2 + x + 3}{x - 1} = \dfrac{x^3 - 2x + 5}{x - 1}$.

The domains of $S(x)$, $D(x)$ and $P(x)$ are $(-\infty, 1) \cup (1, +\infty)$.

$D(x) = x^2 + x - 2 - \dfrac{x+3}{x-1} = \dfrac{x^3 + x^2 - 2x - x^2 - x + 2 - x - 3}{x - 1} = \dfrac{x^3 - 4x - 1}{x - 1}$

$P(x) = (x+2)(x-1) \cdot \dfrac{x+3}{x-1} = (x+2)(x+3) = x^2 + 5x + 6$

$Q(x) = (x+2)(x-1) \cdot \dfrac{(x-1)}{(x+3)} = \dfrac{(x-1)^2(x+2)}{(x+3)}$.

The domain of $Q(x)$ is $(-\infty, -3) \cup (-3, 1) \cup (1, +\infty)$

15. $(f \circ g)(x) = f(g(x)) = f(2x+3) = (2x+3)^2 + 2(2x+3) - 3 = 4x^2 + 16x + 12.$

$(g \circ f)(x) = g(f(x)) = g(x^2 + 2x - 3) = 2(x^2 + 2x - 3) + 3 = 2x^2 + 4x - 3.$

17. $f(x) = 2x + 3$ is a linear function. The graph is a line with slope 2 and y intercept 3. Two points on the line are

x	0	-1
$f(x)$	3	1

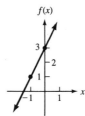

19. $h(x) = -3x^2 + 24x - 33$ is a quadratic function with graph a concave down parabola. The vertex occurs when $x = \dfrac{-24}{-6} = 4.$

Points on the graph include

x	3	4	5
$h(x)$	12	15	12

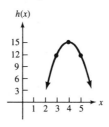

21.

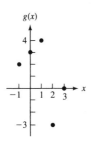

23. Let $h(x) = 2x + 3$ and $g(x) = x^4$. Then $(g \circ h)(x) = g(2x+3) = (2x+3)^4 = f(x).$

25. (a) The graph of g is shown.

By the horizontal line test g is one-to-one.

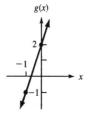

(b) Interchanging x and y $x = 3y + 2$ $3y = x - 2$ $y = \frac{1}{3}x - \frac{2}{3} = g^{-1}(x)$

(c)

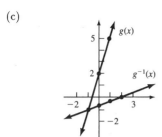

x	-1	0	1
$g(x)$	-1	2	5

x	0	1	2	-1
$g^{-1}(x)$	$-\frac{2}{3}$	$-\frac{1}{3}$	0	-1

27. (a) The slope is $\dfrac{4-2}{3-(-1)} = \dfrac{2}{4} = \dfrac{1}{2}$. The equation is $(y-2) = \frac{1}{2}(x-(-1))$

$y - 2 = \frac{1}{2}(x+1)$ $y = \frac{1}{2}x + \frac{1}{2} + 2$ $y = \frac{1}{2}x + \frac{5}{2}$

(b) If $2x + 3y = 5$, $3y = 5 - 2x$, $y = \frac{5}{3} - \frac{2}{3}x$. The slope of the line is $-\frac{2}{3}$.
The equation is $y - 5 = -\frac{2}{3}(x-2)$.

(c) If $x + 2y = 5$, $2y = 5 - x$, $y = \frac{5}{2} - \frac{x}{2}$. The slope of the desired line is

$-\dfrac{1}{-\frac{1}{2}} = 2$. The equation is $y - 3 = 2(x+1)$ $y = 2x + 5$.

(d) The line is vertical and has equation $x = 1$.

(e) The line is horizontal and has equation $y = 4$.

29.   Some points on the graph are

x	-2	-1	0	1	2
$f(x)$	9	3	1	$\frac{1}{3}$	$\frac{1}{9}$

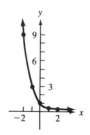

31. $A = 700\left(1 + \frac{.06}{12}\right)^{36} = 700(1.005)^{36} = \$837.68.$ Hence the interest earned is $137.68.

33. (a) $\log 6 = \log 2 + \log 3 = .3010 + .4771 = .7781$

(b) $\log 8 = \log 2^3 = 3 \log 2 = 3(.3010) = .9030$

(c) $\log \sqrt{3} = \frac{1}{2} \log 3 = \frac{1}{2}(.4771) = .2385$

(d) $\log \frac{4}{27} = \log 4 - \log 27 = 2 \log 2 - 3 \log 3 = 2(.3010) - 3(.4771) = -.8293$

(e) $\log 250 = \log (2 \cdot 125) = \log 2 + \log 5^3 = \log 2 + 3 \log 5 = .3010 + 3 \log \frac{10}{2}$

$\quad = .3010 + 3 (\log 10 - \log 2) = .3010 + 3 (1 - .3010) = .3010 + 3 - .9030 = 2.398$

(f) $\log_2 10 = \frac{\log 10}{\log 2} = \frac{1}{.3010} = 3.3222$

35. $4P = P\left(1 + \frac{.06}{12}\right)^t$ where t is the number of months needed. $4 = (1.005)^t$

$\ln 4 = t \ln(1.005). \quad \frac{\ln 4}{\ln(1.005)} = t = 277.95$ months ~ 23.2 years.

37. (a) $\frac{20(21)(41)}{6} = 2870$

(b) $5(30) = 150$

(c) $\sum_{i=1}^{100} \left(\frac{1}{i} - \frac{1}{(i+1)}\right) = \left(1 - \frac{1}{2}\right) + \left(\frac{1}{2} - \frac{1}{3}\right) + \left(\frac{1}{3} - \frac{1}{4}\right) + \cdots + \left(\frac{1}{100} - \frac{1}{101}\right)$

$\quad = 1 - \frac{1}{101} = \frac{100}{101}$

39.

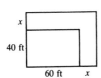

Let x be the width of the strips added.

new area $= 4800 = (60 + x)(40 + x)$ $4800 = 2400 + 100x + x^2$

$0 = x^2 + 100x - 2400 \quad x = \frac{-100 \pm \sqrt{10000 + 9600}}{2} = \frac{-100 \pm \sqrt{19600}}{2}$

$\quad = \frac{-100 \pm 140}{2} = -120\,\text{ft. or } 20\,\text{ft.}$ The width of the strip must be 20 feet.

41.

Quantity sold q	Price p
40	$ 90
80	$ 70

(a) The slope is $\frac{90 - 70}{40 - 80} = -\frac{1}{2}.$

The equation is $p - 90 = -\frac{1}{2}(q - 40)$ or $p = -\frac{1}{2}q + 110$

(b) $R(q) = qP(q) = q\left(-\frac{1}{2}q + 110\right).$ $R(q) = -\frac{1}{2}q^2 + 110q$

(c) Profit $= R(q) - C(q) = -\frac{1}{2}q^2 + 110q - (60q + 250)$. Profit $= -\frac{1}{2}q^2 + 50q - 250$.

The profit is a maximum when $q = \dfrac{-50}{2(-.5)} = 50$ racquets.

The price per racquet is $-\dfrac{50}{2} + 110 = \$\,85$.

The maximum profit is $-\frac{1}{2}(50)^2 + 50(50) - 250 = \$\,1,000$.

43.

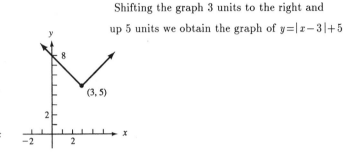

Shifting the graph 3 units to the right and up 5 units we obtain the graph of $y = |x - 3| + 5$

45.

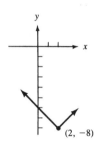

The graph is shifted 2 units to the right and down 8 units. $f(x) = |x - 2| - 8$

CHAPTER 2

THE DERIVATIVE

EXERCISE SET 2.1   RATE OF CHANGE

1. $\$5,600 - \$3,200 = \$2,400$

3. $\Delta x = 8-2 = 6 \qquad \Delta y = f(8) - f(2) = -38 - (-8) = -30$

5. $\Delta x = -2-4 = -6 \qquad \Delta y = f(-2)-f(4) = -13-11 = -24$

7. $\Delta x = 7-2 = 5 \qquad \Delta y = f(7)-f(2) = \sqrt{9}-\sqrt{4} = 3-2 = 1$

9. $\Delta x = e^3-e \sim 17.37 \qquad \Delta y = f(e^3)-f(e) = 5 \ln e^3-5 \ln e = 15-5 = 10$

11. $\Delta x = -1-2 = -3 \qquad \Delta y = f(-1)-f(2) = -3^{-1} + 3^2 = -1/3 + 9 = 26/3$

13. $\Delta t = 6-3 = 3$ seconds $\qquad \Delta d = 16(6^2)-16(3^2) = 576-144 = 432$ feet

15. $\Delta t = 4.01-4 = .01$ seconds $\qquad \Delta d = 16(4.01)^2-16(4^2) = 1.2816$ feet

17. $\Delta t = 5-1 = 4$ seconds, $\Delta d = 16(5^2)-16(1)^2 = 384$ feet

$\Delta d/\Delta t = 384/4 = 96$ feet/second

19. $\Delta t = 5.1-5 = .1$ seconds $\qquad \Delta d = 16(5.1)^2-16(5^2) = 16.16$ feet

$\Delta d/\Delta t = 16.16/.1 = 161.6$ feet/second

21. $\Delta t = 3+h-3 = h$ seconds $\qquad \Delta d = 16(3+h)^2-16(3^2) = 16(6h+h^2)$ feet

$\Delta d/\Delta t = 16(6h+h^2)/h = 16(6+h)$ feet/second

23. $\Delta z = 3-(-2) = 5 \qquad \Delta w = g(3)-g(-2) = 0-(-5) = 5$

$\Delta w/\Delta z = 5/5 = 1$

25. $\Delta z = 2+a-2 = a$

$\Delta w = f(2+a)-f(2) = (2+a)^2+3(2+a)-1-9 = 4+4a+a^2+6+3a-10 = 7a+a^2$

$\Delta w/\Delta z = (7a+a^2)/a = 7+a$

27. $\Delta C = C(800) - C(600) = 19{,}600 - 17{,}200 = \$2{,}400$

$\Delta R = R(800) - R(600) = 28{,}800 - 25{,}200 = \$3{,}600$

$P(q) = R(q) - C(q) = 60q - .03q^2 - (10000 + 12q) = 48q - .03q^2 - 10{,}000$

$\Delta P = P(800) - P(600) = 9{,}200 - 8000 = \$1{,}200$

$\Delta C / \Delta q = 2400/200 = \$12/\text{doll}$
$\Delta R / \Delta q = 3600/200 = \$18/\text{doll}$
$\Delta P / \Delta q = 1200/200 = \$6/\text{doll}$

29. $\Delta t = 7 - 5 = 2$ weeks $\qquad \Delta w = 80(1 - \frac{10}{49}) - 80(1 - \frac{10}{25}) = 80(\frac{10}{25} - \frac{10}{49}) \sim 15.67$

$\Delta w / \Delta t \sim 7.835$ words/week

31. $\Delta t = 3 - 1 = 2$ seconds $\qquad \Delta d = 7557 - 2551 = 5006$ feet

$\Delta d / \Delta t = 5006/2 = 2503$ feet/second

33. (a) The points on the line are $(2, 11)$ and $(4, 29)$. The slope is
$(29 - 11)/(4 - 2) = 18/2 = 9$.

(b) $\dfrac{f(2+h) - f(2)}{2+h-2} = \dfrac{(2+h)^2 + 3(2+h) + 1 - 11}{h} = \dfrac{4 + 4h + h^2 + 6 + 3h - 10}{h} = 7 + h$

35. (a) The points on the line are $(7, 3)$ and $(23, 5)$. The slope is $(5-3)/(23-7) = 2/16 = 1/8$.

(b) $\dfrac{f(7+h) - f(7)}{h} = \dfrac{\sqrt{7+h+2} - \sqrt{7+2}}{h} = \dfrac{\sqrt{9+h} - 3}{h} = \dfrac{\sqrt{9+h} - 3}{h} \cdot \dfrac{\sqrt{9+h} + 3}{\sqrt{9+h} + 3} =$

$\dfrac{9+h-9}{h(\sqrt{9+h} + 3)} = \dfrac{1}{\sqrt{9+h} + 3}$.

37. (a) The points on the line are $(-1, -\frac{1}{2})$ and $(1, 3)$. The slope is
$(3 - (-\frac{1}{2}))/(1 - (-1)) = 3.5/2 = 1.75$

(b) $\dfrac{f(-1+h) - f(-1)}{-1+h-(-1)} = \dfrac{h - 1 + 2^{h-1} - (-1 + \frac{1}{2})}{h} = \dfrac{h + 2^{h-1} + \frac{1}{2}}{h}$.

EXERCISE SET 2.2 INTUITIVE DESCRIPTION OF LIMIT

1.

x	2.1	2.01	1.9	1.99
$5x-2$	8.5	8.05	7.5	7.95

$\lim\limits_{x \to 2}(5x - 2) = 5(2) - 2 = 8$

3.

x	.1	.01	$-.1$	$-.01$
$2^x + 5$	6.072	6.007	5.933	5.993

$\lim\limits_{x \to 0}(2^x + 5) = 2^0 + 5 = 1 + 5 = 6$

5.

x	2.1	2.01	1.9	1.99
$(3x+2)^{5/3}$	34.02	32.20	30.03	31.80

$$\lim_{x \to 2}(3x+2)^{5/3} = (3 \cdot 2+2)^{5/3} = 8^{5/3} = 2^5 = 32$$

7.

x	2.01	2.001	1.99	1.999
x^2+6x+7	23.100	23.010	22.900	22.990

$$\lim_{x \to 2}(x^2+6x+7) = 4+12+7 = 23$$

9.

x	1.1	1.01	.9	.99
$\frac{2x+1}{3x-2}$	2.462	2.932	4	3.072

$$\lim_{x \to 1}\frac{(2x+1)}{(3x-2)} = \frac{2+1}{3-2} = 3$$

11.

y	2.1	2.01	1.9	1.99
$f(y)$	18.32	175.81	-16.69	-174.19

$\lim_{y \to 2}\frac{y^2+y+1}{y^2-4}$ is undefined since $\lim_{y \to 2}y^2+y+1 = 7$ and $\lim_{y \to 2}(y^2-4) = 4-4 = 0$.

13.

u	3.1	3.01	2.9	2.99
$f(u)$	.023	.0023	$-.024$	$-.0023$

$$\lim_{u \to 3}\frac{u^2-9}{u^2+5u+2} = \frac{9-9}{9+15+2} = \frac{0}{26} = 0$$

15.

x	.55	.51	.5001	.499
$f(x)$	-3.78	-5.58	-5.996	-6.042

$$\lim_{x \to .5}[(6x+3)(8x-5)] = [(6(.5)+3)(8(.5)-5)] = 6(-1) = -6$$

17.

x	4.1	4.01	3.9	3.99
$f(x)$	14.49	141.77	-13.78	-141.07

Since $\lim\limits_{x\to 4}\sqrt{x-2}=\sqrt{2}$ and $\lim\limits_{x\to 4}(x-4)=0$, $\lim\limits_{x\to 4}\dfrac{\sqrt{x-2}}{x-4}$ does not exist.

19.

x	$-.9$	$-.99$	-1.1	-1.01
$f(x)$	.248	.2498	.2516	.2502

$$\lim_{x\to -1}\frac{\sqrt{5+x}-2}{x+1}=\lim_{x\to -1}\frac{\sqrt{5+x}-2}{x+1}\cdot\frac{\sqrt{5+x}+2}{\sqrt{5+x}+2}=\lim_{x\to -1}\frac{1}{\sqrt{5+x}+2}=\frac{1}{\sqrt{4}+2}=\frac{1}{4}.$$

21.

x	5.1	5.001	4.9	4.99
$f(x)$	-3.017	-3.00017	-2.983	-2.99833

$$\lim_{x\to 5}\frac{5-x}{\sqrt{2x-1}-3}=\lim_{x\to 5}\frac{5-x}{\sqrt{2x-1}-3}\cdot\frac{\sqrt{2x-1}+3}{\sqrt{2x-1}+3}=\lim_{x\to 5}\frac{(5-x)(\sqrt{2x-1}+3)}{2x-10}=$$

$$=\lim_{x\to 5}\frac{-\sqrt{2x-1}-3}{2}=\frac{-3-\sqrt{10-1}}{2}=-3$$

23.

x	1.1	1.01	.9	.99
$f(x)$	-1.495	-1.499	-1.505	-1.501

$$\lim_{x\to 1}\frac{\sqrt{x+3}-2}{3-\sqrt{x+8}}\cdot\frac{\sqrt{x+3}+2}{\sqrt{x+3}+2}\cdot\frac{3+\sqrt{x+8}}{3+\sqrt{x+8}}=\lim_{x\to 1}\frac{(x-1)(3+\sqrt{x+8})}{(1-x)(\sqrt{x+3}+2)}=\frac{-(3+\sqrt{9})}{\sqrt{4}+2}=-1.5$$

25.

x	4.1	4.01	3.9	3.99
$f(x)$	.0774	.0780	.0789	.0782

$$\lim_{x\to 4}\frac{x-\sqrt{3x+4}}{(x-4)(x+4)}\cdot\frac{x+\sqrt{3x+4}}{x+\sqrt{3x+4}}=\lim_{x\to 4}\frac{x^2-3x-4}{(x-4)(x+4)(x+\sqrt{3x+4})}$$

$$=\lim_{x\to 4}\frac{(x-4)(x+1)}{(x-4)(x+4)(x+\sqrt{3x+4})}=\frac{5}{8(4+\sqrt{16})}=\frac{5}{64}=.078125$$

27.

x	4.1	4.01	3.9	3.99
$f(x)$	.2353	.2368	.2388	.2372

$(6x+3)-27 = (\sqrt[3]{6x+3}-3)\left((\sqrt[3]{6x+3})^2+3\sqrt[3]{6x+3}+9\right)$

$x^3-(3x+52) = (x-\sqrt[3]{3x+52})\left(x^2+x\sqrt[3]{3x+52}+(\sqrt[3]{3x+52})^2\right)$

Furthermore $(6x+3)-27 = 6x-24 = 6(x-4)$ and $x^3-3x-52 = (x-4)(x^2+4x+13)$.

$\lim\limits_{x\to 4}\dfrac{(6x+3)^{1/3}-3}{x-(3x+52)^{1/3}}$

$= \lim\limits_{x\to 4}\dfrac{6(x-4)}{(\sqrt[3]{6x+3})^2+3\sqrt[3]{6x+3}+9} \div \dfrac{(x-4)(x^2+4x+13)}{x^2+x\sqrt[3]{3x+52}+(3x+52)^{2/3}}$

$= \lim\limits_{x\to 4}\dfrac{6(x^2+x\sqrt[3]{3x+52}+(3x+52)^{2/3})}{\left((\sqrt[3]{6x+3})^2+3\sqrt[3]{6x+3}+9\right)(x^2+4x+13)} = \dfrac{6(16+4(4)+16)}{(9+9+9)(16+16+13)}$

$= \dfrac{6(48)}{27(45)} = \dfrac{32}{135} \sim .2370$

29. $\lim\limits_{h\to 0}\dfrac{(3+h)^2+2(3+h)+5-20}{h} = \lim\limits_{h\to 0}\dfrac{9+6h+h^2+6+2h-15}{h} = \lim\limits_{h\to 0}\dfrac{8h+h^2}{h} =$

$= \lim\limits_{h\to 0}\dfrac{h(8+h)}{h} = \lim\limits_{h\to 0}(8+h) = 8$

31. The average speed is $\dfrac{\Delta d}{\Delta t} = \dfrac{\left(-16(4+h)^2+192(4+h)\right)-512}{(4+h)-h}$

$= \dfrac{-256-128h-16h^2+768+192h-512}{h} = \dfrac{-16h^2+64h}{h} = 64-16h$ ft/sec.

The instantaneous speed is $\lim\limits_{h\to 0}\dfrac{\Delta d}{\Delta t} = \lim\limits_{h\to 0}(64-16h) = 64$ ft/sec.

33. (a) $f(2) = 4+2+3 = 9$.

(b) $\dfrac{f(2+h)-9}{(2+h)-2} = \dfrac{(2+h)^2+(2+h)+3-9}{h} = \dfrac{4+4h+h^2+2+h-6}{h} = \dfrac{5h+h^2}{h} = 5+h$

(c) $\lim\limits_{h\to 0}(5+h) = 5$

(d) $y-9 = m(x-2)$
$y = m(x-2)+9 = 5(x-2)+9$
$y = 5x-1$

(e) $f(x) = x^2+x+3$ has a parabola as its graph with vertex when $x = -\frac{1}{2}$. Some points on the graph are

(continued)

(problem #33, continued)

x	-1	$-\frac{1}{2}$	0	1	2	3
$f(x)$	3	2.75	3	5	9	15

Points on the graph of $y = 5x-1$ include $(1, 4)$, $(2, 9)$ and $(3, 14)$.

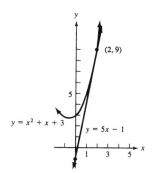

35. $\displaystyle\lim_{h\to 0}\frac{f(4+h)-f(4)}{h} = \lim_{h\to 0}\frac{-3(4+h)+2-(-10)}{h} = \lim_{h\to 0}\frac{-12-3h+2+10}{h}$

$\displaystyle = \lim_{h\to 0}\frac{-3h}{h} = -3.$

37. $\displaystyle\lim_{h\to 0}\frac{f(4+h)-f(4)}{h} = \lim_{h\to 0}\frac{2(4+h)^2+3(4+h)+1-45}{h} = \lim_{h\to 0}\frac{32+16h+2h^2+12+3h-44}{h}$

$\displaystyle = \lim_{h\to 0}\frac{h(19+2h)}{h} = 19.$

39. $\displaystyle\lim_{h\to 0}\frac{15/(3+h)-5}{h} = \lim_{h\to 0}\frac{15-5(3+h)}{h(3+h)} = \lim_{h\to 0}\frac{-5h}{h(3+h)} = -5/3$

EXERCISE SET 2.3 CONTINUITY AND ONE-SIDED LIMITS

1. f is not continuous at -4 since $f(-4)$ is not defined.

3. $\displaystyle\lim_{t\to 1}\frac{t^2+t-2}{t^2-3t+2} = \lim_{t\to 1}\frac{(t+2)(t-1)}{(t-2)(t-1)} = \frac{3}{-1} = -3 \neq g(1) = 2.$ The function is not continuous at 1.

5. $\displaystyle\lim_{x\to 2}\frac{x^2+5x-14}{x^2+3x-10} = \lim_{x\to 2}\frac{(x+7)(x-2)}{(x+5)(x-2)} = \frac{9}{7} = f(2).$ The function is continuous at 2.

7. $\displaystyle\lim_{s\to 9}\frac{\sqrt{s}-3}{s-9}\cdot\frac{\sqrt{s}+3}{\sqrt{s}+3} = \lim_{s\to 9}\frac{1}{\sqrt{s}+3} = \frac{1}{6} \neq h(9) = 0.$ The function is not continuous at 9.

9. $\lim\limits_{t\to-2}\dfrac{t^2+3t+2}{t^2+4t+4} = \lim\limits_{t\to-2}\dfrac{(t+2)(t+1)}{(t+2)(t+2)} = \lim\limits_{t\to-2}\dfrac{(t+1)}{(t+2)}$, which does not exist.

The function is not continuous at -2.

11. $f(x) = [x/3], \ -6 \le x \le 6$

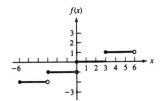

$\lim\limits_{x\to-3^+} f(x) = -1$

$\lim\limits_{x\to-3^-} f(x) = -2.$

13.

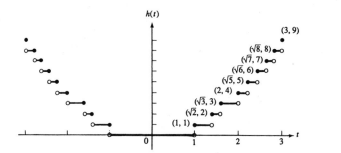

$\lim\limits_{t\to-2^-} h(t) = 4$

$\lim\limits_{t\to-2^+} h(t) = 3.$

15.

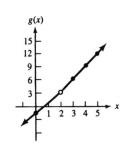

$\lim\limits_{x\to2^+} g(x) = \lim\limits_{x\to2^+} (3x-3) = 3$

$\lim\limits_{x\to2^-} g(x) = \lim\limits_{x\to2^-} (2x-1) = 3$

17.

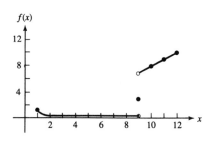

$\lim\limits_{x\to9^+} f(x) = \lim\limits_{x\to9^+} (x-2) = 7$

$\lim\limits_{x\to9^-} f(x) = \lim\limits_{x\to9^-} \dfrac{1}{\sqrt{x}+3} = \dfrac{1}{6}$

19. (a) -2 (b) -2 (c) -4

21. (a) 1   (b) 3   (c) 3

23. $\lim_{x \to a} (-3x^4 + 3x^2 - 8x + 4) = \lim_{x \to a} (-3x^4) + \lim_{x \to a} (3x^2) - \lim_{x \to a} (8x) + \lim_{x \to a} 4 =$
$\lim_{x \to a} (-3) \lim_{x \to a} x^4 + \lim_{x \to a} 3 \lim_{x \to a} x^2 - 8a + 4 = -3a^4 + 3a^2 - 8a + 4 = f(a).$

Hence p is continuous.

25.

x	-5	-4	-3.5	-2.5	-2	-1
y	-1	-2	-4	4	2	1

The graph is discontinuous when $x = -3$.

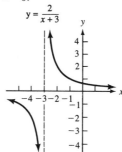

$y = \dfrac{2}{x+3}$

27. Let $p(x) = x^3 - 4x^2 - 5x + 14.$

$p(4) = 64 - 64 - 20 + 14 = -6$
$p(5) = 125 - 100 - 25 + 14 = 0$

Since p is continuous and $p(4)$ and $p(5)$ have opposite signs the equation $x^3 - 4x^2 - 5x + 14 = 0$ has a solution between 4 and 5.

$p(4.5) = 1.625 \qquad p(4.40) = -.256 \qquad p(4.44) = .474$

The solution is approximately 4.4

29. $\lim_{x \to a} p(x) = \lim_{x \to a} (a_0 x^n + a_1 x^{n-1} + \cdots + a_{n-1} x + a_n)$
$= \lim_{x \to a} a_0 x^n + \lim_{x \to a} a_1 x^{n-1} + \cdots + \lim_{x \to a} a_{n-1} x + \lim_{x \to a} a_n$
$= a_0 \lim_{x \to a} x^n + a_1 \lim_{x \to a} x^{n-1} + \cdots + a_{n-1} \lim_{x \to a} x + a_n$
$= a_0 a^n + a_1 a^{n-1} + \cdots + a_{n-1} a + a_n = p(a).$

Thus $p(x)$ is continuous.

EXERCISE SET 2.4   THE DERIVATIVE

1. $f(2+h) = 3(2+h)+2 = 8+3h$
    $f(2) = 3(2)+2 = 8$
    $f(2+h)-f(2) = 8+3h-8 = 3h$

    $\dfrac{f(2+h)-f(2)}{h} = \dfrac{3h}{h} = 3$

    $f'(2) = \lim\limits_{h\to 0} \dfrac{f(2+h)-f(2)}{h} = \lim\limits_{h\to 0} 3 = 3$

3. $f(1+h) = (1+h)^3+(1+h)+1 = 1+3h+3h^2+h^3+h+2 = h^3+3h^2+4h+3$
    $f(1) = 1+1+1 = 3$
    $f(1+h)-f(1) = h^3+3h^2+4h+3-3 = h^3+3h^2+4h$

    $\dfrac{f(1+h)-f(1)}{h} = \dfrac{h^3+3h^2+4h}{h} = h^2+3h+4$

    $f'(1) = \lim\limits_{h\to 0} h^2+3h+4 = 4.$

5. $f'(4) = \lim\limits_{h\to 0} \dfrac{\sqrt{4+h}-\sqrt{4}}{h} = \lim\limits_{h\to 0} \dfrac{\sqrt{4+h}-2}{h} \cdot \dfrac{\sqrt{4+h}+2}{\sqrt{4+h}+2} = \lim\limits_{h\to 0} \dfrac{4+h-4}{h(\sqrt{4+h}+2)}$

    $= \lim\limits_{h\to 0} \dfrac{1}{\sqrt{4+h}+2} = \dfrac{1}{4}$

7. $f'(8) = \lim\limits_{h\to 0} \dfrac{\dfrac{3}{\sqrt{8+h+1}} - \dfrac{3}{\sqrt{9}}}{h} = \lim\limits_{h\to 0} \dfrac{9-3\sqrt{9+h}}{3h\sqrt{9+h}} \cdot \dfrac{9+3\sqrt{9+h}}{9+3\sqrt{9+h}}$

    $= \lim\limits_{h\to 0} \dfrac{81-9(9+h)}{3h\sqrt{9+h}(9+3\sqrt{9+h})} = \lim\limits_{h\to 0} \dfrac{-9h}{9h\sqrt{9+h}(3+\sqrt{9+h})} = \lim\limits_{h\to 0} \dfrac{-1}{\sqrt{9+h}(3+\sqrt{9+h})}$

    $= \dfrac{-1}{3(6)} = -1/18$

9. $\dfrac{dy}{dx} = \lim\limits_{h\to 0} \dfrac{g(x+h)-g(x)}{h} = \lim\limits_{h\to 0} \dfrac{5(x+h)+3-(5x+3)}{h} = \lim\limits_{h\to 0} \dfrac{5h}{h} = 5.$

11. $\dfrac{dy}{dx} = \lim\limits_{h\to 0} \dfrac{(x+h)^2+5(x+h)+13-(x^2+5x+13)}{h}$

    $= \lim\limits_{h\to 0} \dfrac{x^2+2xh+h^2+5x+5h+13-x^2-5x-13}{h} = \lim\limits_{h\to 0} \dfrac{2xh+h^2+5h}{h}$

    $= \lim\limits_{h\to 0} (2x+h+5) = 2x+5.$

13. $\dfrac{dy}{dx} = \lim\limits_{h\to 0} \dfrac{\sqrt{3(x+h)+1}-\sqrt{3x+1}}{h} \cdot \dfrac{\sqrt{3(x+h)+1}+\sqrt{3x+1}}{\sqrt{3(x+h)+1}+\sqrt{3x+1}}$

    $= \lim\limits_{h\to 0} \dfrac{3(x+h)+1-(3x+1)}{h(\sqrt{3x+3h+1}+\sqrt{3x+1})} = \lim\limits_{h\to 0} \dfrac{3h}{h(\sqrt{3x+3h+1}+\sqrt{3x+1})} = \dfrac{3}{2\sqrt{3x+1}}$

15. $\dfrac{dy}{dx} = \lim\limits_{h \to 0} \dfrac{\dfrac{2}{\sqrt{x+h+5}} - \dfrac{2}{\sqrt{x+5}}}{h} = \lim\limits_{h \to 0} \dfrac{2(\sqrt{x+5}-\sqrt{x+h+5})}{h(\sqrt{x+h+5})\sqrt{x+5}}$

$= \lim\limits_{h \to 0} \dfrac{2\Big(x+5-(x+h+5)\Big)}{h\sqrt{x+h+5}\sqrt{x+5}(\sqrt{x+5}+\sqrt{x+h+5})}$

$= \lim\limits_{h \to 0} \dfrac{-2}{\sqrt{x+h+5}\sqrt{x+5}(\sqrt{x+5}+\sqrt{x+h+5})} = \lim\limits_{h \to 0} \dfrac{-2}{(x+5)(2\sqrt{x+5})} = -1(x+5)^{-3/2}$

17. (a) $f(2) = 4+10-3 = 11.$

(b)
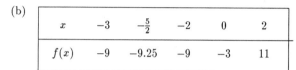

x	-3	$-\frac{5}{2}$	-2	0	2
$f(x)$	-9	-9.25	-9	-3	11

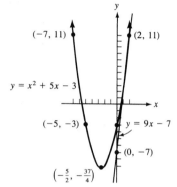

(c) $f'(2) = \lim\limits_{h \to 0} \dfrac{(2+h)^2+5(2+h)-3-11}{h} = \lim\limits_{h \to 0} \dfrac{4+4h+h^2+10+5h-14}{h} = \lim\limits_{h \to 0} \dfrac{h(9+h)}{h} = 9$

(d) $y-11 = 9(x-2)$
$y = 9x-7$

19. (a) $g(2) = \sqrt[3]{4+4} = 2.$

(b)

x	0	1	2	3
$g(x)$	1.59	1.71	2	2.35

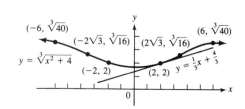

(continued)

(problem #19, continued)

(c) $g'(2) = \lim_{h \to 0} \dfrac{\sqrt[3]{(2+h)^2+4}-2}{h}$

$(2+h)^2+4-8 = h^2+4h$

$= \left(\sqrt[3]{(2+h)^2+4}-2\right)\left(\left(\sqrt[3]{(2+h)^2+4}\ \right)^2+2\sqrt[3]{(2+h)^2+4}\ +4\right).$

Multiplying the numerator and denominator of the difference quotient by the second factor we obtain

$g'(2) = \lim_{h \to 0} \dfrac{(2+h)^2+4-8}{h\left(\left((2+h)^2+4\right)^{2/3}+2\left((2+h)^2+4\right)^{1/3}+4\right)}$

$= \lim_{h \to 0} \dfrac{h(h+4)}{h\left(\left((2+h)^2+4\right)^{2/3}+2\left((2+h)^2+4\right)^{1/3}+4\right)} = \dfrac{4}{4+2(2)+4} = \dfrac{1}{3}.$

(d) $y-2 = \frac{1}{3}(x-2)$

$y = \frac{1}{3}x+\frac{4}{3}$

21. $s'(4) = \lim_{h \to 0} \dfrac{s(4+h)-s(4)}{h} = \lim_{h \to 0} \dfrac{(4+h)^2+3(4+h)+7-35}{h}$

$= \lim_{h \to 0} \dfrac{16+8h+h^2+12+3h+7-35}{h} = \lim_{h \to 0} \dfrac{h(11+h)}{h} = 11 \text{ ft/second}$

23. $s'(2) = \lim_{h \to 0} \dfrac{16(2+h)^2+64(2+h)+70-262}{h} = \lim_{h \to 0} \dfrac{64+64h+16h^2+128+64h-192}{h}$

$= \lim_{h \to 0} \dfrac{h(128+16h)}{h} = 128 \text{ feet/second}$

25. $s'(4) = \lim_{h \to 0} \dfrac{\dfrac{120}{\sqrt{9+h}} - 40}{h} = \lim_{h \to 0} \dfrac{40(3-\sqrt{9+h})}{h\sqrt{9+h}} \cdot \dfrac{3+\sqrt{9+h}}{3+\sqrt{9+h}} = \lim_{h \to 0} \dfrac{40(9-9-h)}{h\sqrt{9+h}(3+\sqrt{9+h})}$

$= \dfrac{-40}{3(6)} = \dfrac{-20}{9} \text{ ft/sec.}$

27. (a) $\lim_{x \to 2^-}(x-2) = 0$

(b) $\lim_{x \to 2^+}(2-x) = 0$

(c) $f(2) = 2-2 = 0$

(d) Yes

(e) $\lim_{h \to 0^-} \dfrac{(2+h-2)-0}{h} = 1$

(f) $\lim_{h \to 0^+} \dfrac{2-(2+h)-0}{h} = -1$

(g) No

29. (a) $\lim\limits_{x \to 2^-}(x^3+x+1) = 8+2+1 = 11$

(b) $\lim\limits_{x \to 2^+}(x^2+3x+1) = 4+6+1 = 11$

(c) $f(2) = 4+6+1 = 11$

(d) Yes

(e) $\lim\limits_{h \to 0^-}\dfrac{(2+h)^3+(2+h)+1-11}{h} = \lim\limits_{h \to 0^-}\dfrac{8+12h+6h^2+h^3+2+h-10}{h}$

$= \lim\limits_{h \to 0^-}\dfrac{h(13+6h+h^2)}{h} = 13$

(f) $\lim\limits_{h \to 0^+}\dfrac{(2+h)^2+3(2+h)+1-11}{h} = \lim\limits_{h \to 0^+}\dfrac{4+4h+h^2+6+3h-10}{h} = \lim\limits_{h \to 0^+}\dfrac{h(7+h)}{h} = 7$

(g) No

EXERCISE SET 2.5   BASIC DIFFERENTIATION FORMULAS

1. $f'(x) = 0$ since 6 is a constant.

3. $h'(x) = 0$ since $\log_7 3$ is a constant.

5. $f'(t) = D_t(2t^2)+D_t(3t)+D_t(1) = 2D_t(t^2)+3D_t(t^1)+0 = 2 \cdot 2t^{2-1}+3 \cdot 1 \cdot t^0 = 4t+3$

7. $h'(x) = 3D_xx^4+10D_xx^2-6D_xx+D_x3 = 3 \cdot 4x^3+10 \cdot 2x^1-6 \cdot 1+0 = 12x^3+20x-6$

9. $g(x) = \sqrt{x^3} = x^{3/2}$ $g'(x) = \frac{3}{2}x^{1/2}$

11. $f(s) = (s^2+1)^3 = (s^2)^3+3(s^2)^2+3s^2+1 = s^6+3s^4+3s^2+1$ $f'(s) = 6s^5+12s^3+6s$

13. $h(t) = (t^2+1)(3t^3+2) = 3t^5+3t^3+2t^2+2$

$h'(t) = 3(5t^4)+3(3t^2)+2(2t)+0 = 15t^4+9t^2+4t$

15. $f(x) = \frac{5}{2}x^{4/3-2/5}+\frac{3}{2}x^{2/3-2/5}+\frac{5}{2}x^{-2/5} = \frac{5}{2}x^{14/15}+\frac{3}{2}x^{4/15}+\frac{5}{2}x^{-2/5}$

$f'(x) = \frac{5}{2} \cdot \frac{14}{15}x^{-1/15}+\frac{3}{2} \cdot \frac{4}{15}x^{-11/15}+\frac{5}{2}(-\frac{2}{5})x^{-7/5} = (7/3)x^{-1/15}+(2/5)x^{-11/15}-x^{-7/5}$

17. $h(u) = \dfrac{4}{\sqrt[5]{u^3}} = \dfrac{4}{u^{3/5}} = 4u^{-3/5}$ $h'(u) = 4(-3/5)u^{-8/5} = -2.4u^{-1.6}$

19. $g'(x) = -2x+3$. The slope of the tangent line at $(-2, -6)$ is $g'(-2) = 4+3 = 7$. The equation of the tangent line is $y-(-6) = 7(x+2)$ or $y = 7x+8$.

21. $g(x) = x^{3/2}+5x^{1/2}-10x^{-1/2}$ $g'(x) = \frac{3}{2}\sqrt{x}+\frac{5}{2\sqrt{x}}+\frac{5}{x\sqrt{x}}$.

The slope of the tangent line at $(4, 13)$ is $g'(4) = \frac{3}{2}(2)+\frac{5}{2(2)}+\frac{5}{4(2)} = 3+\frac{5}{4}+\frac{5}{8} = \frac{39}{8}$.

The equation of the line is $y-13 = \frac{39}{8}(x-4)$ or $8y-104 = 39x-156$ or $39x-8y-52 = 0$.

23. $s'(t) = 10t+6$ $s'(2) = 20+6 = 26$. The instantaneous velocity is 26 ft/sec.

25. $s = 5t^{2/3}$. $s'(t) = \frac{10}{3}t^{-1/3}$. $s'(8) = \frac{10}{6} = \frac{5}{3}$. The instantaneous velocity is $\frac{5}{3}$ ft/sec.

27. $s = t^{3/2}+7t^{1/2}-2t^{-1/2}$ $s' = \frac{3}{2}t^{1/2}+\frac{7}{2}t^{-1/2}+t^{-3/2}$

$s'(4) = \frac{3}{2}\cdot 2+\frac{7}{4}+\frac{1}{4} = 3+1.75+.125 = 4.875$ ft/sec

29. (a) $C'(q) = 150-4q+.03q^2$

(b) $C'(100) = 150-400+.03(10000) = -250+300 = \$50$

(c) $C(100) = 20000+150(100)-2(10000)+.01(100)^3$

$C(99) = 20000+150(99)-2(99^2)+.01(99)^3$

The cost of producing the 100th set is
$C(100)-C(99) = 150-2(10000-99^2)+.01(100^3-99^3) = -248+297.01 = \$49.01$

31. (a) $s'(t) = -32t+v_0$. The instantaneous velocity is zero when $-32t+384 = 0$ or
$t = \frac{384}{32} = 12$ seconds

(b) $s(12) = -16(144)+384(12) = 2304$ feet.

(c) $s'(t) = -32t+384 = 96$ when $32t = 288$ or $t = \frac{288}{32} = 9$ seconds.

(d) $s(t) = -16t^2+384t = -16t(t-24) = 0$ when $t = 0$ seconds or $t = 24$ seconds.

33. (a) The instantaneous velocity is zero when $-32t+1664 = 0$ or $t = \frac{1664}{32} = 52$ seconds.

(b) $s(52) = -16(52)^2+1664(52) = 43,264$ feet.

(c) $-32t+1664 = 96$ when $t = \frac{1568}{32} = 49$ seconds.

(d) The object reaches the ground when $t = 0$ or $t = 104$ seconds since the time the object falls equals the time the object rises.

35. $y' = 6x^2+6x-36 = 0$ when $6(x^2+x-6) = 6(x+3)(x-2) = 0$. The tangent lines at the points $(2, -34)$ and $(-3, 91)$ are parallel to the x-axis.

37. $y' = 4x^3-32 = 0$ if $x^3 = 8$ or $x = 2$. The tangent line at the point $(2, -33)$ has slope zero and thus is parallel to the x-axis.

39. The line $3x+5y = 7$ has slope $-3/5$. $D_x(x^2-5x+2) = 2x-5 = -\frac{3}{5}$ if $2x = 5-.6 = 4.4$, or $x = 2.2$. The point of tangency is $(2.2, -4.16)$

41. (a) $f(3) = 27 + 18 - 2 = 43$

(b) $\dfrac{f(x) - f(3)}{x - 3} = \dfrac{3x^2 + 6x - 2 - 43}{x - 3} = \dfrac{3x^2 + 6x - 45}{x - 3} = \dfrac{3(x^2 + 2x - 15)}{x - 3} =$

$\dfrac{3(x + 5)(x - 3)}{x - 3} = 3(x + 5)$ if $x \neq 3$. Hence $h(x) = 3(x + 5)$.

(c) $f'(x) = 6x + 6$

(d) $h(3) = 24$ and $f'(3) = 24$

1. $D_x[5 \cdot 3x] = D_x[15x] = 15$ $f'(x) = 0.$ $g'(x) = 3.$ $f'(x)g'(x) \neq 15$

3. $D_x[3x^2 \cdot 5x^4] = D_x[15x^6] = 15 \cdot 6x^5 = 90x^5$ $f'(x)g'(x) = 6x(20x^3) = 120x^4 \neq 90x^5$

5. $D_x\dfrac{f(x)}{g(x)} = D_x\left(\dfrac{x^3}{x^2}\right) = D_x(x) = 1.$

$\dfrac{f'(x)}{g'(x)} = \dfrac{3x^2}{2x} = \dfrac{3x}{2} \neq D_x\dfrac{f(x)}{g(x)}$

7. $D_x\dfrac{f(x)}{g(x)} = D_x\dfrac{2x^2}{4x^5} = D_x[\tfrac{1}{2}x^{-3}] = \dfrac{-3}{2x^4}.$

$\dfrac{f'(x)}{g'(x)} = \dfrac{4x}{20x^4} = \dfrac{1}{5x^3} \neq \dfrac{-3}{2x^4}.$

9. $y' = (x^2+11x+3)D_x(x^4+5)+(x^4+5)D_x(x^2+11x+3) = (x^2+11x+3)(4x^3)+(x^4+5)(2x+11)$

$\quad = 4x^5+44x^4+12x^3+2x^5+10x+11x^4+55 = 6x^5+55x^4+12x^3+10x+55.$

Also $y = (x^2+11x+3)(x^4+5) = x^6+11x^5+3x^4+5x^2+55x+15$

and $y' = 6x^5+55x^4+12x^3+10x+55$

11. $y' = (5x^3+3x+8)(8x^3)+(2x^4+5)(15x^2+3) = 40x^6+24x^4+64x^3+30x^6+75x^2+6x^4+15$

$\quad = 70x^6+30x^4+64x^3+75x^2+15$

Alternately,

$y = (5x^3+3x+8)(2x^4+5) = 10x^7+6x^5+16x^4+25x^3+15x+40$ and

$y' = 70x^6+30x^4+64x^3+75x^2+15$

13. $y' = (x^2+3x-1)(\frac{1}{2\sqrt{x}})+(\sqrt{x}+3)(2x+3) = \dfrac{x^2}{2\sqrt{x}}+\dfrac{3x}{2\sqrt{x}}-\dfrac{1}{2\sqrt{x}}+2x^{3/2}+6x+3\sqrt{x}+9$

$\quad = \tfrac{1}{2}x^{3/2}+\tfrac{3}{2}\sqrt{x}-\dfrac{1}{2\sqrt{x}}+2x^{3/2}+6x+3\sqrt{x}+9 = \tfrac{5}{2}x^{3/2}+\tfrac{9}{2}\sqrt{x}-\dfrac{1}{2\sqrt{x}}+6x+9.$

Alternately,

$y = (x^2+3x-1)(\sqrt{x}+3) = x^{5/2}+3x^{3/2}-x^{1/2}+3x^2+9x-3$

$y' = \tfrac{5}{2}x^{3/2}+\tfrac{9}{2}x^{1/2}-\dfrac{1}{2\sqrt{x}}+6x+9.$

15. $y' = \dfrac{(x^4+3)D_x(x^2+1)-(x^2+1)D_x(x^4+3)}{(x^4+3)^2}$

$\quad = \dfrac{(x^4+3)(2x)-(x^2+1)(4x^3)}{(x^4+3)^2} = \dfrac{2x^5+6x-4x^5-4x^3}{(x^4+3)^2} = \dfrac{-2x^5-4x^3+6x}{(x^4+3)^2}$

17. $y' = \dfrac{(3x^2+10)(10x^4)-(2x^5-4)(6x)}{(3x^2+10)^2} = \dfrac{30x^6+100x^2-12x^6+24x}{(3x^2+10)^2} = \dfrac{18x^6+100x^4+24x}{(3x^2+10)^2}$

19. $y' = \dfrac{(x^{1/2}-2)D_x(x^{3/2}+3)-(x^{3/2}+3)D_x(x^{1/2}-2)}{(\sqrt{x}-2)^2}$

$= \dfrac{(x^{1/2}-2)(\frac{3}{2}x^{1/2})-(x^{3/2}+3)(\frac{1}{2}x^{-1/2})}{(\sqrt{x}-2)^2} = \dfrac{\frac{3}{2}x-3\sqrt{x}-\frac{1}{2}x-\frac{3}{2\sqrt{x}}}{(\sqrt{x}-2)^2} =$

$= (x-3\sqrt{x}-\dfrac{3}{2\sqrt{x}})(\sqrt{x}-2)^{-2}$

21. (a) $C'(q) = 20+60000D_q\left(\dfrac{1}{\sqrt{q}+10}\right) = 20+\dfrac{60000}{(\sqrt{q}+10)^2}(-1)D_q(\sqrt{q}+10)$

$= 20-\dfrac{60000}{(\sqrt{q}+10)^2}\cdot\dfrac{1}{2\sqrt{q}} = 20-\dfrac{30000}{\sqrt{q}(\sqrt{q}+10)^2}.$

(b) $C'(100) = 20-\dfrac{30000}{10(20)^2} = 20-\dfrac{3000}{400} = 12.50$

The approximate cost of producing the 101th item is \$12.50.

23. (a) $P(q) = R(q)-C(q) = \dfrac{200q^2+500q}{q+2}-\dfrac{100q^2+600q}{q+3} = \dfrac{100q(2q+5)}{q+2}-\dfrac{100q(q+6)}{q+3}$

$= 100q\left[\dfrac{(2q+5)(q+3)-(q+6)(q+2)}{(q+2)(q+3)}\right] = 100q\dfrac{\left[2q^2+11q+15-q^2-8q-12\right]}{(q+2)(q+3)}$

$= 100q\left[q^2+3q+3\right]/((q+2)(q+3))$

(b) $C'(q) = \dfrac{(q+3)(200q+600)-(100q^2+600q)}{(q+3)^2}$

(c) $R'(q) = \dfrac{(q+2)(400q+500)-(200q^2+500q)}{(q+2)^2}$

(d) $P'(q) = R'(q)-C'(q)$. The results of parts (b) and (c) may now be used to find $P'(q)$, or the result of part (a) may be differentiated.

25. (a) $P(q) = R(q)-C(q) = \dfrac{1}{1250}(-q^4+100q^3-q^2+100q)-1000-50q-\dfrac{30,000q}{q^2+500},\ 0\le q\le 100$

(b) $C'(q) = 50+\dfrac{30,000(-q^2+500)}{(q^2+500)^2},\ 0<q<100$

(c) $R'(q) = \dfrac{1}{1250}(-4q^3+300q^2-2q+100),\ 0<q<100$

(d) $P'(q) = R'(q)-C'(q)$ where $C'(q)$ and $R'(q)$ were found in parts (b) and (c) respectively.

27. **(a)** $D_x\dfrac{1}{x^{10}} = \dfrac{x^{10}D_x(1)-1\cdot D_x x^{10}}{(x^{10})^2} = \dfrac{-10x^9}{x^{20}} = \dfrac{-10}{x^{11}} = -10x^{-11}$

(b) $D_x x^{-n} = D_x\left(\dfrac{1}{x^n}\right) = \dfrac{x^n D_x(1)-1\cdot D_x(x^n)}{(x^n)^2} = \dfrac{0-nx^{n-1}}{x^{2n}} = -nx^{n-1-2n}$

$= -nx^{-n-1}$

EXERCISE SET 2.7 THE CHAIN RULE

1. $f'(x) = 40(x^2+6x+3)^{39}D_x(x^2+6x+3) = 40(2x+6)(x^2+6x+3)^{39}$

3. $h(x) = 5(x^2+3x+1)^{-13}$.

$h'(x) = -13(5)(x^2+3x+1)^{-14}D_x(x^2+3x+1) = -65(2x+3)(x^2+3x+1)^{-14}$

5. $g(u) = (u^2+5u+1)^{5/4}$.

$g'(u) = \frac{5}{4}(u^2+5u+1)^{1/4}D_u(u^2+5u+1) = \frac{5}{4}(2u+5)\sqrt[4]{u^2+5u+1}$.

7. $f'(x) = (x^3+1)^4 D_x(x^4+3x^2+2)^3+(x^4+3x^2+2)^3 D_x(x^3+1)^4$

$= (x^3+1)^4\cdot 3(x^4+3x^2+2)^2 D_x(x^4+3x^2+2)+(x^4+3x^2+2)^3(4)(x^3+1)^3(3x^2)$

$= 3(x^3+1)^4(x^4+3x^2+2)^2(4x^3+6x)+(x^4+3x^2+2)^3(12x^2)(x^3+1)^3$

9. $h'(u) = \dfrac{(u^3+2u^2+3)^5 D_u(u^2+1)^4-(u^2+1)^4 D_u(u^3+2u^2+3)^5}{(u^3+2u^2+3)^{10}}$

$= \dfrac{(u^3+2u^2+3)^5(4)(u^2+1)^3(2u)-(u^2+1)^4(5)(u^3+2u^2+3)^4(3u^2+4u)}{(u^3+2u^2+3)^{10}}$

11. $f'(t) = \dfrac{(t^3+4)^2 D_t(t^2+1)^{1/3}-(t^2+1)^{1/3}D_t(t^3+4)^2}{(t^3+4)^4}$

$= \dfrac{(t^3+4)^2\cdot\frac{1}{3}(t^2+1)^{-2/3}(2t)-(t^2+1)^{1/3}(2)(t^3+4)(3t^2)}{(t^3+4)^4}$

13. $h'(t) = D_t(1+t)^{1/3} = \frac{1}{3}(1+t)^{-2/3}D_t(1+t) = \frac{1}{3}(1+t)^{-2/3}$

15. $g'(t) = 12\left(\dfrac{t^2+3}{t^4+1}\right)^{11}D_t\left(\dfrac{t^2+3}{t^4+1}\right)$

$= 12\left(\dfrac{t^2+3}{t^4+1}\right)^{11}\cdot\dfrac{(t^4+1)(2t)-(t^2+3)(4t^3)}{(t^4+1)^2}$

$= \dfrac{12(t^2+3)^{11}}{(t^4+1)^{13}}(2t^5+2t-4t^5-12t^3) = \dfrac{12(t^2+3)^{11}(-2t^5+2t-12t^3)}{(t^4+1)^{13}}$

17. $f'(s) = 20\left(\frac{3s+1}{\sqrt{s+2}}\right)^{19} D_s\left[\frac{3s+1}{(s+2)^{1/2}}\right] = 20\left(\frac{3s+1}{\sqrt{s+2}}\right)^{19} \cdot \frac{3\sqrt{s+2}-(3s+1)\cdot\frac{1}{2}(s+2)^{-1/2}\cdot 1}{s+2}$

19. $h'(y) = 17\left(y^2+\frac{1}{\sqrt{y}}\right)^{16} D_y(y^2+y^{-1/2}) = 17\left(y^2+\frac{1}{\sqrt{y}}\right)^{16}(2y - \frac{1}{2}y^{-3/2})$

21. $y' = 5(x^2+1)^4(2x)$. Thus the slope of the line at $(1, 32)$ is $5(2)^4(2) = 160$. The equation of the line is
$$y-32 = 160(x-1) \qquad \text{or} \qquad y = 160x-128$$

23. $y' = \dfrac{(x^3+x+29)^5 \cdot 4(x^2+1)^3 \cdot 2x-(x^2+1)^4(5)(x^3+x+29)^4(3x^2+1)}{(x^3+x+29)^{10}}$.

When $x = -3$, $y' = \dfrac{(-1)(8)(-3)(10)^3-10^4(5)(-1)^4(28)}{(-1)^{10}} = 24(10^3)-140(10)^4$

$= 10^3(24-1400) = 10^3(-1376) = -1,376,000$ which is the slope at $(-3, -10,000)$. The equation of the tangent line is

$\qquad y = -1,376,000(x+3) - 10,000.$

25. $y = (x^3+x^2+x+1)^3(x^2+1)^{-1/2}$

$y' = 3(x^3+x^2+x+1)^2(3x^2+2x+1)(x^2+1)^{-1/2} - \frac{1}{2}(2x)(x^2+1)^{-3/2}(x^3+x^2+x+1)^3$.

When $x = -1$, $y' = 3(0)-(-1)(2)^{-3/2}(0) = 0$.

The slope of the tangent line is zero and its equation is $y = 0$.

27. (a) $y = -(169-x^2)^{1/2}$

$\qquad y' = -\frac{1}{2}(169-x^2)^{-1/2}(-2x) = \dfrac{x}{\sqrt{169-x^2}}$.

The slope of the tangent line at $(12, -5)$ is $\dfrac{12}{\sqrt{25}} = 12/5$.

(b) The slope is $\dfrac{-5-0}{12-0} = -5/12$.

(c) Since $\frac{12}{5}\left(-\frac{5}{12}\right) = -1$ the lines of parts (a) and (b) are perpendicular.

29. $d' = 32t-81 \cdot \frac{1}{5}(t^2+18)^{-4/5}(2t)$. When $t = 15$ seconds, the instantaneous velocity is

$32(15) - \frac{81}{5}(243)^{-4/5}(30) = 32(15) - \dfrac{81(6)}{81} = 32(15)-6 = 474$ ft/sec.

31. (a) $C'(q) = 30 - \frac{1}{5}(q^2+625)^{-4/5}(2q)$

(b) $C'(50) = 30 - \frac{1}{5}(3125)^{-4/5}(100) = 30-20(.0016) = 29.968.$

The approximate cost of producing the 50th item is $29.968.

33. (a) $C'(q) = \left(10 + \frac{q^2}{q^3+50}\right)^3 \cdot 1 + q \cdot 3\left(10 + \frac{q^2}{q^3+50}\right)^2 D_q\left(\frac{q^2}{q^3+50}\right)$

$$= \left(10 + \frac{q^2}{q^3+50}\right)^3 + 3q\left(10 + \frac{q^2}{q^3+50}\right)^2\left[\frac{2q(q^3+50)-q^2(3q^2)}{(q^3+50)^2}\right]$$

(b) $C'(10) = (10 + \frac{100}{1050})^3 + 30(10 + \frac{100}{1050})^2\left[\frac{20(1050)-3(10,000)}{(1050)^2}\right]$

$$= 1028.84 + 3057.4[-.008163] = 1003.88.$$

The approximate cost of producing the tenth item is $1003.88.

35. $P'(x) = -60D_x(x^2+1)^{-.1} = -60(-.1)(x^2+1)^{-1.1}(2x).$

$P'(12) = 6(145)^{-1.1}(24) = .60375$ items/week.

37. $h'(z) = \frac{1}{z+1}$, $y = f(t) = h(t^4+3)$. Let $u = t^4+3$ so that $y = h(u)$.

$$f'(t) = \frac{dy}{dt} = \frac{dy}{du} \cdot \frac{du}{dt} = h'(u)(4t^3) = h'(t^4+3) \cdot (4t^3) = \frac{1}{t^4+4} \cdot 4t^3 = \frac{4t^3}{t^4+4}.$$

EXERCISE SET 2.8   IMPLICIT DIFFERENTIATION

1. Let $a = 1$, $b = 5$, $c = -x^2$. $y = \frac{-5 \pm \sqrt{25+4x^2}}{2}$.

$$f_1(x) = \frac{-5+\sqrt{25+4x^2}}{2} \quad \text{and} \quad f_2(x) = \frac{-5-\sqrt{25+4x^2}}{2}.$$

3. $y^2 = 4x^2$ $y = \pm 2x$ $f_1(x) = 2x,$ $f_2(x) = -2x$

5.
Let $y = f(x)$

$$D_x(x^3+x^2y+5) = D_x(3x^2y^2-7y+6x)$$

$$D_x(x^3+x^2f(x)+5) = D_x(3x^2[f(x)]^2-7f(x)+6x)$$

$$3x^2+D_x(x^2f(x))+0 = D_x(3x^2[f(x)]^2)-7f'(x)+6$$

$$3x^2+2xf(x)+x^2f'(x) = 6x[f(x)]^2+3x^2 \cdot 2f(x)f'(x)-7f'(x)+6$$

$$x^2f'(x)-6x^2f(x)f'(x)+7f'(x) = 6x[f(x)]^2+6-3x^2-2xf(x)$$

$$(x^2-6x^2f(x)+7)f'(x) = 6x[f(x)]^2+6-3x^2-2xf(x)$$

$$D_xy = f'(x) = \frac{6xy^2+6-3x^2-2xy}{x^2-6x^2y+7}$$

7.

$$D_x(5x^2y^4+6x) = D_x(8x^3+7y^2-10)$$

$$y^4D_x(5x^2)+5x^2D_xy^4+6 = 24x^2+D_x(7y^2)-0$$

$$y^4(10x)+5x^2(4y^3)\frac{dy}{dx}+6 = 24x^2+14y\frac{dy}{dx}$$

$$(20x^2y^3-14y)\frac{dy}{dx} = 24x^2+10xy^4-6$$

$$D_xy = \frac{dy}{dx} = \frac{24x^2-10xy^4-6}{20x^2y^3-14y}$$

9. Note $\log_5 3$, e^6 and π^2 are constants.

$$D_x((\log_5 3)x^4)+D_x(6y^7) = e^6D_x(x^2y^4)+D_x(\pi^2)$$

$$4x^3\log_5 3+42y^6D_xy = e^6(2xy^4+4x^2y^3D_xy)+0$$

$$42y^6D_xy-4x^2y^3e^6D_xy = 2xy^4e^6-4x^3\log_5 3$$

$$D_xy = \frac{2xy^4e^6-4x^3\log_5 3}{42y^6-4x^2y^3e^6}$$

11.

$$xy^{3/4}+y^2x^{3/7} = x^2y+7$$

$$D_x(xy^{3/4}+y^2x^{3/7}) = D_x(x^2y+7)$$

$$y^{3/4}+\tfrac{3}{4}xy^{-1/4}D_xy+2yx^{3/7}D_xy+\tfrac{3}{7}x^{-4/7}y^2 = 2xy+x^2D_xy$$

$$\tfrac{3}{4}xy^{-1/4}D_xy+2yx^{3/7}D_xy-x^2D_xy = 2xy-\tfrac{3}{7}x^{-4/7}y^2-y^{3/4}$$

$$D_xy = \frac{2xy-\tfrac{3}{7}x^{-4/7}y^2-y^{3/4}}{\tfrac{3}{4}xy^{-1/4}+2yx^{3/7}-x^2}$$

13. (a) $2^2+3(2^2)(-2) = 4+3(4)(-2) = 4-24 = -20$

Also $5(2^2)(-2)^3+140 = 20(-8)+140 = -160+140 = -20$.

(b)
$$D_x(x^2+3x^2y) = D_x(5x^2y^3+140)$$

$$2x+6xy+3x^2D_xy = 10xy^3+15x^2y^2D_x(y).$$

When $x = 2$ and $y = -2$,

$$4-24+12D_xy = -160+240D_xy \text{ and}$$

$228D_xy = 140$ or $D_xy = \frac{140}{228} = \frac{35}{57}$ which is the slope of the tangent line.

The equation of the tangent line is $y+2 = \frac{35}{57}(x-2)$ or $y = \frac{35}{57}x - \frac{184}{57}$.

15. (a) $\sqrt[3]{8}+1 = 3$

Also $\sqrt{4}+3-2 = 2+3-2 = 3$

(b)
$$D_x\left((x^2+4)^{1/3}+y^2\right) = D_x\left((y^2+3)^{1/2}+3y-2\right)$$

$$\tfrac{1}{3}(x^2+4)^{-2/3}(2x)+2yD_xy = \tfrac{1}{2}(y^2+3)^{-1/2}(2y)D_xy+3D_xy.$$

When $x = 2$ and $y = 1$,

$$\tfrac{1}{3}(8)^{-2/3}(4) + 2D_xy = \tfrac{1}{2}(4)^{-1/2}(2)D_xy + 3D_xy$$

$$\tfrac{4}{3} \cdot \tfrac{1}{4} + 2D_xy = \tfrac{1}{2}D_xy + 3D_xy$$

$D_xy = \tfrac{1}{3} \cdot \tfrac{2}{3} = 2/9$ which is the slope of the tangent line at $(2, 1)$.

The equation of the tangent line is $y - 1 = \tfrac{2}{9}(x - 2)$ or $9y - 9 = 2x - 4$ or $2x - 9y = -5$.

17.

$$y^2 + 3xy - 4 = 0$$

$$D_x(y^2 + 3xy - 4) = 0$$

$$2\,yy' + 3y + 3xy' = 0$$

$$y' = \frac{-3y}{2y + 3x}$$

but $y = \dfrac{-3x \pm \sqrt{9x^2 + 16}}{2}$, so

$$y' = \frac{-3y}{2y + 3x} = -3\left(\frac{-3x \pm \sqrt{9x^2 + 16}}{2}\right)\frac{1}{-3x \pm \sqrt{9x^2 + 16} + 3x} = -\frac{3}{2}\left(\frac{-3x \pm \sqrt{9x^2 + 16}}{\pm\sqrt{9x^2 + 16}}\right) =$$

$$-\frac{3}{2}\left(\pm\frac{3x}{\sqrt{9x^2 + 16}} + 1\right) = \pm\frac{9x}{2\sqrt{9x^2 + 16}} - \frac{3}{2} \qquad \text{or} \qquad -\frac{3}{2} \pm \frac{9x}{2\sqrt{9x^2 + 16}}.$$

If $f_1(x) = \dfrac{-3x + \sqrt{9x^2 + 16}}{2}$ and $f_2(x) = \dfrac{-3x - \sqrt{9x^2 + 16}}{2}$,

$$f_1(x) = \frac{-3}{2} + \frac{9x}{2\sqrt{9x^2 + 16}} \qquad \text{and} \qquad f_2(x) = \frac{-3}{2} - \frac{9x}{2\sqrt{9x^2 + 16}}.$$

The results are consistent.

19. $y^3 - 2xy^2 - 5x^2y + 6x^3 = 0$

$$D_x(y^3 - 2xy^2 - 5x^2y + 6x^3) = 3y^2y' - 2y^2 - 4xyy' - 10xy - 5x^2y' + 18x^2 = 0$$

$$y' = \frac{2y^2 + 10xy - 18x^2}{3y^2 - 4xy - 5x^2}.$$

$$y^3 - 2xy^2 - 5x^2y + 6x^3 = (y - x)(y - 3x)(y + 2x) = 0$$

If $y = x$, $y = 3x$, $y = -2x$,

$f_1(x) = x,$ $\qquad f_2(x) = 3x,$ $\qquad f_3(x) = -2x$ $\qquad$ and

$f_1'(x) = 1,$ $\qquad f_2'(x) = 3,$ $\qquad f_3'(x) = -2.$

However, if $y = x$,

$$y' = \frac{2x^2 + 10x^2 - 18x^2}{3x^2 - 4x^2 - 5x^2} = \frac{-6x^2}{-6x^2} = 1.$$

If $y = 3x$,

$$y' = \frac{18x^2 + 30x^2 - 18x^2}{27x^2 - 12x^2 - 5x^2} = 3.$$

If $y = -2x$,

$$y' = \frac{8x^2 - 20x^2 - 18x^2}{12x^2 + 8x^2 - 5x^2} = \frac{-30x^2}{15x^2} = -2.$$

The results are consistent.

EXERCISE SET 2.9   DERIVATIVES OF EXPONENTIAL AND LOGARITHMIC FUNCTIONS

1. $\frac{dy}{dx} = e^{x^5}D_x(x^5) = 5x^4 e^{x^5}$

3. $dy/dx = e^{2x^2+5x+1}D_x(2x^2+5x+1) = (4x+5)e^{2x^2+5x+1}$

5. $dy/dx = e^{(x^2+5x+1)^2}D_x(x^2+5x+1)^2 = e^{(x^2+5x+1)^2}(2)(x^2+5x+1)^1 D_x(x^2+5x+1)$

   $= 2(x^2+5x+1)(2x+5)e^{(x^2+5x+1)^2}$

7. $\frac{dy}{dx} = (x^2+3x+7)D_x(e^{x^2+3}) + e^{x^2+3}D_x(x^2+3x+7)$

   $= (x^2+3x+7)e^{x^2+3}D_x(x^2+3) + e^{x^2+3}(2x+3)$

   $= (x^2+3x+7)(2x)e^{x^2+3} + (2x+3)e^{x^2+3} = e^{x^2+3}(2x^3+6x^2+16x+3)$

9. $dy/dx = (x^2+7)^{1/2}D_x e^{x^2+7} + e^{x^2+7}D_x(x^2+7)^{1/2}$

   $= (x^2+7)^{1/2}(2x)e^{x^2+7} + e^{x^2+7}\cdot\frac{1}{2}(x^2+7)^{-1/2}(2x) = e^{x^2+7}\left(2x\sqrt{x^2+7} + \frac{x}{\sqrt{x^2+7}}\right)$

11. $dy/dx = \frac{x^2 D_x e^{5x} - e^{5x}D_x(x^2)}{(x^2)^2} = \frac{x^2 \cdot 5e^{5x} - e^{5x}\cdot 2x}{x^4}$

    $= \frac{e^{5x}(5x-2)}{x^3}$

13. $dy/dx = \frac{e^{5x}D_x(x^4+2) - (x^4+2)D_x e^{5x}}{(e^{5x})^2}$

    $= \frac{4x^3 e^{5x} - 5(x^4+2)e^{5x}}{e^{10x}} = \frac{e^{5x}(4x^3-5x^4-10)}{e^{10x}} = \frac{4x^3-5x^4-10}{e^{5x}}$

15. $dy/dx = \frac{e^{3x+2}\cdot\frac{1}{2}(x^2+1)^{-1/2}2x - 3e^{3x+2}\sqrt{x^2+1}}{(e^{3x+2})^2} = \frac{x/\sqrt{x^2+1} - 3\sqrt{x^2+1}}{e^{3x+2}}$

17. $dy/dx = \frac{D_x(x^6+x^2+1)}{x^6+x^2+1} = \frac{6x^5+2x}{x^6+x^2+1}$

19. $dy/dx = \dfrac{D_x(x^5+3x^2-6)}{x^5+3x^2-6} = \dfrac{5x^4+6x}{x^5+3x^2-6}$

21. $dy/dx = (x^4+6x+1)D_x\ln|x^3-1|+\ln|x^3-1|D_x(x^4+6x+1)$

$$= (x^4+6x+1)\dfrac{D_x(x^3-1)}{x^3-1}+\ln|x^3-1|(4x^3+6)$$

$$= \dfrac{(x^4+6x+1)3x^2}{x^3-1}+(4x^3+6)\ln|x^3-1|$$

23. $\dfrac{dy}{dx} = (x^2+6)^{1/3}D_x\ln(x^2+6)+\ln(x^2+6)D_x(x^2+6)^{1/3}$

$$= (x^2+6)^{1/3}\cdot\dfrac{2x}{x^2+6}+\ln(x^2+6)\cdot\tfrac{1}{3}(x^2+6)^{-2/3}(2x)$$

25. $\dfrac{dy}{dx} = \dfrac{(x^5+2x+1)\cdot\left(\dfrac{20x^3}{5x^4+2}\right)+(5x^4+2)\ln(5x^4+2)}{(x^5+2x+1)^2}$

27. $dy/dx = \dfrac{\ln(x^2+2)D_x\ln(x^4+1)-\ln(x^4+1)D_x\ln(x^2+2)}{\ln^2(x^2+2)}$

$$= \dfrac{\ln(x^2+2)\left(4x^3/(x^4+1)\right)-\ln(x^4+1)\left(2x/(x^2+2)\right)}{\ln^2(x^2+2)}$$

29. $\dfrac{dy}{dx} = \dfrac{D_x(x^2+2)}{x^2+2}+\dfrac{D_x(x^4-3)}{x^4-3}-\left[e^{x^2}D_x(3+\ln|5x|)^2+(3+\ln|5x|)^2D_xe^{x^2}\right]$

$$= \dfrac{2x}{x^2+2}+\dfrac{4x^3}{x^4-3}-\left(e^{x^2}\cdot 2(3+\ln|5x|)\cdot\dfrac{5}{5x}+(3+\ln|5x|)^2\cdot 2xe^{x^2}\right)$$

$$= \dfrac{2x}{x^2+2}+\dfrac{4x^3}{x^4-3}-2(3+\ln|5x|)e^{x^2}\cdot\tfrac{1}{x}-2xe^{x^2}(3+\ln|5x|)^2$$

31. $y = \dfrac{\ln|5x+1|}{\ln(3)} = \dfrac{1}{\ln(3)}\ln|5x+1|$ where $\ln(3)$ is a constant.

$dy/dx = \dfrac{1}{\ln(3)}D_x|5x+1| = \dfrac{5}{(5x+1)\ln 3}$

33. $y = \dfrac{1}{\ln 10}\ln|x^2+5x+2|$ $\qquad dy/dx = \dfrac{1}{\ln 10}\cdot\dfrac{2x+5}{x^2+5x+2} = \dfrac{2x+5}{(x^2+5x+2)\ln 10}$

35. $y = \frac{(x^2+6)}{\ln 6}\ln|x^3+3x+7|$

$\frac{dy}{dx} = \frac{1}{\ln 6}\Big[(x^2+6)D_x\ln|x^3+3x+7|+\ln|x^3+3x+7|D_x(x^2+6)\Big]$

$= \frac{1}{\ln 6}\left[\frac{(x^2+6)(3x^2+3)}{x^3+3x+7}+2x\ln|x^3+3x+7|\right]$

37.
$$D_x(e^{x^2})+D_x(e^{x^3y^2}) = D_x(10x^2y^4)+D_x(11)$$

$$2xe^{x^2}+D_x(x^3y^2)e^{x^3y^2} = 20xy^4+40x^2y^3\frac{dy}{dx}$$

$$2xe^{x^2}+\Big(3x^2y^2+2x^3y\frac{dy}{dx}\Big)e^{x^3y^2} = 20xy^4+40x^2y^3\frac{dy}{dx}$$

$$2x^3ye^{x^3y^2}\frac{dy}{dx}-40x^2y^3\frac{dy}{dx} = 20xy^4-2xe^{x^2}-3x^2y^2e^{x^3y^2}$$

$$\frac{dy}{dx} = \frac{20xy^4-2xe^{x^2}-3x^2y^2e^{x^3y^2}}{2x^3ye^{x^3y^2}-40x^2y^3}$$

39.
$$D_x\ln(x^6y^2)+D_xe^{xy} = D_x3x^2y^3$$

$$\frac{6x^5y^2+2x^6yy'}{x^6y^2}+(y+xy')e^{xy} = 6xy^3+9x^2y^2y'$$

$$6x^5y^2+2x^6yy'+x^6y^3e^{xy}+x^7y^2y'e^{xy} = 6x^7y^5+9x^8y^4y'$$

$$y' = \frac{6x^7y^5-x^6y^3e^{xy}-6x^5y^2}{2x^6y+x^7y^2e^{xy}-9x^8y^4}$$

$$= \frac{6x^2y^4-xy^2e^{xy}-6y}{2x+x^2ye^{xy}-9x^3y^3}$$

41.
$$5x^2y^3+10 = \tfrac{5}{7}(2\ln x+\pi\ln y)+x^2+2$$

$$10xy^3+15x^2y^2y' = \tfrac{5}{7}(\tfrac{2}{x}+\tfrac{\pi y'}{y})+2x$$

$$15x^2y^2y'-\tfrac{5\pi}{7y}y' = \tfrac{10}{7x}+2x-10xy^3$$

$$\frac{dy}{dx} = \frac{\tfrac{10}{7x}+2x-10xy^3}{15x^2y^2-\tfrac{5\pi}{7y}}\cdot\frac{7xy}{7xy} = \frac{10y+14x^2y-70x^2y^4}{105x^3y^3-5\pi x}$$

43.

$$2\ln x + 4\ln y = \tfrac{1}{e}(x+y^2)+2$$

$$\tfrac{2}{x}+\tfrac{4y'}{y} = \tfrac{1}{e}(1+2yy'). \qquad \text{When } x=e \text{ and } y=\sqrt{e}$$

$$\tfrac{2}{e}+\tfrac{4y'}{\sqrt{e}} = \tfrac{1}{e}(1+2\sqrt{e}y')$$

$$2+4\sqrt{e}y' = 1+2\sqrt{e}y'$$

$$4\sqrt{e}y'-2\sqrt{e}y' = -1$$

$$y' = \tfrac{-1}{2\sqrt{e}} \quad \text{which is the slope of the tangent line.}$$

The equation of the tangent line at $(e, \sqrt{e})$ is $y-\sqrt{e} = \tfrac{-1}{2\sqrt{e}}(x-e)$

or $y = \tfrac{-x}{2\sqrt{e}}+\tfrac{\sqrt{e}}{2}+\sqrt{e} = \tfrac{-x}{2\sqrt{e}}+\tfrac{3\sqrt{e}}{2}$

45. If $f(x) > 0$, $D_x|f(x)| = D_x f(x) = f'(x) = f'(x)\cdot 1 = f'(x)\cdot\dfrac{f(x)}{f(x)} = f'(x)\dfrac{|f(x)|}{f(x)}$.

If $f(x) < 0$, $D_x|f(x)| = D_x(-f(x)) = -f'(x) = -f'(x)\cdot 1 = f'(x)\cdot\dfrac{-f(x)}{f(x)} = f'(x)\dfrac{|f(x)|}{f(x)}$

47. $P'(t) = 2000(.2)e^{.2t} = 400e^{.2t}$ $\qquad P'(10) = 400e^2 = 2{,}955.6$ bacteria/hour

49. (a) $C'(q) = 75D_q\left[qe^{-.00002q^2+.002}\right] = 75e^{.002}D_q\left[qe^{-.00002q^2}\right] =$

$$75e^{.002}\left[1e^{-.00002q^2}-.00004q^2e^{-.00002q^2}\right] = 75e^{-.00002q^2}e^{.002}\left[1-.00004q^2\right]$$

(b) $C'(100) = 75e^{-.2}e^{.002}[1-.4] = 36.92.$ \$36.92 is the approximate cost of producing the 100th unit.

51. (a) $C'(q) = .002e^{.002q}(-85{,}000q+60{,}000{,}000)-85{,}000e^{.002q}$

(b) $C'(150) = .002e^{3}(47{,}250{,}000)-85{,}000e^{3} = \$12{,}823.66$ is the approximate cost of producing the 150th unit.

53. $d(t) = 500+\tfrac{1}{2}\ln(t^3+9)$ $\quad v(t) = d'(t) = \tfrac{1}{2}\cdot\dfrac{(3t^2)}{t^3+9}$ $\quad v(6) = \tfrac{1}{2}\cdot\dfrac{3(36)}{225} = \dfrac{18}{75} = \dfrac{6}{25} = .24$ m/min

55. $f'(x) = (2x-1)e^x+(x^2-x-1)e^x = e^x(2x-1+x^2-x-1) = e^x(x^2+x-2) = e^x(x+2)(x-1)$

$f'(x) > 0$ if $x < -2$ or $x > 1.$

x			-2			1		
$x+2$	$-$	$-$	0	$+$	$+$	$+$	$+$	$+$
$x-1$	$-$	$-$	$-$	$-$	$-$	0	$+$	$+$
$(x+2)(x-1)$	$+$	$+$	0	$-$	$-$	0	$+$	$+$

EXERCISE SET 2.10   LOGARITHMIC DIFFERENTIATION

1. $D_x(\ln|y|) = D_x\ln|(x^2+1)(x^3+3x+2)(x^4+x^2+1)|$

$\qquad = D_x[(\ln(x^2+1)+\ln|x^3+3x+2|+\ln(x^4+x^2+1)]$

$\dfrac{y'}{y} = \dfrac{2x}{x^2+1}+\dfrac{3x^2+3}{x^3+3x+2}+\dfrac{4x^3+2x}{x^4+x^2+1}$

$y' = (x^2+1)(x^3+3x+2)(x^4+x^2+1)\Big(\dfrac{2x}{x^2+1}+\dfrac{3x^2+3}{x^3+3x+2}+\dfrac{4x^3+2x}{x^4+x^2+1}\Big)$ whenever $y \neq 0$.

3. $D_x|y| = D_x\ln[(x^3+5x^2+6x+2)^2(x^3+1)^{1/2}]$

$\qquad = D_x[2\ln|x^3+5x^2+6x+2|+\tfrac{1}{2}\ln|x^3+1|]$

$\dfrac{y'}{y} = \dfrac{2(3x^2+10x+6)}{x^3+5x^2+6x+2}+\dfrac{3x^2}{2(x^3+1)}$

$y' = (x^3+5x^2+6x+2)^2\sqrt{x^3+1}\Big(\dfrac{2(3x^2+10x+6)}{x^3+5x^2+6x+2}+\dfrac{3x^2}{2(x^3+1)}\Big)$ $\qquad$ when $y \neq 0$

5. $D_x\ln|y| = D_x[4\ln|3x^5-4x^2+6|+3\ln|x^2-6x+7|-\tfrac{1}{3}\ln(x^2+1)-2|x^3-4x^2+6|]$

$\dfrac{y'}{y} = \dfrac{4(15x^4-8x)}{3x^5-4x^2+6}+\dfrac{3(2x-6)}{x^2-6x+7}-\dfrac{2x}{3(x^2+1)}-\dfrac{2(3x^2-8x)}{x^3-4x^2+6}$

$y' = \dfrac{(3x^5-4x^2+6)^4(x^2-6x+7)^3}{(x^2+1)^{1/3}(x^3-4x^2+6)^2} \times$

$\qquad\qquad\qquad\qquad \left[\dfrac{4(15x^4-8x)}{3x^5-4x^2+6}+\dfrac{3(2x-6)}{x^2-6x+7}-\dfrac{2x}{3(x^2+1)}-\dfrac{2(3x^2-8x)}{x^3-4x^2+6}\right]$

7. $D_x\ln|y| = D_x[\ln e^{x^2}+2\ln|x^3+1|-2\ln(x^4+3)-2\ln|x^3+1|]$

$\dfrac{y'}{y} = 2x - \dfrac{8x^3}{x^4+3}$

$y' = \dfrac{e^{x^2}(x^3+1)^2}{(x^4+3)^2(x^3+1)^2}\Big(2x - \dfrac{8x^3}{x^4+3}\Big) = \dfrac{e^{x^2}}{(x^4+3)^2}\Big(2x - \dfrac{8x^3}{x^4+3}\Big)$

9. $\ln|y| = \ln 5-\ln(x^4+1)-2\ln(x^2+6)-\tfrac{1}{3}\ln|x^3+x^2+x+1|$

$D_x\ln|y| = \dfrac{y'}{y} = 0 - \dfrac{4x^3}{x^4+1} - \dfrac{4x}{x^2+6} - \dfrac{3x^2+2x+1}{3(x^3+x^2+x+1)}$

$$y' = \frac{-5}{(x^4+1)(x^2+6)^2\sqrt[3]{x^3+x^2+x+1}}\left[\frac{4x^3}{x^4+1}+\frac{4x}{x^2+6}+\frac{3x^2+2x+1}{3(x^3+x^2+x+1)}\right]$$

11. $\ln y = \ln(5^{5x^2-6x+10}) = (5x^2-6x+10)\ln 5$

$$D_x\ln y = \frac{y'}{y} = D_x(5x^2-6x+10)\ln 5 = (10x-6)\ln 5$$

$$y' = 5^{5x^2-6x+10}(10x-6)\ln 5$$

13. $\ln y = \ln 4^{\sqrt{x^2+3}} = \sqrt{x^2+3}\ln 4 = (x^2+3)^{1/2}\ln 4$

$$D_x\ln y = \frac{y'}{y} = \tfrac{1}{2}(x^2+3)^{-1/2}(2x)\ln 4 = \frac{x\ln 4}{\sqrt{x^2+3}}$$

$$y' = \frac{4^{\sqrt{x^2+3}}x\ln 4}{\sqrt{x^2+3}}$$

15. $\ln y = \ln(x^2)+\ln(2^x) = 2\ln x+x\ln 2$

$$D_x\ln y = \frac{y'}{y} = \tfrac{2}{x}+\ln 2. \qquad y' = (x^2)(2^x)(\tfrac{2}{x}+\ln 2)$$

17. $\ln y = x^2\ln(e^x+4)$

$$D_x\ln y = \frac{y'}{y} = 2x\ln(e^x+4)+x^2\left(\frac{e^x}{e^x+4}\right)$$

$$y' = (e^x+4)^{x^2}\left(2x\ln(e^x+4)+\frac{x^2e^x}{e^x+4}\right)$$

19. $\ln y = \ln(x^6+3x^2+5)^{e^{3x}} = e^{3x}\ln(x^6+3x^2+5)$

$$\frac{y'}{y} = 3e^{3x}\ln(x^6+3x^2+5)+e^{3x}\cdot\frac{6x^5+6x}{x^6+3x^2+5}$$

$$y' = (x^6+3x^2+5)^{e^{3x}}\left[3e^{3x}\ln(x^6+3x^2+5)+\frac{6e^{3x}(x^5+x)}{x^6+3x^2+5}\right]$$

21. $\ln y = (x^2+6)\ln(x^2+1)^{5x+2} = (x^2+6)(5x+2)\ln(x^2+1) = (5x^3+2x^2+30x+12)\ln(x^2+1)$

$$\frac{y'}{y} = (15x^2+4x+30)\ln(x^2+1)+(5x^3+2x^2+30x+12)\frac{(2x)}{x^2+1}$$

$$y' = \left[(x^2+1)^{5x+2}\right]^{(x^2+6)}\left[(15x^2+4x+30)\ln(x^2+1)+\frac{(2x)(5x^3+2x^2+30x+12)}{x^2+1}\right]$$

23. $\ln y = \ln 108 + \ln x + 2\ln(x^4+1) + \frac{1}{3}\ln(x^2+x+2) - 3\ln(x^2+2) - \frac{1}{2}\ln(x^8+x^4+8x+1)$

$\frac{y'}{y} = \frac{1}{x} + \frac{8x^3}{x^4+1} + \frac{2x+1}{3(x^2+x+2)} - \frac{6x}{x^2+2} - \frac{8x^7+4x^3+8}{2(x^8+x^4+8x+1)}$. When $x=2$ and $y=34$,

$y' = 34(\frac{1}{2} + \frac{64}{17} + \frac{5}{24} - 2 - \frac{532}{289})$

$y' = 17 + 128 + \frac{5(17)}{12} - 68 - \frac{1064}{17} = \frac{77(12)(17) + 5(17)(17) - 1064(12)}{12(17)}$

$= \frac{4385}{204}$, which is the slope of the tangent line.

The equation of the tangent line is $y = \frac{4385}{204}(x-2) + 34$.

25. Let $y = u_1 u_2 u_3 \ldots u_n$

$\ln|y| = \ln|u_1 u_2 \ldots u_n| = \ln|u_1| + \ln|u_2| + \cdots + \ln|u_n| = \ln|f_1(x)| + \ln|f_2(x)| + \cdots + \ln|f_n(x)|$

$D_x y = \frac{y'}{y} = D_x\ln|f_1(x)| + D_x\ln|f_2(x)| + \cdots + D_x\ln|f_n(x)|$

$\quad = \frac{f_1'(x)}{f_1(x)} + \frac{f_2'(x)}{f_2(x)} + \cdots + \frac{f_n'(x)}{f_n(x)}$

$y' = D_x(u_1 u_2 \cdots u_n) = u_1 u_2 \cdots u_n \sum_{i=1}^{n} \frac{D_x(u_i)}{u_i}$

$\quad = \sum_{i=1}^{n} u_1 u_2 \cdots u_n \frac{D_x(u_i)}{u_i}$

27. The correct procedure is

$\ln y = \ln(x^2+1)^{5x} = 5x\ln(x^2+1)$

$D_x\ln y = \frac{y'}{y} = 5\ln(x^2+1) + \frac{5x(2x)}{x^2+1}$

$y' = (x^2+1)^{5x}\left(5\ln(x^2+1) + \frac{10x^2}{x^2+1}\right)$

Adding the two incorrect results,

$10x^2(x^2+1)^{5x-1} + 5(x^2+1)^{5x}[\ln(x^2+1)] = (x^2+1)^{5x}\left(\frac{10x^2}{x^2+1} + 5\ln(x^2+1)\right) = y'$

29. Let $u = (x^4+3)$ and $v = x^2$. By Exercise 28,

$D_x(x^4+3)^{x^2} = x^2(x^4+3)^{x^2-1}(4x^3) + (x^4+3)^{x^2}\ln(x^4+3)(2x)$

EXERCISE SET 2.11   HIGHER ORDER DERIVATIVES

1. $y' = D_x(2x^2+5x+10) = 4x+5$ $y'' = D_x(4x+5) = 4$ $y''' = D_x(4) = 0$

3. $D_x y = D_x(x^5+3x^2-6x+7) = 5x^4+6x-6$

$D_x^2 y = D_x(5x^4+6x-6) = 20x^3+6$

$D_x^3 y = D_x(20x^3+6) = 60x^2$

5. $u = e^{t^4}$

$\dfrac{du}{dt} = e^{t^4}D_t(t^4) = 4t^3e^{t^4}$

$\dfrac{d^2u}{dt^2} = 12t^2e^{t^4} + 4t^3(4t^3)e^{t^4} = 12t^2e^{t^4} + 16t^6e^{t^4} = e^{t^4}(12t^2 + 16t^6)$

$\dfrac{d^3u}{dt^3} = 4t^3e^{t^4}(12t^2 + 16t^6) + e^{t^4}(24t + 96t^5) = e^{t^4}(144t^5 + 64t^9 + 24t)$

7. $t' = u(5e^{5u})+1e^{5u} = 5ue^{5u}+e^{5u} = e^{5u}(5u+1)$

$t'' = 5e^{5u}(5u+1)+e^{5u}(5) = 5e^{5u}(5u+2)$

$t''' = 25e^{5u}(5u+2)+25e^{5u} = 25e^{5u}(5u+3)$

9. $y' = 2x\ln(x^2+1)+\dfrac{(x^2+1)(2x)}{x^2+1} = 2x\ln(x^2+1)+2x = 2x\Big(1+\ln(x^2+1)\Big)$

$y'' = 2\Big(1+\ln(x^2+1)\Big)+2x\Big(\dfrac{2x}{x^2+1}\Big) = 2+2\ln(x^2+1)+\dfrac{4x^2}{x^2+1}$

$y''' = \dfrac{2(2x)}{x^2+1}+\dfrac{8x(x^2+1)-4x^2(2x)}{(x^2+1)^2} = \dfrac{4x}{x^2+1}+\dfrac{8x}{(x^2+1)^2} = \dfrac{4x^3+12x}{(x^2+1)^2}$

11. $z = 2t^{2/3}$ $z' = \tfrac{4}{3}t^{-1/3}$ $z'' = -\tfrac{4}{9}t^{-4/3}$ $z''' = \tfrac{16}{27}t^{-7/3}$

13. $w' = \dfrac{2v(v^4+1)-4v^3(v^2+1)}{(v^4+1)^2} = \dfrac{2v^5+2v-4v^5-4v^3}{(v^4+1)^2} = \dfrac{-2v^5-4v^3+2v}{(v^4+1)^2}$

$w'' = \dfrac{(-10v^4-12v^2+2)(v^4+1)^2-2(4v^3)(v^4+1)(-2v^5-4v^3+2v)}{(v^4+1)^4}$

$= \dfrac{(v^4+1)\big[(-10v^4-12v^2+2)(v^4+1)-8v^3(-2v^5-4v^3+2v)\big]}{(v^4+1)^4}$

$= \dfrac{\big[-10v^8-12v^6+2v^4-10v^4-12v^2+2+16v^8+32v^6-16v^4\big]}{(v^4+1)^3}$

$$= \frac{6v^8+20v^6-24v^4-12v^2+2}{(v^4+1)^3} = (6v^8+20v^6-24v^4-12v^2+2)(v^4+1)^{-3}$$

$$w''' = D_v(6v^8+20v^6-24v^4-12v^2+2)(v^4+1)^{-3}+(6v^8+20v^6-24v^4-12v^2+2)D_v(v^4+1)^{-3}$$
$$= (48v^7+120v^5-96v^3-24v)(v^4+1)^{-3}+(6v^8+20v^6-24v^4-12v^2+2)(-3)(v^4+1)^{-4}(4v^3)$$

15. (a) $v = 6t^2+10t-6$. When $t = 5$ the velocity is $6(25)+50-6 = 194$ cm/sec.

(b) $a = 12t+10$. When $t = 5$, the acceleration is $12(5)+10 = 70$ cm/sec^2.

17. $v = -5e^{-t} + 10t$ and $a = 5e^{-t} + 10$.

When $t = 3$, $v = -5e^{-3} + 30 = 29.75$ cm/sec and $a = 5e^{-3} + 10 = 10.25$ cm/sec^2

19. $D_x(x^3+y^3) = D_x 7$

$3x^2+3y^2y' = 0$ or $x^2+y^2y' = 0$

$y' = \frac{-x^2}{y^2}$. When $x = -1$ and $y = 2$, $y' = -\frac{1}{4}$.

$D_x(x^2+y^2y') = D_x(0)$

$2x+2y(y')^2+y^2y'' = 0$.

When $x = -1$ and $y = 2$, $-2+4(-\frac{1}{4})^2+4y'' = 0$. $4y'' = \frac{7}{4}$ and $y'' = \frac{7}{16}$.

$D_x(2x+2y(y')^2+y^2y'') = 0$

$2+2(y')^3+4yy'y''+2yy'y''+y^2y''' = 0$.

When $x = -1$, $y = 2$, $y' = -\frac{1}{4}$, $y'' = \frac{7}{16}$, and $y''' = \frac{-21}{128}$

21. $D_x(x^2y^3+y^3) = D_x(5x^4-3)$

$2xy^3+3x^2y^2y'+3y^2y' = 20x^3$

$(3x^2y^2+3y^2)y' = 20x^3-2xy^3$

$y' = \frac{20x^3-2xy^3}{3x^2y^2+3y^2}$

$D_x(2xy^3+3x^2y^2y'+3y^2y') = D_x(20x^3)$

$2y^3+6xy^2y'+6xy^2y'+6x^2y(y')^2+3x^2y^2y''+6y(y')^2+3y^2y'' = 60x^2$

$y'' = \frac{60x^2-2y^3-12xy^2y'-6x^2y(y')^2 - 6y(y')^2}{3x^2y^2+3y^2}$

$D_x(2y^3+12xy^2y'+6x^2y(y')^2+3x^2y^2y''+6y(y')^2+3y^2y'') = D_x(60x^2)$

$6y^2y'+12y^2y'+24xy(y')^2+12xy^2y''+12xy(y')^2+6x^2(y')^3+12x^2yy'y''+6xy^2y''$

$\quad +6x^2yy'y''+3x^2y^2y'''+6(y')^3+12y(y')y''+6yy'y''+3y^2y'''=120x$

or $\quad 18y^2y'+36xy(y')^2+18xy^2y''+6x^2(y')^3+18x^2yy'y''+3x^2y^2y'''$

$\quad +6(y')^3+18yy'y''+3y^2y'''=120x$

When $x=1$ and $y=1$

$\quad y'=\dfrac{20-2}{6}=\dfrac{18}{6}=3$

$\quad y''=\dfrac{60-2-12(3)-6(9)-6(9)}{6}=-\dfrac{86}{6}=-\dfrac{43}{3}\quad$ and $\quad y'''=204.$

23. $D_x\ln(x^2+y^2)=D_x(3x^5)$

$\dfrac{2x+2yy'}{x^2+y^2}=15x^4.$ When $x=0$ and $y=1$, $\dfrac{2y'}{1}=0$ and $y'=0$

$2x+2yy'=15x^6+15x^4y^2$

$D_x(2x+2yy')=D_x(15x^6+15x^4y^2)$

$2+2(y')^2+2yy''=90x^5+60x^3y^2+30x^4yy'.$

When $x=0$ and $y=1$, $2+2y''=0$ and $y''=-1.$

$D_x(2+2(y')^2+2yy'')=D_x(90x^5+60x^3y^2+30x^4yy')$

$0+4y'y''+2y'y''+2yy'''$

$\quad\quad =450x^4+180x^2y^2+120x^3yy'+120x^3yy'+30x^4(y')^2+30x^4yy''$

When $x=0$ and $y=1$, $2y'''=0$ and $y'''=0.$

25. (a) $C'(q)=100e^{-.001q}-100(.001)qe^{-.001q}=100e^{-.001q}-.1qe^{-.001q}$

(b) $C'(500)=100e^{-.001(500)}-.1(500)e^{-.001(500)}=50e^{-.5}\sim\$30.33$

(c) $C''(q)=(100)(-.001)e^{-.001q}-.1e^{-.001q}+.1q(.001)e^{-.001q}=(-.2+.0001q)e^{-.001q}$

(d) $C''(500)=(-.2+.05)e^{-.5}\sim-.091.$ When $q=500$ the marginal cost is decreasing at the approximate rate of $9¢$/radio/radio.

27. (a) $C(q)=2000+10q\ln(q^2+10,000)-10q\ln(10,000)$

$\quad C'(q)=0+10\ln(q^2+10,000)+\dfrac{20q^2}{q^2+10,000}-10\ln(10,000)\quad\quad$ (divide)

$\quad\quad =10\ln(q^2+10,000)+20-\dfrac{200,000}{q^2+10,000}-10\ln(10,000)$

(b) $C'(100)=10\ln(20,000)+20-\dfrac{200,000}{20,000}-10\ln(10,000)=10+\ln2=\$10.69$

(c) $C''(q) = \dfrac{20q}{q^2+10,000} + \dfrac{200,000(2q)}{(q^2+10,000)^2} = \dfrac{20q}{q^2+10,000} + \dfrac{400,000q}{(q^2+10,000)^2}$

(d) $C''(100) = \dfrac{2000}{20,000} + \dfrac{40,000,000}{(20,000)^2} = .1+.1 = .2.$

When the 100th racquet is produced the marginal cost is increasing at the rate of $20\cent/\text{racquet}/\text{racquet}$.

29.   Let $\quad u = x^5 \quad$ and $\quad v = e^{5x}.\quad$ Then

$$u' = 5x^4, \qquad\qquad v' = 5e^{5x},$$
$$u'' = 20x^3, \qquad\qquad v'' = 25e^{5x},$$
$$u''' = 60x^2, \quad\text{and}\quad v''' = 125e^{5x}.$$

$$\dfrac{d^3w}{dx^3} = x^5(125e^{5x})+3(5x^4)(25e^{5x})+3(20x^3)(5e^{5x})+60x^2 \cdot e^{5x}$$
$$= e^{5x}(125x^5+375x^4+300x^3+60x^2)$$

31.   Let $u = x^2$ and $v = e^{x^2}$. Then

$$u' = 2x,\ u'' = 2,\ u''' = 0 \text{ and } u^{(4)} = u^{(5)} = 0.$$
$$v' = 2xe^{x^2},\ v'' = 2e^{x^2}+4x^2e^{x^2},$$
$$v''' = 4xe^{x^2}+8xe^{x^2}+8x^3e^{x^2} = 12xe^{x^2}+8x^3e^{x^2},$$
$$v^{(4)}(x) = 12e^{x^2}+24x^2e^{x^2}+24x^2e^{x^2}+16x^4e^{x^2} = 12e^{x^2}+48x^2e^{x^2}+16x^4e^{x^2} \text{ and}$$
$$v^{(5)}(x) = 24xe^{x^2}+96xe^{x^2}+96x^3e^{x^2}+64x^3e^{x^2}+32x^5e^{x^2} = (120x + 160x^3 + 32x^5)e^{x^2}.$$

$$\dfrac{d^5w}{dx^5} = x^2(120x+160x^3+32x^5)e^{x^2}+10x(12+48x^2+16x^4)e^{x^2}+20(12x+8x^3)e^{x^2}$$
$$= e^{x^2}(32x^7+320x^5+760x^3+360x)$$

EXERCISE SET 2.12 CHAPTER REVIEW

1. (a) $\Delta x = 6-3 = 3$
   $\Delta y = 62-20 = 42$
   The average rate of change is $\dfrac{\Delta y}{\Delta x} = \dfrac{42}{3} = 14.$

   (b) $\Delta x = -3-1 = -4$
   $\Delta y = -10-2 = -12$
   The average rate of change is $\dfrac{\Delta y}{\Delta x} = \dfrac{-12}{-4} = 3.$

3. $\dfrac{\Delta A}{\Delta R} = \dfrac{81\pi-36\pi}{9-6} = \dfrac{45\pi}{3} = 15\pi \text{ cm}^2/\text{cm}$

5. (a) $f(2) = 8$, $f(4) = 26$.

   The slope is $\frac{26-8}{4-2} = \frac{18}{2} = 9$

   (b) $f(3) = 16$

   The slope is $\frac{16-8}{3-2} = 8$

   (c) $f(2.1) = 4.41+6.3-2 = 8.71$

   The slope is $\frac{8.71-8}{2.1-1} = \frac{.71}{.1} = 7.1$

   (d) The slope is

   $$f'(2) = \lim_{h\to 0}\frac{f(2+h)-f(2)}{h} = \lim_{h\to 0}\frac{(2+h)^2+3(2+h)-2-8}{h} = \lim_{h\to 0}\frac{4+4h+h^2+6+3h-10}{h}$$

   $$= \lim_{h\to 0}\frac{h(7+h)}{h} = 7$$

   (e) $y - 8 = 7(x-2)$ or $y = 7x - 6$.

7. (a) $\lim_{x\to 3}\frac{(x^2+7x-4)-26}{x-3} = \lim_{x\to 3}\frac{x^2+7x-30}{x-3} = \lim_{x\to 3}\frac{(x+10)(x-3)}{x-3} = 13$.

   (b) $\lim_{x\to -1}\frac{(x^2+7x-4)-(-10)}{x+1} = \lim_{x\to -1}\frac{x^2+7x+6}{x+1} = \lim_{x\to -1}\frac{(x+6)(x+1)}{x+1} = 5$.

   (c) $f'(x) = 2x+7$

   (d) $f'(3) = 6+7 = 13 = \lim_{x\to 3}\frac{f(x)-f(3)}{x-3}$

   $f'(-1) = -2+7 = 5 = \lim_{x\to -1}\frac{f(x)-f(-1)}{x+1}$

9. (a) $\lim_{x\to 1^-}(x+2) = 3$

   (b) $\lim_{x\to 1^+}(-x+3) = 2$

   The function is not continuous at 1.

11. (a) The function f is continuous at c if $f(c)$ exists and $\lim_{x\to c}f(x) = f(c)$.

   (b) The derivative of a function g at b is the value of $\lim_{h\to 0}\frac{g(b+h)-g(b)}{h}$ if g is defined in an open interval containing b and the limit exists.

13. $g'(9) = \lim_{h\to 0}\frac{g(9+h)-g(9)}{h} = \lim_{h\to 0}\frac{\sqrt{9+h}-3}{h}\cdot\frac{\sqrt{9+h}+3}{\sqrt{9+h}+3} = \lim_{h\to 0}\frac{h}{h(\sqrt{9+h}+3)} = \frac{1}{\sqrt{9}+3} = \frac{1}{6}$

15.

$$D_x(x^2y^3+5xy^2+x^2) = D_x(6+5e^y)$$

$$2xy^3+3x^2y^2y'+5y^2+10xyy'+2x = 5y'e^y$$

$$3x^2y^2y' + 10xyy'-5y'e^y = -2xy^3-5y^2-2x$$

$$y' = \frac{-2xy^3-5y^2-2x}{10xy-5e^y+3x^2y^2}$$

17. (a) $\dfrac{dy}{dx} = 4x^3 + 9x^2 - 10x + 9 \qquad \dfrac{d^2y}{dx^2} = 12x^2 + 18x - 10 \qquad \dfrac{d^3y}{dx^3} = 24x + 18.$

When $x = 1$ and $y = 2$, $\dfrac{d^3y}{dx^3} = 24 + 18 = 42.$

(b) $y = (x+1)^{1/2} \qquad y' = \tfrac{1}{2}(x+1)^{-1/2} \qquad y'' = -\tfrac{1}{4}(x+1)^{-3/2}.$

When $x = 8$ and $y = 3$, $y'' = -\tfrac{1}{4}(9)^{-3/2} = -\tfrac{1}{4} \cdot \tfrac{1}{27} = -\tfrac{1}{108}$

(c) $D_x(x^2 - 3xy + y^2) = D_x(11)$

$2x - 3y - 3xy' + 2yy' = 0.$

When $x = -1$ and $y = 2$, $-2 - 6 + 3y' + 4y' = 0.$ $\qquad 7y' = 8 \qquad y' = 8/7.$

$D_x(2x - 3y - 3xy' + 2yy') = 0$

$2 - 3y' - 3y' - 3xy'' + 2(y')^2 + 2yy'' = 0.$

When $x = -1$ and $y = 2$, $2 - 6(\tfrac{8}{7}) + 3y'' + 2(\tfrac{64}{49}) + 4y'' = 0$

$7y'' = \dfrac{48}{7} - \dfrac{14}{7} - \dfrac{128}{49} = \dfrac{34(7) - 128}{49} = \dfrac{110}{49}$

$y'' = \dfrac{110}{343}$

(d) $y = 2e^{x^2} \qquad D_xy = 4xe^{x^2} \qquad D_x^2y = 4e^{x^2} + 8x^2e^{x^2}$

$D_x^3y = 8xe^{x^2} + 16xe^{x^2} + 16x^3e^{x^2}.$ When $x = 0$ and $y = 2$, $D_x^3y = 0.$

(e) $D_xy = \ln(x+1) + 1 \qquad D_x^2y = \dfrac{1}{x+1} \qquad D_x^3y = -\dfrac{1}{(x+1)^2}.$ When $x = e - 1$, $D_x^3y = -\dfrac{1}{e^2}.$

(f) $y' = 2e^{2x} \qquad y'' = 4e^{2x} \qquad y''' = 8e^{2x}$

$y^{(10)} = 2^{10}e^{2x}.$ When $x = 0$, $y^{(10)} = 2^{10} = 1024.$

(g) $y' = 250x^{49} + 930x^{30} + 60x^{11} - 12x + 5.$ y'' is a polynomial of degree 48, y''' is a polynomial of degree 47, $y^{(50)}$ is a constant and $y^{(51)} = 0$ at all points including the point $(0, -3)$.

19. $D_x(x^2e^{y-1} + x^3y^2) = D_x(5x+2)$

$2xe^{y-1} + e^{y-1}x^2y' + 3x^2y^2 + 2x^3yy' = 5.$

When $x = 2$ and $y = 1$, $4 + 4y' + 12 + 16y' = 5.$ $\qquad 20y' = -11 \qquad$ and $\qquad y' = -11/20.$

$D_x(2xe^{y-1} + e^{y-1}x^2y' + 3x^2y^2 + 2x^3yy') = D_x(5)$

$2e^{y-1} + 2xe^{y-1} + 2xe^{y-1}y' + x^2e^{y-1}(y')^2 + x^2e^{y-1}y'' + 6xy^2 + 6x^2yy'$

$\qquad + 6x^2yy' + 2x^3(y')^2 + 2x^3yy'' = 0.$

When $x = 2$ and $y = 1$,

$$2+4+4y'+4(y')^2+4y''+12+24y'+24y'+16(y')^2+16y'' = 0.$$

$$18+52y'+20(y')^2+20y'' = 0$$

$$y'' = -\frac{1}{20}\Big(18+52(-\frac{11}{20})+\frac{20(121)}{400}\Big) = -\frac{1}{20}\Big(\frac{-1820}{400}\Big) = \frac{182}{800} = \frac{91}{400}$$

Similarly, $y''' = .08865$.

21. $f(x) = g(x^2+1) = g\big(h(x)\big)$ where $h(x) = x^2+1$. By the Chain Rule,

$$f'(x) = g'\big(h(x)\big)h'(x) = g'(x^2+1)(2x) = 2x \cdot \frac{5}{\sqrt[3]{(x^2+1)^2}} = 10x(x^2+1)^{-2/3}$$

CHAPTER 3

APPLICATIONS OF DERIVATIVES

EXERCISE SET 3.1 MARGINAL ANALYSIS

1. (a) $P(q) = R(q) - C(q) = -4q^2 + 2400q - 14000 - 13428\sqrt{q}$

   (b) $C'(q) = \dfrac{13428}{2\sqrt{q}} = 6714/\sqrt{q}$

   (c) $R'(q) = -8q + 2400$

   (d) $P'(q) = R'(q) - C'(q) = 2400 - 8q - 6714/\sqrt{q}$

   (e) $C'(144) = 6714/12 = \$559.50$ is the approximate cost of producing the 144th motorcycle.

   (f) $R'(144) = -8(144) + 2400 = \$1248$ is the approximate revenue from the 144th motorcycle.

   (g) $\$1248 - 559.50 = \$688.50$ is the approximate profit on the 144th motorcycle.

3. (a) $P(q) = 100qe^{-.02q} - (-.25q^2 + 25q + 600)$

   (b) $C'(q) = -.5q + 25$

   (c) $R'(q) = 100e^{-.02q} - 2qe^{-.02q}$

   (d) $P'(q) = 100e^{-.02q} - 2qe^{-.02q} - (-.5q + 25)$

   (e) $C'(40) = -20 + 25 = \$5$ is the approximate cost of producing the 40th calculator.

   (f) $R'(40) = 20e^{-.8}$ is the approximate revenue from the 40th calculator.

   (g) $20e^{-.8} - 5 = \$3.99$ is the approximate profit on the 40th calculator.

5. (a) $R(q) - C(q) = -.0125q^2 + 62.5q - 5000 - 1000\sqrt{q}$

   (b) $C'(q) = 500/\sqrt{q}$

   (c) $R'(q) = -.025q + 62.5$

   (d) $P'(q) = -.025q + 62.5 - 500/\sqrt{q}$

(e) $C'(1600) = 500/40 = \$12.50$

(f) $R'(1600) = -.025(1600) + 62.5 = \$22.50$

(g) $P'(1600) = \$22.50 - \$12.50 = \$10.00$

7. (a) $P(q) = -.05q^2 + 150q - 30q - 60000e^{-.001q} = -.05q^2 + 120q - 60000e^{-.001q}$

(b) $C'(q) = 30 + 60000(-.001)e^{-.001q} = 30 - 60e^{-.001q}$

(c) $R'(q) = -.1q + 150$

(d) $P'(q) = -.1q + 150 - 30 + 60e^{-.001q} = -.1q + 120 + 60e^{-.001q}$

(e) $C'(300) = 30 - 60e^{-.3} = -\$14.45$

(f) $R'(300) = -30 + 150 = \$120$

(g) $P'(300) = \$120 + 14.45 = \$134.45$

9. (a) $C(q) = 14000 + 13428\sqrt{q}$

$C'(q) = 6714/\sqrt{q}$

$A(q) = \dfrac{C(q)}{q} = \dfrac{14000 + 13428\sqrt{q}}{q}$

$A'(q) = \dfrac{1}{q}[C'(q) - A(q)] = \dfrac{1}{q}\left[\dfrac{6714}{\sqrt{q}} - \dfrac{14000}{q} - \dfrac{13428}{\sqrt{q}}\right]$

$= \dfrac{1}{q}\left[\dfrac{-6714}{\sqrt{q}} - \dfrac{14000}{q}\right]$

(b) $A'(200) = \dfrac{1}{200}\left[-\dfrac{6714}{\sqrt{200}} - \dfrac{14000}{200}\right] = -2.72$

An increase in production from 199 to 200 motorcycles will decrease the average cost by approximately \$2.72.

11. (a) $A'(q) = \dfrac{1}{q}(C'(q) - A(q)) = \dfrac{1}{q}\left(-.5q + 25 - \dfrac{-.25q^2 + 25q + 600}{q^2}\right)$

$= -.5 + \dfrac{25}{q} + .25 - \dfrac{25}{q} - \dfrac{600}{q^2} = -.25 - \dfrac{600}{q^2}$

(b) $A'(35) = -.25 - \dfrac{600}{(35)^2} = -.7398$

An increase in production of one unit (from 34 to 35) will decrease the average cost of production by approximately \$0.74.

13. (a) $A'(q) = \frac{1}{q}\left(C'(q) - \frac{C(q)}{q}\right) = \frac{1}{q}\left(\frac{500}{\sqrt{q}} - \frac{5000 + 1000\sqrt{q}}{q}\right)$

$= \frac{500}{q\sqrt{q}} - \frac{5000}{q^2} - \frac{1000}{q\sqrt{q}} = -\frac{500}{q\sqrt{q}} - \frac{5000}{q^2}$

(b) $A'(2000) = -\frac{500}{89442.7} - \frac{5000}{4000000} = -.00684$

An increase in production of one unit (from 1999 to 2000) will decrease the average cost of production by approximately $0.0068.

15. (a) $A'(q) = \frac{1}{q}\left(C'(q) - \frac{C(q)}{q}\right) = \frac{1}{q}\left(30 - 60e^{-.001q} - \frac{30q + 60000e^{-.001q}}{q}\right)$

$= \frac{1}{q}\left(30 - 60e^{-.001q} - 30 - \frac{60000e^{-.001q}}{q}\right)$

$= \frac{e^{-.001q}}{q}\left(-60 - \frac{60000}{q}\right)$

(b) $A'(750) = \frac{e^{-.75}}{750}\left(-60 - \frac{60000}{750}\right) = -.0882$

An increase in production of one unit (from 749 to 750) will decrease the average cost of production by approximately 8.82 cents.

17. (a) $R(q) = pq = q(-.001q^2 + 640) = -.001q^3 + 640q$

(b) $R'(q) = -.003q^2 + 640$

(c) $R'(400) = -.003(160000) + 640 = \$120$

The revenue from the 400th unit is approximately $120

19. (a) When $q = 400$, $p = \$50$ and when $q = 480$, $p = \$40$. The slope of the graph is

$\frac{50 - 40}{400 - 480} = -1/8$ and the demand equation is

$p - 50 = -\frac{1}{8}(q - 400)$ or $p = -q/8 + 100$

(b) $R(q) = pq = q(-q/8 + 100) = -q^2/8 + 100q$

(c) $R'(q) = -q/4 + 100$

(d) $R'(350) = \$12.50$. The revenue from the 350th unit is approximately $12.50.

21. $y' = 4x + 5$. The slope of the tangent line at $(-2,4)$ is $4(-2) + 5 = -3$. The equation of the tangent line is $y - 4 = -3(x + 2)$.

23. $y = x^{3/2} - 3x^{1/2} + 6x^{-1/2}$, $y' = (3/2)x^{1/2} - (3/2)x^{-1/2} - 3x^{-3/2}$.

When $x = 4$, $y' = \frac{3}{2}(2) - \frac{3}{2(2)} - \frac{3}{8} = 3 - \frac{3}{4} - \frac{3}{8} = \frac{15}{8}$.

The equation of the tangent line at $(4,5)$ is $y - 5 = \frac{15}{8}(x - 4)$.

25. $y' = 4xe^{x^2}$. When $x = 0$, $y' = 0$. The equation of the tangent line at $(0,2)$ is $y = 2$.

27. $v(t) = 10t + 7$ and $a(t) = 10$. The velocity when $t = 2$ is 27 ft/s and the acceleration is 10 ft/s^2.

29.
$$v(t) = (10t + 6)e^t + (5t^2 + 6t - 4)e^t$$
$$= e^t(5t^2 + 16t + 2)$$

$$a(t) = e^t(5t^2 + 16t + 2) + e^t(10t + 16)$$
$$= e^t(5t^2 + 26t + 18)$$

The velocity when $t = 2$ is $54e^2$ ft/s and the acceleration is $90e^2$ ft/s^2.

EXERCISE SET 3.2 ELASTICITY

1. $\eta = \frac{p}{q}D_p(-p^2 + 400) = \frac{-2p^2}{q}$. When $p = 15$, $q = -225 + 400 = 175$ and

$\eta = -\frac{2(225)}{175} = -2.571$

3. $\eta = \frac{p}{q}D_p(3p^4 - 200p^3 + 6p^2 - 600p + 6,265,000)$

$= \frac{p}{q}(12p^3 - 600p^2 + 12p - 600)$. When $p = 30$,

$q = 3(30)^4 - 200(30)^3 + 6(30)^2 - 600(30) + 6,265,000$

$= 2,430,000 - 5,400,000 + 5400 - 18000 + 6,265,000$

$= 3,282,400$ and $\eta = \frac{30}{3,282,400}(-216,240)$

$= -1.976$

5. $\eta = \frac{p}{q}D_p(p^3 - 4,800p + 128,000) = \frac{p}{q}(3p^2 - 4800)$

When $p = 10$, $q = 1000 - 48,000 + 128,000 = 81,000$ and

$\eta = \frac{10}{81,000}(300 - 4800) = -\frac{4500}{8100} = -.55\overline{5}$

7. $\eta = \frac{p}{q} D_p\left[100(169 - p^2)^{1/2}\right] = \left(\frac{100\,p}{q}\right)\left(\frac{-p}{\sqrt{169 - p^2}}\right) = \frac{-100p^2}{q\sqrt{169 - p^2}}$

When $p = 5$, $q = 100\sqrt{144} = 1200$

$\eta = \frac{-100(25)}{1200(12)} = \frac{-25}{144} = -.1736$

9. The percentage change in price is $\frac{15.3 - 15}{15}(100) = \frac{30}{15} = 2\%$

$\dfrac{\text{percentage change in demand}}{\text{percentage change in price}} \sim \eta = -2.571$

and the demand decreases by approximately $(2.571)(2) = 5.142\%$

11. Since $\eta = -1.976$ by problem 3, $\dfrac{\triangle q}{q} / \dfrac{\triangle p}{p} \sim -1.976$

Thus, the percentage decrease in demand, $\dfrac{\triangle q}{q}(100)$ is approximately $3(-1.976) = -5.93\%$

13. $\eta = -.55\overline{5}$ by problem 5. $\dfrac{\triangle q}{q} / \dfrac{\triangle p}{p} \sim \eta$

Thus, the percentage change increase in price, $\dfrac{\triangle p}{p}(100)$, is approximately

$\dfrac{-2}{\eta} = \dfrac{-2}{-.55\overline{5}} = 3.6\%$

15. Revenue $=$ demand(price)

$R(p) = 100p\sqrt{169 - p^2}$, $0 \le p \le 13$

$R'(p) = 100\sqrt{169 - p^2} - \dfrac{100p^2}{\sqrt{169 - p^2}}$

$\dfrac{pR'(p)}{R(p)} \sim \dfrac{p \triangle R}{R \triangle p} = \dfrac{p}{\triangle p(100)} \cdot \dfrac{100}{R/\triangle R}$

When $p = 5$ and $\triangle p = .2$

$\dfrac{5(1200 - 208.\overline{3}\,)}{6000} \sim .25 \cdot \dfrac{\triangle R}{R}(100)$ and $\dfrac{\triangle R}{R}(100) \sim 3.31\%$

is the approximate percent increase in revenue.

17. $q = -p^2 + 400$, $\dfrac{dq}{dp} = -2p$

(a) When $p = 5$, $q = 375$, $dq/dp = -10$, $\eta = \frac{5}{375}(-10)$

$= -.1\overline{3}$ and the demand is inelastic.

(b) When $p = 10$, $q = 300$, $dq/dp = -20$, $\eta = \frac{10}{300}(-20)$

$= -2/3$ and the demand is inelastic.

(c) When $p = 18$, $q = 76$, $dq/dp = -36$, $\eta = \frac{18}{76}(-36)$

$= -8.526$ and the demand is elastic

19. $q'(p) = 12p^3 - 600p^2 + 12p - 600$

(a) When $p = 10$, $q = 6,089,600$, $q'(10) = -48,480$ and $\eta = \frac{-10}{6,089,600}(48,480)$

$= -.0796$. The demand is inelastic.

(b) When $p = 20$, $q = 5,135,400$, $q'(20) = -144,360$

$\eta = \frac{p}{q}q'(p) = -.5622$. The demand is inelastic.

(c) When $p = 40$, $q = 1,130,600$, $q'(40) = -192,120$ and

$\eta = -6.797$. The demand is elastic.

21. $q'(p) = 3p^2 - 4800$

(a) When $p = 10$, q $= 81,000$, $q'(10) = -4500$, $\eta = \frac{-10(4500)}{81000}$.

$-1 < \eta < 0$ and the demand is inelastic.

(b) When $p = 20$, $q = 40,000$, $q' = 1200 - 4800 = -3600$,

$\eta = \frac{-20(3600)}{40,000} < -1$. The demand is elastic.

(c) When $p = 30$, $q = 11,000$, $q' = 2700 - 4800 = -2100$,

$\eta = \frac{-30(2100)}{11000} < -1$. The demand is elastic.

23. $q'(p) = \frac{-100p}{\sqrt{169 - p^2}}$

(a) When $p = 3$, $q = 1264.9$, $q' = -23.717$, and $\eta = -.05625$.
The demand is inelastic.

(b) When $p = 6$, $q = 1153.3$, $q' = -52.027$ and $\eta = -.271$.
The demand is inelastic.

(c) When $p = 9$, $q = 938.08$, $q' = -95.94$ and $\eta = -.9205$.
The demand is inelastic.

25. $q' = -2p + 5.$ $\eta = \frac{pq'}{q} = \frac{p(-2p+5)}{-p^2 + 5p + 48}$, $3 \le p \le 9$

The demand is elastic when:

$$\frac{p(-2p+5)}{-p^2+5p+48} < -1,$$

$$\frac{-2p^2+5p}{-p^2+5p+48} = \frac{2p^2-5p}{p^2-5p-48} < -1,$$

$$\frac{2p^2-5p}{p^2-5p-48} + \frac{p^2-5p-48}{p^2-5p-48} < 0,$$

$$\frac{3p^2-10p-48}{p^2-5p-48} = \frac{(p-6)(3p+8)}{p^2-5p-48} < 0.$$

$p^2 - 5p - 48 = 0$ if $p = \dfrac{5 \pm \sqrt{217}}{2} \sim \dfrac{5 \pm 14.7}{2}$ and neither solution is between 3 and 9.

$p-6$	$-$	$-$	$-$	$+$	$+$	$+$
$3p+8$	$+$	$+$	$+$	$+$	$+$	$+$
$p^2-5p-48$	$-$	$-$	$-$	$-$	$-$	$-$
$\dfrac{(p-6)(3p+8)}{p^2-5p-48}$	$+$	$+$	$+$	$-$	$-$	$-$

3   6   9

The demand is elastic if $6 < p \le 9$, inelastic if $3 \le p < 6$, and has unit elasticity if $p=6$.

27. $\dfrac{D_x \ln|y|}{D_x \ln|x|} = \dfrac{y'/y}{1/x} = \dfrac{x}{y}f'(x) = \eta$ if $y = f(x)$ and $xy \ne 0.$

29. Not necessarily.

For example, if 100 units are sold when $p = 10$ cents and 150 units are sold when $p = 5$ cents, the percent increase in demand is 50% and the percent decrease in price is 50%. The change in revenue is $100(.1) - 150(.05) = \$2.50$.

EXERCISE SET 3.3 OPTIMIZATION

1. (a) f is increasing on $(-2,2)$ and $(5,7)$.

    (b) f is decreasing on $(-4,-2)$ and $(2,5)$.

    (c)   Horizontal tangents occur at the points $(-2,1)$, $(2,5)$, and $(5,-1)$.

    (d)   Local maxima occur when $x = -4$, 2 and 7.

    (e)   Local minima occur when $x = -2$ and 5.

(f)   The absolute maximum is 5 which occurs when $x = 2$.

(g)   The absolute minimum is -1 which occurs when $x = 5$.

3. (a) f is increasing on $(-4, -2)$ and $(1,3)$.

(b) f is decreasing on $(-5, -4)$, $(-2,1)$, and $(3,5)$.

(c)   Horizontal tangents occur at the points $(-4, -3)$, $(-2,2)$, $(1, -1)$, and $(3,2)$.

(d)   Local maxima occur when $x = -5$, $x = -2$, and $x = 3$.

(e)   Local minima occur when $x = -4$, $x = 1$, and $x = 5$.

(f)   The absolute maximum is 2, which occurs when $x = -5$, $x = -2$, and $x = 3$.

(g)   The absolute minimum is -3, which occurs when $x = -4$.

5. $f(x) = x^2 - 6x + 3$

$f'(x) = 2x - 6$

(a) $2x - 6 > 0$ if $x > 3$. f is increasing on $(3, +\infty)$

(b) $2x - 6 < 0$ if $x < 3$. f is decreasing on $(-\infty, 3)$

(c)
        $f(3) = 9 - 18 + 3 = -6$
(d)

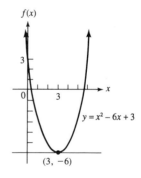

7. $f(x) = x^2 - 16$, $f'(x) = 2x$.

(a) $2x > 0$ if $x > 0$. $f(x)$ is increasing on $(0, +\infty)$.

(b) f is decreasing on $(-\infty, 0)$

(c)
        $f(0) = -16$
(d)

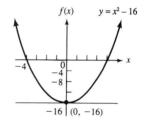

9. $f(x) = 2x^3 - 3x^2 - 12x + 10$

$f'(x) = 6x^2 - 6x - 12 = 6(x^2 - x - 2) = 6(x - 2)(x + 1)$

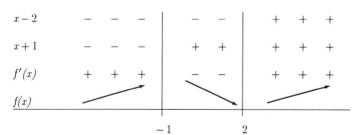

(a) f is increasing on $(-\infty, -1)$ and $(2, +\infty)$

(b) f is decreasing on $(-1,2)$

x	$f(x)$
-1	$+17$
$+2$	-10

(c)

(d)

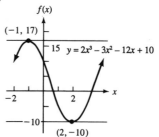

11. $f(x) = 2x^3 + 3x^2 - 72x + 100$

$f'(x) = 6x^2 + 6x - 72 = 6(x^2 + x - 12) = 6(x+4)(x-3)$

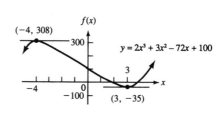

$x+4$	$-$ $-$	$+$ $+$ $+$	$+$ $+$
$x-3$	$-$ $-$	$-$ $-$ $-$	$+$ $+$
$f'(x)$	$+$ $+$	$-$ $-$ $-$	$+$ $+$
$f(x)$	↗	↘	↗
x	-4	3	

(a) f is increasing on $(-\infty, -4)$ and $(3, +\infty)$

(b) f is decreasing on $(-4,3)$

x	$f(x)$
-4	308
$+3$	-35

(c)

(d)

13. $f(x) = x^4 - 32x + 45$

$f'(x) = 4x^3 - 32 = 4(x^3 - 8)$

(a) $4(x^3 - 8) > 0$ if $x^3 > 8$ or $x > 2$. f is increasing on $(2, +\infty)$

(b) f is decreasing on $(-\infty, 2)$

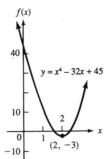

x	$f(x)$
0	45
2	-3
1	14

$y = x^4 - 32x + 45$

(2, −3)

(c)

(d)

15. $g'(x) = 2xe^{x^2}$

(a) $g'(x) > 0$ if $x > 0$, since e^{x^2} is positive for all x. That is $g'(x) > 0$ on $(0, +\infty)$.

(b) $g'(x) < 0$ on $(-\infty, 0)$

(c) $g'(x) = 0$ if $x = 0$

(d) $g'(x)$ is defined for all real numbers x.

17. $g'(x) = 2e^{x/2} + (2x + 4) \cdot \frac{1}{2}e^{x/2} = e^{x/2}(x + 4)$

(a) $e^{x/2}(x + 4) > 0$ if $x + 4 > 0$ or $x > -4$. $g'(x) > 0$ on $(-4, +\infty)$.

(b) $(-\infty, -4)$

(c) $e^{x/2}(x + 4) = 0$ if $x = -4$.

(d) $g'(x)$ is defined for all real numbers x.

19. $g'(x) = \ln|x + 1| + \frac{x+1}{x+1} = 1 + \ln|x + 1|$

(a) $1 + \ln|x + 1| > 0$ if $\ln|x + 1| > -1$.   This occurs if $|x + 1| > \frac{1}{e}$.   Hence, $x + 1 > \frac{1}{e}$ or $x + 1 < -\frac{1}{e}$. $g'(x) > 0$ on $(-\infty, -\frac{1}{e} - 1)$ and $(\frac{1}{e} - 1, +\infty)$

(b) $g'(x) < 0$ on $(-\frac{1}{e} - 1, -1)$ and $(-1, \frac{1}{e} - 1)$

(c) $g'(x) = 0$ if $x = -\frac{1}{e} - 1$ or $x = \frac{1}{e} - 1$

(d) The domain of g is $(-\infty, -1) \cup (-1, +\infty)$ and $g'(x)$ is defined at all points in the domain.

21. $g'(x) = \frac{1}{3}x^{-2/3} = \dfrac{1}{3\sqrt[3]{x^2}}$

    (a) $g'(x) > 0$ on $(-\infty,0)$ and $(0,+\infty)$ since $\sqrt[3]{x^2}$ is never negative.

    (b)   None

    (c)   None. A fraction can only be zero if its numerator is zero.

    (d) $g'(x)$ is undefined if $x = 0$ and $x = 0$ is in the domain of g.

23. $g'(x) = \dfrac{2(3x-4) - 3(2x+3)}{(3x-4)^2} = \dfrac{-17}{(3x-4)^2}$

    (a) $g'(x)$ is never positive, since the denominator is positive and the numerator negative.

    (b) $g'(x)$ is negative on $(-\infty, 4/3)$ and $(4/3, +\infty)$

    (c) $g'(x) \neq 0$, since the numerator is never zero.

    (d)   None. The domain of g is the same as the domain of g'.

25. f is increasing when f' is positive. f' is positive on the intervals $(-3,2),(5,\infty)$, and $(-\infty,-5)$.

27. The tangent is horizontal when $f'(x) = 0$. $f'(x) = 0$ when $x = -3, -2$, or 5.

29. $f'' < 0$ when f' is decreasing. f' is decreasing on $(-5, -4)$ and $(-1,4)$

31. $P'(x) = -3q^2 + 18q - 15 = -3(q^2 - 6q + 5) = -3(q-5)(q-1)$

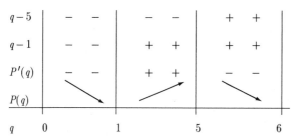

    (a)   The profit is decreasing on $(5,6)$ and $(0,1)$

    (b)   The profit is increasing on $(1,5)$

    (c)   A loss occurs if $P(q) = -q^3 + 9q^2 - 15q - 9 < 0$. $P(0) = -9$, $P(1) = -16$, $P(5) = 16$ and $P(6) = 9$. Thus, the company operates at a loss over $(0,1)$ but not over the entire interval $(1,5)$. It operates at a gain over $(5,6)$.

33. (a) $R(q) = qp = q(-q^2 + 48) = -q^3 + 48q$

(b) $R'(q) = -3q^2 + 48 > 0$ if $48 > 3q^2$ or $q^2 < 16$. The revenue function is increasing if $0 < q < 4$.

(c) The revenue function is decreasing if $q^2 > 16$ and $0 \le q \le 6$. Thus, it is decreasing if $4 < q < 6$.

35. (a) $P'(q) = -e^{-.1q} + (10 - q)(-.1)e^{-.1q}$

$\quad = e^{-.1q}(-1 - 1 + .1q) = e^{-.1q}(.1q - 2) < 0$ if $.1q - 2 < 0$ or $q < 20$.

Hence, P is decreasing for $0 \le q \le 10$.

(b) $R(q) = pq = (10 - q)e^{-.1q}q = (10q - q^2)e^{-.1q}$

(c) $R'(q) = (10 - 2q)e^{-.1q} + (10q - q^2)(-.1)e^{-.1q}$

$\quad = e^{-.1q}(10 - 2q - q + .1q^2) = e^{-.1q}(10 - 3q + .1q^2)$.

$.1q^2 - 3q + 10 = 0$ if $q^2 - 30q + 100 = 0$ or $q = \dfrac{30 \pm \sqrt{900 - 400}}{2}$

$\quad = 15 \pm 5\sqrt{5}$. $\quad 15 - 5\sqrt{5}\epsilon(0, 10)$ and $R'(q) < 0$ if $15 - 5\sqrt{5} < q < 10$.

R is decreasing on $(15 - 5\sqrt{5}, 10)$ and is increasing on $(0, 15 - 5\sqrt{5})$

37. (a) $f'(x) = 7x^6 + 10x^4 + 9x^2 + 5 \ge 5$ since x^6, x^4 and x^2 are nonnegative for all x. Thus, $f'(x) \ge 5 > 0$.

(b) Since $f'(x) > 0$ for all x, f is strictly increasing and any horizontal line can only intersect the graph of f in, at most, one point. Thus, by the horizontal line test, f is one $-$ to $-$ one.

EXERCISE SET 3.4 THE MEAN VALUE THEOREM

1. (a) $f'(x) = 2x$

$\quad f'(c) = 2c = \dfrac{f(7) - f(3)}{7 - 3} = \dfrac{49 - 9}{4}. \quad c = 5.$

(b)

3. (a) $f'(x) = \frac{1}{2\sqrt{x}}$

$f'(c) = \frac{1}{2\sqrt{c}} = \frac{f(9) - (4)}{9 - 4} = \frac{3 - 2}{5} = \frac{1}{5}$. $\quad \frac{5}{2} = \sqrt{c}$ and $c = \frac{25}{4}$

(b)

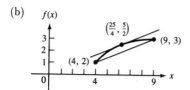

5. (a) $f'(x) = -\frac{1}{x^2}$

$f'(c) = -\frac{1}{c^2} = \frac{f(9) - f(4)}{9 - 4} = \frac{\frac{1}{9} - \frac{1}{4}}{5} = -\frac{1}{36}$. $\quad c^2 = 36$ and $c = 6$.

(b)

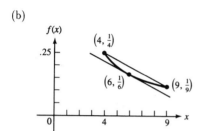

7. (a) $f'(x) = \frac{1}{3\sqrt[3]{x^2}}$

$f'(c) = \frac{1}{3\sqrt[3]{c^2}} = \frac{f(8) - f(1)}{(8 - 1)} = \frac{2 - 1}{7} = \frac{1}{7}$.

$\frac{7}{3} = \sqrt[3]{c^2}$, $\frac{343}{27} = c^2$, $c = \sqrt{\frac{343}{27}}$. $\quad c \sim 3.56$

(b)

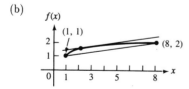

9. (a) $f(x) = (x+2)^3 + 3$

$$f'(x) = 3(x+2)^2$$

$$\frac{f(-2+\sqrt{3}) - f(-2-\sqrt{3})}{-2+\sqrt{3}+2+\sqrt{3}} = \frac{3\sqrt{3}+3+3\sqrt{3}-3}{2\sqrt{3}}$$

$$= \frac{6\sqrt{3}}{2\sqrt{3}} = 3$$

$f'(c) = 3(c+2)^2 = 3$ if $(c+2)^2 = 1$.

$c = -1, \ c = -3$

Both -1 and -3 are in the interval $\left[-2-\sqrt{3}, \ -2+\sqrt{3}\right]$

(b)

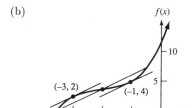

11. (a) Suppose $a < b$. Then $b - a > 0$ and $b^3 - a^3 = (b-a)(a^2 + ab + b^2) > 0$

since $a^2 + ab + b^2 > 0$ by Exercise 10. Thus, $b^3 > a^3$ and

$f(x) = x^3$ is strictly increasing.

(b) $f'(x) = 3x^2$; $f'(0) = 0$.

13. (a) The average velocity is $\dfrac{62 - 10}{6 - 2} = \dfrac{52}{4} = 13$ ft/s.

(b) $d' = 2t + 5 = 13$ if $2t = 8$ or $t = 4$ and $4\epsilon[2,6]$

EXERCISE 3.5 THE FIRST DERIVATIVE TEST

1. $f'(x) = 2x + 5 = 0$ if $x = -\dfrac{5}{2}$, $-\dfrac{5}{2}$ is the only critical value.

3. $g'(x) = 4x^{-1/5} = \dfrac{4}{5\sqrt{x}}$. Since $g'(x)$ is never zero, the only critical value occurs when $g'(x)$

is undefined, which occurs when $x = 0$. Since 0 is in the domain of g, 0 is a critical value.

5. $h'(x) = -\dfrac{4}{3}(x+1)^{-5/3} = \dfrac{-4}{3 \cdot \sqrt[3]{(x+1)^5}}$. $h'(x)$ is never zero and $h'(x)$ is undefined

if $x = -1$. However, -1 is not in the domain of h, so there are no critical values.

7. $f'(x) = 6x^2 + 18x - 108 = 6(x^2 + 3x - 18) = 6(x + 6)(x - 3) = 0$ if $x = -6$ or $x = 3$.
-6 and 3 are the critical values.

9. $g'(x) = 2xe^{3x} + 3x^2 e^{3x} = xe^{3x}(2 + 3x)$. $g'(x) = 0$ if $x = 0$ or $2 + 3x = 0$.
The critical values are 0 and $-2/3$.

11. $h'(x) = 3x^2 + 12x + 9 = 3(x^2 + 4x + 3) = 3(x + 1)(x + 3) = 0$ if $x = -1$ or $x = -3$.
The critical value is -1 (-3 is not in the domain of h).

13. $f(x) = \ln|x + 1| + 1 = 0$ if $\ln|x + 1| = -1$. $e^{-1} = \frac{1}{e} = |x + 1|$.
$x + 1 = \frac{1}{e}$ or $x + 1 = -\frac{1}{e}$.   The critical values are $\frac{1}{e} - 1$ and $-\frac{1}{e} - 1$.

15. $f'(x) = -2x + 6 = 0$ if $x = 3$. $x = 3$ is the only critical value.   There is a local maximum at $x = 3$.

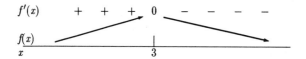

17. $g'(x) = 3x^2 + 12x - 36 = 3(x^2 + 4x - 12) = 3(x + 6)(x - 2)$.
-6 and 2 are the critical values, but $-6 \notin [-4, 4]$.

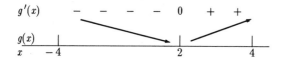

There is a local minimum at $x = 2$ and local maxima at $x = -4$ and $x = 4$.

19. $h'(x) = 3x^2 + 6x - 24 = 3(x^2 + 2x - 8) = 3(x + 4)(x - 2)$
2 and -4 are the critical values, but 2 is not in $[-5, 1]$

There is a local maximum at $x = -4$. There are local minima at $x = -5$ and $x = 1$.

21. $f'(t) = 4e^{3t} + 12te^{3t} = 4e^{3t}(1 + 3t)$. $t = -\frac{1}{3}$ is the only critical value.

$f'(t)$ $-$ $-$ 0 $+$ $+$

$f(t)$

t $-1/3$

There is a local minimum at $t = -1/3$.

23. $g'(x) = \ln|x+1| + \frac{x}{x+1}$.   The critical value is 0.

$g'(1) > 0$ and $g'(-1/2) < 0$.   There is a local minimum at $x = 0$.

25. $h(t) = \frac{t^2 + t + 2}{t - 1} = t + 2 + \frac{4}{t - 1}$ by division.

$h'(t) = 1 - \frac{4}{(t-1)^2} = 0$ if $(t-1)^2 = 4$. $t = 3$ and $t = -1$ are critical values.

$h'(t)$ $+$ $+$ 0 $-$ $-$ U $-$ $-$ 0 $+$ $+$

$h(t)$

t -1 1 3

There is a local minimum when $t = 3$.   There is a local maximum at $t = -1$.

27. $f'(x) = (2x + 3)^3\sqrt{x-2} + (x^2 + 3x + 11)$. $\dfrac{1}{3(x-2)^{2/3}}$

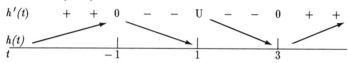

$$= \frac{1}{3(x-2)^{2/3}} \cdot \left[3(2x+3)(x-2) + x^2 + 3x + 11\right]$$

$$= \frac{3(2x^2 - x - 6) + x^2 + 3x + 11}{3(x-2)^{2/3}} = \frac{7x^2 - 7}{3(x-2)^{2/3}} = \frac{7(x-1)(x+1)}{3(x-2)^{2/3}}$$

Critical points are $1, -1$, and 2

$(x-1)$ $-$ $-$ $-$ $-$ $-$ $-$ 0 $+$ $+$ $+$ $+$ $+$ $+$

$(x+1)$ $-$ $-$ $-$ 0 $+$ $+$ $+$ $+$ $+$ $+$ $+$ $+$ $+$

$(x-2)^{2/3}$ $+$ $+$ $+$ $+$ $+$ $+$ $+$ $+$ $+$ 0 $+$ $+$ $+$

$f'(x)$ $+$ $+$ $+$ 0 $-$ $-$ 0 $+$ $+$ U $+$ $+$ $+$

$f(x)$

x -1 1 2

A local maximum occurs at $x = -1$.

A local minimum occurs at $x = 1$.

29. $f'(x) = 2x + 4 = 0$ if $x = -2$. However, -2 is not in the interval. Evaluating $f(x)$ at the end points, $f(-1) = 1$, $f(2) = 16$.

There is an absolute minimum of 1 when $x = -1$ and an absolute maximum of 16 when $x = 2$.

31. $g'(x) = 6x^2 + 18x - 108 = 6(x^2 + 3x - 18) = 6(x + 6)(x - 3) = 0$

if $x = -6$ or $x = 3$. 3 is the only critical point.

$g(3) = -159$
$g(-4) = 478$
$g(5) = -35$

The absolute maximum is $g(-4) = 478$

The absolute minimum is $g(3) = -159$

33. $h'(x) = 2xe^{3x} + 3x^2 e^{3x} = xe^{3x}(2 + 3x) = 0$ if $x = 0$ of $x = -2/3$

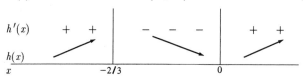

There is an absolute minimum of $h(0) = 0$, since $h(x)$ is positive unless $x = 0$.

There is no absolute maximum as $h(-2/3) < h(100)$.

35. $f'(x) = e^{-x/4} - \frac{x}{4}e^{-x/4} = e^{-x/4}(1 - x/4) = 0$ if $x = 4$.

$f(4) = 4e^{-1} \sim 1.47$

$f(-1) = -e^{1/4}$

$f(6) = 6e^{-1.5} \sim 1.34$

The absolute minimum is $f(-1) = -e^{-1/4}$ and the absolute maximum is $f(4) \sim 1.47$.

37. From Exercise 27, f has local extrema at $x = 1$ and $x = -1$.

$f(1) = -15$ $\qquad\qquad$ $f(-2) = -9\sqrt[3]{4}$

$f(-1) = -9\sqrt[3]{3}$ $\qquad\qquad$ $f(3) = 29$

The absolute maximum is $f(3) = 29$ and the absolute minimum is $f(1) = -15$.

39. Suppose $f'(x_1) < 0$. Then, there is an interval $(x_1 - h_1, x_1 + h)$ such that $\dfrac{f(x) - f(x_1)}{x - x_1}$

is negative for $x \in (x_1 - h, x + h)$. If $x_1 - h < s < x_1$, $s - x_1$ is negative, $f(s) - f(x_1)$ must be

positive, and $f(s) > f(x_1)$. If $x_1 < t < x_1 + h$, $t - x_1$ is positive, $f(t) - f(x_1)$ must be
negative, and

$f(t) < f(x_1)$. Therefore, $f(t) < f(x_1) < f(s)$ when $x_1 - h < s < x_1 < t < x_1 + h$.

41. (a) The average variable cost is $A(q) = \dfrac{V(q)}{q}$. $A(q)$ can have a minimum only at a

critical point, which occurs when $A'(q) = \dfrac{qV'(q) - V(q)}{q^2} = 0$ or $V'(q) = \dfrac{V(q)}{q} = A(q)$.

Since $C(q) = F + V(q)$, $C'(q) = V'(q)$.

If the average variable cost is a minimum at q_0, $C'(q_0) = V'(q_0) = V(q_0)/q_0$. Thus,
at q_0, the marginal cost and the average variable cost are equal.

(b) The average total cost is $\dfrac{F + V(q)}{q}$ which has derivative $\dfrac{-F + qV'(q) - V(q)}{q^2}$.

This derivative is zero at q_0 only if $\dfrac{F + V(q_0)}{q_0} = V'(q_0) = C'(q_0)$

EXERCISE SET 3.6 THE SECOND DERIVATIVE TEST

1. $f'(x) = 2x + 5 = 0$ if $x = -5/2$

$f''(x) = 2$, $f''(-5/2) > 0$. Hence there is a local minimum at $x = -5/2$.

3. $g'(x) = 4/\sqrt[5]{x}$. 0 is a critical value but $g'(0)$ does not exist, so the Second Derivative Test
cannot be used for finding extrema.

5. $h'(x) = (-2/3)(x+1)^{-5/3}$. There are no critical values and therefore no local extrema.

7. $f'(x) = 6x^2 + 18x - 108 = 6(x^2 + 3x - 18) = 6(x + 6)(x - 3) = 0$ if $x = 3$ or $x = -6$.

$f''(x) = 12x + 18$. $f''(3) = 36 + 18 > 0$ and $f''(-6) = 12(-6) + 18$.

There is a local minimum at $x = 3$ and a local maximum at $x = -6$.

9. $g'(x) = xe^{3x}(2 + 3x) = 0$ if $x = 0$ or $x = -2/3$.

$g''(x) = D_x(2xe^{3x} + 3x^2 e^{3x}) = 2e^{3x} + 6xe^{3x} + 6xe^{3x} + 9x^2 e^{3x}$

$= e^{3x}(2 + 12x + 9x^2)$.

$g'(0) = 2$ and $g'(-2/3) = (1/e^2)(2 - 8 + 4) < 0$. There is a local minimum when $x = 0$
and a local maximum when $x = -2/3$.

11. $h'(x) = 3x^2 + 12x + 9 = 3(x^2 + 4x + 3) = 3(x + 3)(x + 1)$

The critical value in $[-2,2]$ is -1. $h''(x) = 6x + 12$. $h''(-1) = 6$.

There is a local minimum when $x = -1$.

13. $f'(x) = \ln|x+1| + 1 = 0$ if $x = -1 \pm 1/e$

$f''(x) = \frac{1}{x+1}$, $f''(-1+1/e) = e > 0$, $f''(-1-1/e) = -e < 0$.

There is a local minimum when $x = -1+1/e$ and a local maximum when $x = -1-1/e$.

15. $f'(x) = -2x+6 = 0$ if $x = 3$

$f''(x) = -2$. $f''(3) = -2$. There is a local maximum when $x = 3$.

17. $g'(x) = 3x^2 + 12x - 36 = 3(x^2 + 4x - 12) = 3(x+6)(x-2) = 0$ if $x = -6$ or 2.

2 is the only critical value in $[-4,4]$. $g''(x) = 6x + 12$. $g''(2) > 0$. There is a local minimum at $x = 2$.

19. $h'(x) = 3x^2 + 6x - 24 = 3(x^2 + 2x - 8) = 3(x+4)(x-2) = 0$ if $x = -4$ or $x = 2$.

The only critical value in $[-5,1]$ is -4.
$h''(x) = 6x + 6$. $h''(-4) < 0$.

There is a local maximum when $x = -4$.

21. $f'(t) = 4e^{3t} + 12te^{3t} = 4e^{3t}(1+3t) = 0$ if $t = -1/3$,

$f''(t) = 12e^{3t}(1+3t) + 12e^{3t} = 12e^{3t}(2+3t)$

$f''(-1/3) > 0$. There is a local minimum of $-4/(3e)$ when $t = -1/3$.

23. $g'(x) = \ln|x+1| + \frac{x}{x+1}$. $g''(x) = \frac{1}{x+1} + \frac{1}{(x+1)^2}$. 0 is the critical value.

$g''(0) > 0$. A local minimum occurs when $x = 0$.

25. $h'(t) = \dfrac{(2t+1)(t-1)^2 - (t^2+t+2)}{(t-1)^2} = \dfrac{t^2 - 2t - 3}{(t-1)^2} = \dfrac{(t-3)(t+1)}{(t-1)^2}$

3 and -1 are the critical values.

$h''(t) = \dfrac{(2t-2)(t-1)^2 - 2(t-1)(t^2 - 2t - 3)}{(t-1)^4} = \dfrac{2(t-1)^2 - 2(t^2 - 2t - 3)}{(t-1)^3}$

$= \dfrac{8}{(t-1)^3}$. $h''(3) > 0$ and $h''(-1) < 0$.

There is a local maximum when $t = -1$ and a local minimum when $t = 3$.

27. $f'(x) = (2x+3)(x-2)^{1/3} + (x^2 + 3x + 11)(x-2)^{-2/3}(1/3)$

$= \dfrac{3(2x+3)(x-2) + (x^2 + 3x + 11)}{3(x-2)^{2/3}} = \dfrac{7(x-1)(x+1)}{3(x-2)^{2/3}}$

$x = 1$, -1 and 2 are critical values but only 1 and -1 are in the domain of f'.

$f''(x) = \frac{7}{3}\left[2x(x-2)^{2/3} - (x^2-1)\frac{2}{3}(x-2)^{-1/3}\right] \cdot \frac{1}{(x-2)^{4/3}}$

$f''(1) = \frac{7}{3}\left[\frac{2}{1}\right] = \frac{14}{3}$. There is a local minimum when $x = 1$.

$f''(-1) = \frac{7}{3}\left[-2(-3)^{2/3}/(-3)^{4/3}\right] < 0$. There is a local maximum when $x = -1$.

29. $f'(x) = 2x + 4 = 0$ if $x = -2$. There are no critical values in $[-1,2]$ so the Second Derivative Test cannot be used.

31. $g'(x) = 6x^2 + 18x - 108 = 6(x^2 + 3x - 18) = 6(x+6)(x-3) = 0$ if $x = -6$ or $x = 3$.

The only critical value in $[-4,5]$ is 3.

$g''(x) = 12x + 18$. $g''(3) > 0$. There is a local minimum when $x = 3$.

33. $h'(x) = 2xe^{3x} + 3x^2e^{3x} = xe^{3x}(2 + 3x) = 0$ if $x = 0$ or $-2/3$.

$h''(x) = 2e^{3x} + 6xe^{3x} + 6xe^{3x} + 9x^2e^{3x} = e^{3x}(2 + 12x + 9x^2)$

$h''(0) > 0$ and $h''(-2/3) = \frac{1}{e^2}(2 - 8 + 4) < 0$.

There is a local minimum when $x = 0$ and a local maximum when $x = -2/3$.

35. $f'(x) = e^{-x/4} - \frac{x}{4}e^{-x/4} = e^{-x/4}(1 - \frac{x}{4}) = 0$ if $x = 4$.

$f''(x) = -\frac{1}{4}e^{-x/4}(1 - x/4) - \frac{1}{4}e^{-x/4} = -\frac{1}{2}e^{-x/4} + \frac{xe^{-x/4}}{16}$

$f''(4) = -\frac{1}{2}e^{-1} + \frac{1}{4}e^{-1} < 0$. There is a local maximum when $x = 4$.

37. From Exercise 27, the critical points are 1 and -1. These critical points are in the interval $[-2, 3]$. As in Exercise 27, there is a local minimum at $x = 1$ and a local maximum at $x = -1$.

39. (a)

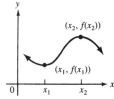
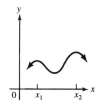

f is increasing on $[x_1, x_2]$ with critical values at x_1 and x_2.

If the graph were to ever decrease between x_1 and x_2, as shown above,

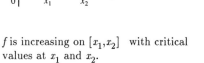

a local minimum would occur between x_1 and x_2, since f is continuous on $[x_1, x_2]$. However, there cannot be such a local minimum as there are no critical values between x_1 and x_2.

(b)

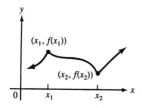

f is decreasing on $[x_1, x_2]$

EXERCISE SET 3.7 APPLICATIONS

1.

$2x + 2y = 400$, $x + y = 200$, $y = 200 - x$

The area is $xy = x(200 - x) = 200x - x^2$

Let $A(x) = 200x - x^2$. $A'(x) = 200 - 2x = 0$ if $x = 100$. $A''(x) = -2$. $A''(100) = -2$.

The area is maximized when $x = y = 100$ feet.

3.

Volume $= l^2 h = 64$, $h = \dfrac{64}{l^2}$

cost $=$ cost of bottom $+$ cost of the 4 sides

$= l^2(6) + 4(lh)(1.5) = 6l^2 + 6l(\dfrac{64}{l^2})$

$C(l) = 6l^2 + \dfrac{384}{l}$

$C'(l) = 12l - \dfrac{384}{l^2} = 0$ if $12l^3 = 384$, $l^3 = 32$ or $l = \sqrt[3]{32}$ feet.

$C''(l) = 12 + \dfrac{2(384)}{l^3} > 0$ when $l = \sqrt[3]{32}$

The cost is minimized when $l = \sqrt[3]{32}$ and $h = \dfrac{64}{(32)^{2/3}} = \dfrac{16}{4^{2/3}} = \dfrac{4^2}{4^{2/3}} = 4^{4/3}$ feet.

5.

$xy = 1200$, $x = 1200/y$

cost $= (2x + y)9 + 15y = 18x + 24y$

$C(y) = 18(1200)/y + 24y$, $C'(y) = -\dfrac{21600}{y^2} + 24 = 0$ if $y^2 = 900$ or $y = \pm 30$.

$C''(y) = \dfrac{43200}{y^3}$. $C''(30) > 0$. When $y = 30$, $x = 40$. The cost is a minimum when the

expensive side is 30 feet long, the opposite side is 30 feet long and the other 2 sides are 40 feet in length.

7.

Volume $= 7xy = 1764$

$$y = \frac{1764}{7x} = \frac{252}{x}$$

cost = cost of bottom + cost of front and back
+ cost of sides $= xy(20) + 2(7x)8 + 2(7y)14$

$C(x) = 20x(\frac{252}{x}) + 112x + 196(252)/x$, $0 < x$.

$C'(x) = 112 - \dfrac{196(252)}{x^2} = 0$ if $x^2 = 441$ or $x = 21$.

$C''(x) = \dfrac{2(196)(252)}{x^3}$, $C''(21) > 0$. The cost is minimized when the bottom is 21 ft by 12

ft, the front and back are 21 ft by 7 ft and the sides are 12 ft by 7 ft.

9.

cost = cost of ends + cost of front

$= 2x(8) + 5y = 320$. $5y = 320 - 16x$,

$$y = \frac{320 - 16x}{5}$$

Area $= xy = x(\dfrac{320 - 16x}{5}) = 64x - \dfrac{16x^2}{5} = A(x)$

$A'(x) = 64 - \dfrac{32x}{5} = 0$ if $x = 10$. $A''(x) = -\dfrac{32}{5}$.

The area is maximized when the ends are 10 feet, and the front is 32 feet.

11.

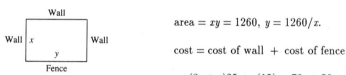

area $= xy = 1260$, $y = 1260/x$.

cost = cost of wall + cost of fence

$= (2x + y)35 + y(15) = 70x + 50y$.

$C(x) = 70x + \dfrac{1260(50)}{x}$. $C'(x) = 70 - \dfrac{1260(50)}{x^2} = 0$ if $x^2 = 900$ or $x = 30$

$C''(x) = 2(1260)(50)/x^3$. $C''(30) > 0$. The minimum cost occurs when the rectangle is 30 ft by 42 ft with the fence on the 42 ft side.

13. Let t be the number of weeks until picking. The income is maximized when the revenue is greatest.

$R(t)$ = number of pounds (price/lb).

$R(t) = (96 + 4t)(90 - 3t) = 90(96) + 72t - 12t^2$.

$R'(t) = 72 - 24t = 0$ if $t = 3$

$R''(3) < 0$. The maximum income occurs when $t = 3$ weeks. The revenue per tree is $87.48.

15. Profit = income − cost = number of cans sold (price per can) − cost
If x is the price in dollars,

$$P(x) = x(-x^2 + 144) - 5.5(-x^2 + 144) = -x^3 + 5.5x^2 + 144x - 792$$
$$P'(x) = -3x^2 + 11x + 144 = -(3x^2 - 11x - 144) = -(3x + 16)(x - 9) = 0$$

if $x = 9$ or $-16/3$. $P''(x) = -6x + 11$. $P''(9) < 0$.

The maximum profit occurs if the price per can is \$9.

17. Cost = cost above ground + cost under water.

$$C(x) = 300(30 - x) + 500\sqrt{256 + x^2} \text{ where } x \text{ is shown in the diagram.}$$
$$C'(x) = -300 + 500x/\sqrt{256 + x^2} = 0 \text{ if } 3/5 = \frac{x}{\sqrt{256 + x^2}} \cdot \quad 9/25 = \frac{x^2}{256 + x^2},$$

$256(9) + 9x^2 = 25x^2$, $16x^2 = 256(9)$, $x^2 = (16)(9)$. $x = \pm 12$, and $x\epsilon[0, 30]$.

$C(0) = 9000 + 500(16) = \$17{,}000$

$C(30) = 500(34) = \$17{,}000$

$C(12) = 300(18) + 500(20) = \$15{,}400$. Thus, the cost is minimized when $x = 12$ miles and the total minimum cost is \$15,400.

19. Revenue = Revenue from cattle + Revenue from grain A.
Let x be the number of acres in grain B, $0 \le x \le 500$.

$$R(x) = 500(300 + 3x - .003x^2)(1.20) + (500 - x)240(3)$$
$$= 600(300 + 3x - .003x^2) + 720(500 - x)$$
$$= 540{,}000 + 1080x - 1.8x^2$$

$R'(x) = 1080 - 3.6x = 0$ if $x = 300$. $R''(300)$ is negative and the revenue is maximized when 200 acres are planted in Grain A and 300 acres in Grain B.

21. Let x be the number over 40 that take the tour.

Profit = Income − expense = $(x + 40)(300 - 3x) - 2000 - 120(x + 40)$
$$= 300x + 12{,}000 - 3x^2 - 120x - 2000 - 120x - 4800$$
$$R(x) = 5200 + 60x - 3x^2$$
$$R'(x) = 60 - 60x = 0 \text{ when } x = 10$$
$R''(x) = -6$. The maximum profit occurs when $40 + 10 = 50$ people take the tour.

23. Let x be the speed where $x\epsilon[30, 55]$

Cost of trip = cost of driver + cost of gas
$$= 10(\text{number of hours}) + (3 + \frac{x^2}{600})(\text{number of hours})(1.25)$$
distance = speed $\cdot$ time

$\frac{600}{x}$ = number of hours

$C(x) = 10(\frac{600}{x}) + (3 + \frac{x^2}{600})(\frac{600}{x})(1.25) = \frac{600}{x} + (\frac{3}{x} + \frac{x}{600})750$

$C(x) = \frac{8250}{x} + \frac{5}{4}x.$

$C'(x) = \frac{-8250}{x^2} + \frac{5}{4} = 0$ if $x^2 = 6600$, $x = 10\sqrt{66}$ which is not in the domain.

The most economical speed occurs when $x = 55$ mph.

25. $C(x) = 4.9(\frac{600}{x}) + (3 + \frac{x^2}{600})(\frac{600}{x})(4.20)$

$\qquad = \frac{2940}{x} + (\frac{3}{x} + \frac{x}{600})(2520)$

$\qquad = \frac{10500}{x} + 4.2x$

$C'(x) = -\frac{10500}{x^2} + 4.2 = 0$ if $x = 50$

$C''(x) = \frac{21000}{x^3}$, $C''(50) > 0$. The most economical speed is 50 mph.

27. $\qquad y = x^2 + 2x + 3$

$\qquad y' = 2x + 2$

The tangent line at $(2,11)$ has slope $2(2) + 2 = 6$.

The slope of the line through $(14,9)$ and $(2,11)$ is $\frac{11-9}{2-14} = \frac{2}{-12} = \frac{-1}{6}$.
Since $6(-\frac{1}{6}) = -1$, the lines are perpendicular.

29. (a) The absolute maximum occurs at the point $(3/2, 5)$.

(b) 1 is the largest integer less than $3/2$.

(c) $f(1) = 4$ and $f(2) = 2$.

(d) If $k = 4$, $f(4) > f(1)$ and $f(4) > f(2)$

31. If f has a local maximum at the critical value c, f is increasing on $[a,c]$ and decreasing on $[c,b]$. If c_1 is the largest integer less than c, $f(c_1) \geq f(k)$ for all integers k in $[a,c]$ and $f(c_1 + 1) \geq f(k)$ for all integers k in $[c,b]$. Thus, either $f(c_1)$ or $f(c_1 + 1)$ is greater than or equal to $f(k)$, where k is any integer in $[a,b]$. The explanation is similar if a local minimum occurs at the critical value.

33. Profit = Income − Expense

$\quad$ = (number of computers sold)price − expense

$P(q) = q(-.00005q^2 - .3q + 8000) - (4500000 + 500q)$

$\quad = -.00005q^3 - .3q^2 + 8000q - 4500000 - 500q$

$\quad = -.00005q^3 - .3q^2 + 7500q - 4500000, \quad 0 \leq q \leq 9000.$

$P'(q) = -.00015q^2 - .6q + 7500 = 0$ if $q = -2000 \pm 500\sqrt{216}$.

$c = 5348.46$ is the approximate critical value.

$P''(q) = -.0003q - .6$

$P''(c) < 0.$ $P(5348) = 19381733.6 > P(5349)$

The maximum profit is realized when $q = 5348$.

35. The demand equation is $p = -4q + 600$.

$P(q) = R(q) - C(q) = q(-4q + 600) - (5000 + 2000 \ln(q+1)),\ 0 \le q \le 100$

$P'(q) = -8q + 600 - \dfrac{2000}{q+1} = 0$

$q^2 - 74q + 175 = 0$

$q = \dfrac{74 \pm 69.1}{2}$

$q \sim 71.55$ or 2.45

$C''(71.55) < 0.$ The absolute maximum occurs at 71 or 72.

$P(72) > (71)$.

The maximum occurs when $q = 72$.

37. $P(q) = R(q) - C(q) = q(-.000025q^2 - .25q + 15000) - 22{,}500{,}000 - 1500q$

$\qquad = -.000025q^3 - .25q^2 + 13{,}500q + 22{,}500{,}000$

$P'(q) = -.000075q^2 - .5q + 13500 = 0,$ if $q = \dfrac{.5 \pm \sqrt{4.3}}{-.00015}.$

The critical number is approximately 10490.96.

$P(10{,}490) < P(10{,}491)$

The maximum profit occurs when $q = 10{,}491$.

39. The average cost $A(q) = \dfrac{C(q)}{q} = \dfrac{.0005q^3 + 3q + 8}{q}$

$= .0005q^2 + 3 + \dfrac{8}{q}.$ $A'(q) = .001q - 8/q^2 = 0$ if $q^3 = 8000.$ $q = 20.$

$A''(q) = .001 + 16/q^3.$ $A''(20)$ is positive. Thus, the minimum for A occurs when $q = 20$.

41. The average cost per floor $A(x) = \dfrac{\text{total cost}}{\text{number of floors}}.$ Let x be the number of floors.

The cost is $343{,}000 + 15{,}000x + 7000x^2$.

$A(x) = \dfrac{343{,}000 + 15{,}000x + 7000x^2}{x} = \dfrac{343{,}000}{x} + 15{,}000 + 7000x.$

$A'(x) = -\dfrac{343{,}000}{x^2} + 7{,}000 = 0$ if $\dfrac{343{,}000}{7{,}000} = x^2 = 49$

The only critical value is 7.

$A''(x) > 0$ when $x = 7.$ A minimum occurs when $x = 7$.

The minimum average cost occurs when 7 floors are constructed.

43. Let x be the distance indicated on the diagram below.
distance $=$ rate(time)
time $=$ time rowing $+$ time jogging

$$T(x) = \tfrac{1}{3}\sqrt{x^2 + 256} + (30 - x)\tfrac{1}{5}, \quad 0 \le x \le 30.$$

$$T'(x) = \frac{x}{3\sqrt{x^2 + 256}} - \frac{1}{5} = 0 \quad \text{if } 5x = 3\sqrt{x^2 + 256}, \quad 25x^2 = 9(x^2 + 256)$$

$16x^2 = 9(256), \quad x = 12.$

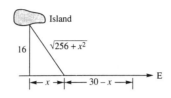

$T(0) = \dfrac{16}{3} + \dfrac{30}{5} = 11\tfrac{1}{3}$ hours

$T(12) = \dfrac{20}{3} + \dfrac{18}{5} = 10.27$ hours

$T(30) = \tfrac{1}{3}(34) = 11\tfrac{1}{3}$ hours

The time is minimized if the man arrives on the mainland 18 miles west of the point to which he is going.

45. (a) The light is greatest when the window area is largest.

window area $=$ area of semi-circle $+$ area of rectangle $= \tfrac{1}{2}\pi(\tfrac{d}{2})^2 + xd.$

$2x + 2d + \dfrac{\pi d}{2} = 20. \quad x = (20 - 2d - \dfrac{\pi d}{2})\tfrac{1}{2} = 10 - d - \dfrac{\pi d}{4}$

The window area $A(d) = \dfrac{\pi}{8}d^2 + 10d - d^2 - \dfrac{\pi}{4}d^2 = 10d - (1 + \tfrac{\pi}{8})d^2.$

$A'(d) = 10 - 2(1 + \tfrac{\pi}{8})d = 0 \quad \text{if } d = \dfrac{-5}{1 + \pi/8} \sim 3.59 \text{ feet.}$

} xft.

dft.

The maximum light enters when $d \sim 3.59$ feet and $x \sim 3.59$ feet.

(b) The amount of light entering is greatest when $\tfrac{3}{4}$(area of semicircle) $+$ area of rectangle is maximized.

$$f(d) = \tfrac{3}{4}(\tfrac{\pi}{8}d^2) + 10d - d^2 - \tfrac{\pi}{4}d^2$$

$$f'(d) = \dfrac{3\pi d}{16} + 10 - 2d - \tfrac{\pi}{2}d = 0$$

$$10 = 2d + \tfrac{\pi}{2}d - \dfrac{3\pi}{16}d = d(2 + \tfrac{5\pi}{16})$$

$$d = \dfrac{10}{2 + \dfrac{5\pi}{16}} \sim 3.35 \text{ feet.} \quad \text{The maximum occurs when } d \sim 3.35 \text{ feet and } x \sim 4.0 \text{ feet.}$$

47. Let x be the number of cases ordered. The number of orders/year is $\dfrac{1620}{x}$. The average number of cases in inventory is $x/2$.

Total cost $=$ cost of placing orders $+$ cost of maintaining inventory.

$C(x) = 66(\dfrac{1620}{x}) + 1.65(x/2).$

$C'(x) = \dfrac{-66(1620)}{x^2} + \dfrac{1.65}{2} = 0 \quad \text{if } x^2 = \dfrac{66(1620)(2)}{1.65} = 129,600 \text{ and } x = 360.$

$C''(x) > 0$ when $x = 360.$ The minimum cost occurs when 360 cases are ordered.

49. Cost = employee salaries + loss in profit.

$$C(x) = 5.5x + \frac{792}{x+1}, \quad x \geq 0.$$

$$C(x) = 5.5 - \frac{792}{(x+1)^2} = 0 \text{ if } (x+1)^2 = \frac{792}{5.5} = 144, \quad x = 11.$$

$C''(11) > 0$. The minimum cost occurs when 11 employees work during the noon hour.

EXERCISE SET 3.8 CURVE SKETCHING

1. The graph is a parabola with vertex when $x = \dfrac{-6}{2} = -3$

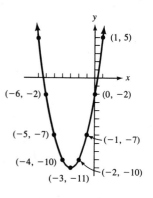

x	$f(x)$
0	-2
-3	-11
-6	-2

$$f(x) = x^2 + 6x - 2$$

The x-intercepts occur when $x = \dfrac{-6 \pm \sqrt{36+8}}{2} = -3 \pm \sqrt{11}$

3. The graph is a parabola with vertex at $x = \dfrac{12}{4} = 3$.

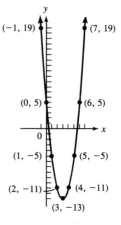

x	$f(x)$
3	-13
0	5
6	5

$$y = h(x) = 2x^2 - 12x + 5$$

The x-intercepts occur when $x = \dfrac{12 \pm \sqrt{104}}{4} = 3 \pm \frac{1}{2}\sqrt{21}$

5. $g(x) = x^3 + 3x^2 - 24x + 5$

$g'(x) = 3x^2 + 6x - 24 = 3(x^2 + 2x - 8) = 3(x+4)(x-2)$

$g''(x) = 6x + 6 = 6(x+1)$

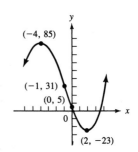

x		-4		-1		2	
$x+4$	$-$	0	$+$	$+$	$+$	$+$	$+$
$x-2$	$-$	$-$	$-$	$-$	$-$	0	$+$
$x+1$	$-$	$-$	$-$	0	$+$	$+$	$+$
g'	$+$	0	$-$	$-$	$-$	0	$+$
g''	$-$	$-$	$-$	0	$+$	$+$	$+$

x	$g(x)$	
-4	85	(local maximum)
2	-23	(local minimum)
-1	31	(point of inflection)

7. $f'(x) = -3x^2 - 6x + 9 = -3(x^2 + 2x - 3) = -3(x+3)(x-1)$

   $f''(x) = -6x - 6$

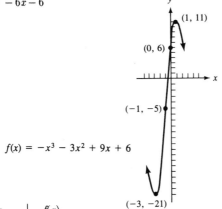

$(1, 11)$

$(0, 6)$

$(-1, -5)$

$f(x) = -x^3 - 3x^2 + 9x + 6$

$(-3, -21)$

x	$f(x)$	
-3	-21	local minimum since $f''(-3) > 0$
1	11	local maximum since $f''(1) < 0$
-1	-5	the direction of concavity changes

9. $h(x) = x^4 - 4x + 3$

   $h'(x) = 4x^3 - 4 = 4(x^3 - 1) = 0$ if $x = 1.$

   $h''(x) = 12x^2. \quad f''(1) > 0.$

   h'' never changes signs.

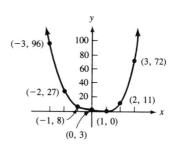

$(-3, 96)$ $(3, 72)$

$(-2, 27)$ $(2, 11)$

$(-1, 8)$ $(1, 0)$

$(0, 3)$

x	$h(x)$	
1	0	local minimum
0	3	
2	11	

11. (a) $\displaystyle\lim_{x \to +\infty} \frac{x^2 + 3x - 2}{3x^2 - 5x + 10} \cdot \frac{1/x^2}{1/x^2} = \lim_{x \to +\infty} \frac{1 + \frac{3}{x} - 2/x^2}{3 - \frac{5}{x} + \frac{10}{x^2}} = \frac{1 + 0 - 0}{3 - 5 + 10} = 1/3$

   (b) $\displaystyle\lim_{x \to -\infty} \frac{1 + 3/x - 2/x^2}{3 - 5/x + 10/x^2} = \frac{1 + 0 - 0}{3 - 0 + 0} = 1/3$

13. $f(x) = \dfrac{-x^2 + 7x + 4}{x^2 + 8x - 6} \cdot \dfrac{1/x^2}{1/x^2} = \dfrac{-1 + 7/x + 4/x^2}{1 + 8/x - 6/x^2}$

(a) $\displaystyle\lim_{x \to \infty} f(x) = \dfrac{-1 + 0 + 0}{1 + 0 - 0} = -1.$

(b) $\displaystyle\lim_{x \to -\infty} f(x) = \dfrac{-1 + 0 + 0}{1 + 0 - 0} = -1$

15. $f(x) = \dfrac{4x^3 - 2x^2 + 5x + 1}{2x^3 + 6x - 7} \cdot \dfrac{1/x^3}{1/x^3} = \dfrac{4 - 2/x + 5/x^2 + 1/x^3}{2 + 6/x^2 - 7/x^3}$

(a) $\displaystyle\lim_{x \to \infty} f(x) = \dfrac{4 - 0 + 0 + 0}{2 + 0 - 0} = 2$

(b) $\displaystyle\lim_{x \to -\infty} f(x) = \dfrac{4 - 0 + 0 + 0}{2 + 0 - 0} = 2$

17. $f(x) = \dfrac{4x^2 - 5x + 7}{6x^3 + 3x^2 - 5} \cdot \dfrac{1/x^2}{1/x^2} = \dfrac{4 - 5/x + 7/x^2}{6x + 3 - 5/x^2}$

(a) $\displaystyle\lim_{x \to +\infty} f(x) = 0$ since the numerator $\to 4$ and the denominator becomes large.

(b) $\displaystyle\lim_{x \to -\infty} f(x) = 0$ since the numerator $\to 4$ and the denominator becomes negative without bound.

19. (a) As $x \to -3^-$, $x + 3 \to 0$ and $x + 3$ is negative.

Thus, $\displaystyle\lim_{x \to -3^-} \dfrac{-2}{x + 3} = +\infty.$

(b) As $x \to -3^+$, $x + 3 \to 0$ and $x + 3$ is positive.

Thus, $\displaystyle\lim_{x \to -3^+} \dfrac{-2}{x + 3} = -\infty$

21. $f(x) = \dfrac{5 - 3x}{x + 2}.$ If x is near -2, $5 - 3x$ is near 11. The denominator is near 0 and negative

if $x < -2$, and near 0 and positive if $x > -2$.

Thus, $\displaystyle\lim_{x \to 2^-} f(x) = -\infty$ and $\displaystyle\lim_{x \to 2^+} f(x) = +\infty$

23. If x is near 1, $x + 2$ is near 3 and $(x - 1)^2$ is near 0.

Thus, $\displaystyle\lim_{x \to 1^+} f(x) = \lim_{x \to 1} f(x) = +\infty.$

25. If x is near -1, $x + 8$ is near 7 and $x - 5$ is near -6. If $x < -1$, $x + 1$ is near zero and negative. Thus, $\displaystyle\lim_{x \to -1^-} f(x) = +\infty.$ However, if $x > -1$, $x + 8$ is near 9, $x - 5$ is near

-6 and $x + 1$ is near 0 and positive.

$\displaystyle\lim_{x \to -1^+} f(x) = -\infty.$

If x is near 5, $x+8$ is near 13, and $x+1$ is near 6.   If $x < 5$, $x-5$ is near zero and is negative.   If $x > 5$, x-5 is near zero and positive.

Thus $\lim_{x\to 5^-} f(x) = -\infty$ and $\lim_{x\to 5^+} f(x) = +\infty$

27. If x is near 2, $5 - x$ is near 3, $2x + 3$ is near 7, and $7 - x$ is near 5.   If $x < 2$, $2 - x$ is positive and near 0, but if $x > 2$, $2 - x$ is negative and near zero.

$\lim_{x\to 2^-} f(x) = +\infty$ and $\lim_{x\to 2^+} f(x) = -\infty$.

If x is near 7, $5 - x$ is near -2, $2x + 3$ is near 17, and $2 - x$ is near -5.   If $x < 7$, $7 - x$ is positive and near zero.   If $x > 7$, $7 - x$ is negative and near zero.

$\lim_{x\to 7^-} f(x) = +\infty$

$\lim_{x\to 7^+} f(x) = -\infty$

29. $g'(x) = 4x^3 - 24x^2 + 48x - 32$

$\quad = 4(x^3 - 6x^2 + 12x - 8)$

$\quad = 4(x - 2)(x^2 - 4x + 4) = 4(x - 2)^3 = 0$

only if $x = 2$.   2 is the only critical value of g.

$g''(x) = 12x^2 - 48x + 48 = 12(x^2 - 4x + 4) = 12(x - 2)^2$
g' changes from negative to positive at $x = 2$.   Thus, there is a local minimum at $x = 2$.
There is no point of inflection as g'' is never negative.

x	$g(x)$
2	0
0	16
4	16

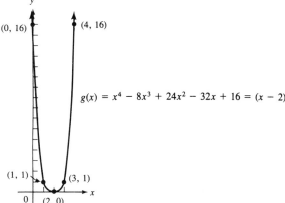

$g(x) = x^4 - 8x^3 + 24x^2 - 32x + 16 = (x - 2)^4$

31. $g(x) = x^8 + 6x^4 - 4x + 2$

$g'(x) = 8x^7 + 24x^3 - 4$

$g''(x) = 56x^6 + 72x^2 = 8x^2(7x^4 + 9)$

The second derivative never changes sign and there is no point of inflection.

33. $f(x) = x^4 - 4x^3 - 48x^2 + 36x + 3$

$f'(x) = 4x^3 - 12x^2 - 96x + 36$

$f''(x) = 12x^2 - 24x - 96 = 12(x^2 - 2x - 8) = 12(x - 4)(x + 2)$

$f''(x)$ changes sign at $x = 4$ and $x = -2$.

The points of inflection are $(-2, f(-2))$ and $(4, f(4))$

35. $h(x) = x^{17/9}$

$h'(x) = \frac{17}{9} x^{8/9}$

$h''(x) = \frac{17}{9} \cdot \frac{8}{9} \cdot \frac{1}{x^{1/9}}$

The second derivative changes sign when $x = 0$. $(0,0)$ is a point of inflection.

37. $f(x) = x^4 + x$

(a) $f'(x) = 4x^3 + 1$, $f''(x) = 12x^2$, $f''(0) = 0$.

(b) $f''(x)$ is never negative so there are no points of inflection.

(c) $f'(x)$ does not change sign at $x = 0$, so $(0,0)$ is not an extremum.

(d)

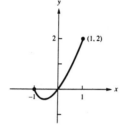

EXERCISE 3.9 MORE ON CURVE SKETCHING

1. $f(x) = \frac{x-3}{x+2}$. $\lim\limits_{x \to -2^-} f(x) = +\infty$. $\lim\limits_{x \to -2^+} f(x) = -\infty$.

$\lim\limits_{x \to +\infty} f(x) = \lim\limits_{x \to -\infty} f(x) = \lim\limits_{x \to \infty} \frac{1 - 3/x}{1 + 2/x} = 1$

The y-intercept is $(0, -3/2)$.   The x-intercept is $(3,0)$.

$f'(x) = \frac{x + 2 - (x-3)}{(x+2)^2} = \frac{5}{(x+2)^2}$.   The graph is increasing on $(-\infty, -2)$ and $(-2, +\infty)$

as $f'(x)$ is never negative, $f''(x) = \frac{-10}{(x+2)^3}$. $f'' < 0$ on $(2, +\infty)$ and $f'' > 0$ on

$(-\infty, -2)$.   The graph is concave down on $(-2, +\infty)$ and concave up on $(-\infty, -2)$

x	$f(x)$
3	0
0	$-3/2$
-1.9	-49
-2.1	51

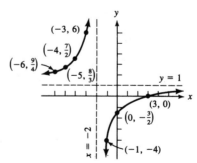

3. $h(x) = \dfrac{(x-3)(x-4)}{(x-1)(x-6)} = \dfrac{x^2 - 7x + 12}{x^2 - 7x + 6} = 1 + \dfrac{6}{x^2 - 7x + 6}$

$h'(x) = \dfrac{-6}{(x^2 - 7x + 6)^2}(2x - 7)$

h is increasing if $2x - 7 < 0$ or $x < 3.5$ and $x \neq 1$.

At other points in the domain h is decreasing.

$\displaystyle\lim_{x\to\infty} h(x) = \lim_{x\to\infty} \dfrac{1 - 7/x + 12/x^2}{1 - 7/x + 6/x^2} = 1$

Also, $\displaystyle\lim_{x\to\infty} h(x) = 1$

h is discontinuous at $x = 1$ and $x = 6$.

$\displaystyle\lim_{h\to 1^+} h(x) = -\infty \qquad\qquad \lim_{h\to 1^-} h(x) = \infty$

$\displaystyle\lim_{h\to 6^+} h(x) = \infty \qquad\qquad \lim_{h\to 6^-} h(x) = -\infty$

x	$h(x)$
3	0
4	0
3.5	.04
-2	5/4
0	2
7	2

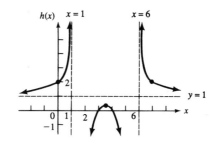

5. $g(x) = \dfrac{(x-4)(x-1)}{(x-3)^2}$

$\displaystyle\lim_{x\to\infty} g(x) = \lim_{x\to -\infty} g(x) = 1$

g is discontinuous when $x = 3$ and $\displaystyle\lim_{x\to 3^+} g(x) = \lim_{x\to 3^-} g(x) = -\infty$

$g'(x) = \dfrac{(2x-5)(x-3)^2 - (x^2 - 5x + 4)2(x-3)}{(x-3)^4} = \dfrac{(2x-5)(x-3) - 2x^2 + 10x - 8}{(x-3)^3}$

$= \dfrac{7 - x}{(x-3)^3} \cdot g'(x) > 0$ if $3 < x < 7$ and the function is increasing. If $x < 3$ or $x > 7$ the

function is decreasing.

x	$g(x)$
4	0
1	0
2	-2
5	1
0	4/9
7	1.125

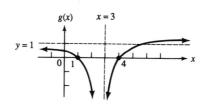

7. $f(x) = \dfrac{(x-6)(x-3)}{x-2} = \dfrac{x^2 - 9x + 18}{x-2} = x - 7 + \dfrac{4}{x-2}$

$f'(x) = 1 - \dfrac{4}{(x-2)^2}$ $f''(x) = \dfrac{8}{(x-2)^3}$

$f'(x) > 0$ if $x > 4$ or $x < 0$. $f'(x) < 0$ if $0 < x < 2$ or $2 < x < 4$. $f''(x) > 0$ if $x > 2$ and $f''(x) < 0$ if $x < 2$.   The function has a local maximum at $x = 4$, a local minimum at $x = 0$, is concave downward on $(-\infty, 2)$ and concave upward on $(2, \infty)$.

The function is discontinuous when $x = 2$, and $\lim\limits_{x \to 2^+} f(x) = \infty$ while $\lim\limits_{x \to 2^-} f(x) = -\infty$.

x	$f(x)$
0	-9
1	-10
3	0
6	0
4	-1
-2	-10

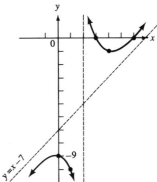

9. $h(x) = \dfrac{(x-1)(x+7)}{x^2 + 1} = \dfrac{x^2 + 6x - 7}{x^2 + 1}$

h is continuous and has x-intercepts of 1 and -7.

$\lim\limits_{x \to +\infty} h(x) = \lim\limits_{x \to -\infty} h(x) = 1.$

$h'(x) = \dfrac{(2x+6)(x^2+1) - (x^2 + 6x - 7)2x}{(x^2+1)^2} = \dfrac{-6x^2 + 16x + 6}{(x^2+1)^2}$

$= \dfrac{-2(3x^2 - 8x - 3)}{(x^2+1)^2} = \dfrac{-3(x-3)(3x+1)}{(x^2+1)^2}$

h has critical points at 3 and $-1/3$ with a local maximum at $x = 3$ and a local minimum at $x = -1/3$.

$\lim\limits_{x \to \infty} h(x) = \lim\limits_{x \to -\infty} h(x) = 1$

x	$h(x)$
1	0
-7	0
0	-7
3	2
4	1.94
-1	-6
$-1/3$	-8
-2	-3

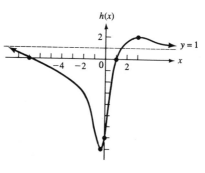

11. $h(x) = \dfrac{(x+3)(2x+1)}{(x-2)(x+5)}$.   The y-intercept is $-3/10$.

The x-intercepts are -3 and $-1/2$.

$$\lim_{x \to +\infty} h(x) = \lim_{x \to -\infty} h(x) = \lim_{x \to \infty} \frac{(1+3/x)(2+1/x)}{(1-2/x)(1+5/x)} = 2$$

$$\lim_{x \to 2^-} h(x) = -\infty \qquad \lim_{x \to 2^+} h(x) = +\infty$$

$$\lim_{x \to -5^-} h(x) = +\infty \qquad \lim_{x \to -5^+} h(x) = -\infty$$

$$
\begin{aligned}
h(x) &= \frac{2x^2 + 7x + 3}{(x-2)(x+5)} = \frac{2x+11}{(x+5)} + \frac{25}{(x-2)(x+5)} \\
&= 2 + \frac{1}{x+5} + \frac{25}{x^2 + 3x - 10} = 2 + (x+5)^{-1} + 25(x^2 + 3x - 10)
\end{aligned}
$$

$$
\begin{aligned}
h'(x) &= \frac{-1}{(x+5)^2} - \frac{25(2x+3)}{(x+5)^2(x-2)^2} = \frac{-(x-2)^2 - 50x - 75}{(x+5)^2(x-2)^2} \\
&= \frac{-x^2 - 46x - 79}{(x+5)^2(x-2)^2}
\end{aligned}
$$

The critical points occur if

$$x^2 + 46x + 79 = 0$$

$$x = \frac{-46 \pm \sqrt{46^2 - 4(79)}}{2}$$

$$\sim \frac{-46 \pm 42.426}{2}$$

$x = -1.787$ or $x = -44.2$

$h' < 0$ if $x < -44.2$
or $-1.787 < x < 2$ or $2 < x$
$h' > 0$ if $-44.2 < x < -5$
or $-5 < x < -1.787$

---

13. $f(x) = xe^{-3x}$. $f(x) = 0$ if $x = 0$. $(0,0)$ is the x-intercept and the y-intercept.   If $x > 0$, $f(x) > 0$ and the graph is in Quadrant I.   If $x < 0$, $f(x) < 0$ and the graph is in Quadrant III.

The function is neither even nor odd. $\displaystyle\lim_{x \to +\infty} f(x) = 0$. $\displaystyle\lim_{x \to -\infty} f(x) = -\infty$.   The x-axis is a horizontal asymptote to the right.

$$f'(x) = e^{-3x} - 3xe^{-3x} = e^{-3x}(1 - 3x) = 0 \text{ if } x = 1/3$$
$$f''(x) = -3e^{-3x} - 3e^{-3x} + 9xe^{-3x} = e^{-3x}(9x - 6).$$
$f''(\tfrac{1}{3}) < 0$.   A local maximum occurs when $x = 1/3$.
$f''(x) > 0$ on $(2/3, +\infty)$. $f''(x) < 0$ on $(-\infty, +2/3)$

We locate some pertinent points.

x	$f(x)$
0	0
1/3	.12
2/3	.09
-1	-20
1	.05

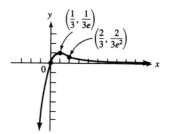

15. $h(x) = (x+1)\ln|x+1|$. The domain is $\{x | x \neq -1\}$. The function is neither odd nor even.

$\lim_{x \to +\infty} h(x) = +\infty$, $\lim_{x \to -\infty} h(x) = -\infty$. $\lim_{x \to -1} h(x) = 0$

$h'(x) = \ln|x+1| + 1 = 0$ if $x = -1 \pm \frac{1}{e}$.

$h''(x) = \frac{1}{x+1}$, $h''(-1 - \frac{1}{e}) < 0$, $h''(-1 + \frac{1}{e}) > 0$.

There is a local maximum when $x = -1 - 1/e \sim -1.37$ and a local minimum when $x = -1 + 1/e \sim -.63$. There are no points of inflection but the graph is concave upward if $x > -1$ and concave downward if $x < -1$. Some relevant points on the graph follow:

x	$h(x)$
$-1 - 1/e$	.36
$-1 + 1/e$	$-.36$
0	0
-2	0
1	1.4
$-.9$	.23
$-.99$	$-.05$

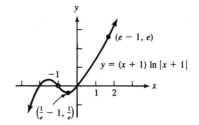

17.

$$\lim_{x \to \infty} \left| \frac{x^2 + 5x - 6}{x+1} - (x+5) \right|$$

$$= \lim_{x \to \infty} \left| \frac{x^2 + 5x - 6 - (x+5)(x+1)}{x+1} \right|$$

$$= \lim_{x \to \infty} \left| \frac{x^2 + 5x - 6 - x^2 - 6x - 5}{x+1} \right|$$

$$= \lim_{x \to \infty} \left| \frac{-x - 11}{x+1} \right| = \lim_{x \to \infty} \left| \frac{-1 - 11/x}{1 + 1/x} \right| = 1.$$

19. (a) $\frac{5.0001}{2.0001} - \frac{5}{2} = -.000074996$

- 144 -

(b) $\dfrac{9000000.0001}{2.0001} - \dfrac{9000000}{2} \doteq -224.99$

The first result is near zero, while the second is not.

21.　$y = 50x^{7/5}$

$y' = 70x^{2/5}$

$y'' = 28x^{-3/5}$

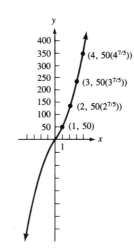

y is increasing and continuous at all x and the graph is concave upward if $x > 0$ and concave downward if $x < 0$. It is symmetric about the origin.

x	$f(x)$
0	0
$1/32$	$.39$
$-1/32$	$-.32$
1	50
-1	-50

EXERCISE SET 3.10 DIFFERENTIALS

In Exercises 1-7 the formulas $dy = f'(x)dx$ and $\triangle y = f(x + \triangle x) - f(x)$ are used.

1.　$f(x) = x^2 + 6x - 3.$　$f'(x) = 2x + 6.$
　(a)　$\triangle y = f(4.2) - f(4) = 39.84 - 37 = 2.84$
　(b)　$dy = f'(4)dx = (14)(.2) = 2.8$
　(c)　$\triangle y = 2.84 \doteq 2.8 = dy$

3.　$f(x) = x^3 + x^2 - 5x + 1.$　$f'(x) = 3x^2 + 2x - 5$
　(a)　$\triangle y = f(3.1) - f(3) = 24.901 - 22 = 2.901$
　(b)　$dy = f'(3)dx = 28(.1) = 2.8$
　(c)　$dy = 2.8 \doteq \triangle y = 2.901$

5.　$f(x) = \sqrt{x}.$　$f'(x) = 1/(2\sqrt{x})$
　(a)　$\triangle y = f(225.3001) - f(225) = 15.01 - 15 = .01$
　(b)　$dy = f'(225)dx = \dfrac{1}{30}(.3001) = .01000\overline{3}$
　(c)　$\triangle y = .01 \doteq .01000\overline{3} = dy$

7. $f(x) = \ln x.$ $f'(x) = 1/x$

   (a) $\Delta y = f(73) - f(75) = -.027028672$

   (b) $dy = f'(75)(-2) = -2/75 = -.02\overline{6}$

   (c) $\Delta y = -.027028672 \doteq -.026 = dy$

9. $f(x) = \sqrt[3]{x}.$ $f'(x) = \dfrac{1}{3x^{2/3}}$

   $f(515) \doteq f(512) + f'(512)(3) = 8 + 1/64 = 8.01562$

11. $f(x) = \sqrt[5]{x}.$ $f'(x) = (1/5)x^{-4/5}$

   $f(245) \doteq f(243) + f'(243)(2) = 3 + .4/81 = 3.004938272$

13. $f(x) = x^6.$ $f'(x) = 6x^5$

   $f(9.9) \doteq f(10) + f'(10)(-.1) = 940,000$

15. $|\Delta r| \le 30(.03) = .9$ feet

   (a) $C = 2\pi r.$ $\dfrac{dC}{dr} = 2\pi.$ $dC = 2\pi \, dr$

      If $r = 30$, $|dr| \le .03(30) = .9$

      $|dC| \le 2\pi(.9) = 1.8\pi$ ft is an approximation of the maximum possible error in circumference. The maximum relative error is approximately $\dfrac{1.8\pi}{2\pi(30)} = .03$ and the maximum percentage error is approximately 3%.

   (b) $A = \pi r^2.$ $\dfrac{dA}{dr} = 2\pi r.$ $dA = 2\pi r \, dr.$

      $|dA| = 2\pi r |dr| \le 60\pi(.9) = 54\pi$ is an approximation of the maximum possible error in area. The maximum relative error is approximately $\dfrac{54\pi}{\pi(30)^2} = .06$ and the maximum percentage error approximately 6%.

17. $V = (4/3)\pi r^3 = (4/3)\pi(\tfrac{d}{2})^3 = (1/6)\pi d^3$

   $|\Delta d| \le .5$

   $\Delta V \doteq dV = (1/2)\pi d^2 \, dd$

   $|\Delta V| \doteq |dV| \le (1/2)\pi(50)^2|.5| = 625\pi$

   The approximate maximum relative error is $\dfrac{625\pi}{(\pi/6)(50)^3} = .03$

   The approximate maximum percentage error is 3%

19. $\Delta r = .0225.$ $V = (4/3)\pi r^3$

   $\Delta V \doteq 4\pi r^2 \, \Delta r = 4\pi(1)^2(.0225) = .28$ in^3

21. Let q be the number of PC's manufactured and sold.

$|\triangle q| \le 6000(.03) = 180$

$P = -.00005q^3 - .3q^2 + 7500q - 4,500,000$

$\triangle P \doteq dP = (-.00015q^2 - .6q + 7500)\,dq$

$|\triangle P| \doteq |dP| \le |-1500||180| = 270,000$

The approximate maximum relative error is $\dfrac{270,000}{18,900,000} = .0143$.

The approximate maximum percentage error is 1.43%.

23. $|\triangle q| \le 10,000(.04) = 400$

$\triangle P \doteq dP = (-.000075q^2 - .5q + 13500)\,dq$

$|\triangle P| \doteq |dP| \le 1000(400) = 400,000$

The approximate maximum relative error is $\dfrac{400,000}{62,500,000} = .0064$

The approximate maximum percentage error is .64%.

EXERCISE SET 3.11 RELATED RATES

1.

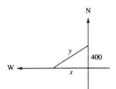

x is the distance between the sports car and the intersection and y the distance between the officer and the sports car at time t. Want $\dfrac{dx}{dt}$ when $y = 500$.

$y^2 = x^2 + (400)^2 \qquad 2y\dfrac{dy}{dt} = 2x\dfrac{dx}{dt}. \qquad \dfrac{y\,dy}{x\,dt} = \dfrac{dx}{dt}$

When $y = 500$, $x = 300$ and $\dfrac{dx}{dt} = \dfrac{500}{300}(66) = 110$ ft/s $= 75$ mph

3.

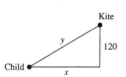

With x and y as labeled in the diagram, we are given $\dfrac{dy}{dt} = 2.5$, and must find $\dfrac{dx}{dt}$ when $y = 130$.

$x^2 + 120^2 = y^2. \qquad 2x\dfrac{dx}{dt} = 2y\dfrac{dy}{dt}.$ When $y = 130$, $x = 50$

$\dfrac{dx}{dt} = \dfrac{y\,dy}{x\,dt} = \dfrac{130}{50}(2.5) = 6.5$ ft/s.

5.

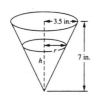

$$V = (1/3)\pi r^2 \qquad \frac{dV}{dt} = -.75$$

$$\frac{3.5}{7} = \frac{r}{h} \quad \cdot \quad r = \frac{h}{2}. \quad \text{We want } \frac{dh}{dt} \text{ when } h = 4.$$

$$V = (1/3)\pi(\tfrac{h}{2})^2 h = \tfrac{1}{12}\pi h^3$$

$\frac{dV}{dt} = (1/12)\pi \cdot 3h^2\frac{dh}{dt}$. When $h = 4$, $-.75 = \pi(4)\frac{dh}{dt}$ and $\frac{dh}{dt} = \frac{-.75}{4\pi} \doteq -.0597$ in./s.

7.

$\frac{dx}{dt} = 80$ mi/hr. y is the distance of the ball

from 3rd base.

Find $\frac{dy}{dt}$ when $x = 120$.

$x^2 + 90^2 = y^2$. $2x\frac{dx}{dt} = 2y\frac{dy}{dt}$. $\frac{dy}{dt} = \frac{x\,dx}{y\,dt}$. When $x = 120$, $y = 150$ and $\frac{dy}{dt} = \frac{120}{150} \cdot 80 = 64$ mph

9.

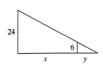

y is the length of the shadow. $\frac{24}{x+y} = \frac{6}{y}$

$$\frac{dx}{dt} = 5 \text{ mph}$$

$4y = x + y$, $3y = x$. $3\frac{dy}{dt} = \frac{dx}{dt}$. $\frac{dy}{dt} = \frac{1}{3}\frac{dx}{dt} = \frac{5}{3}$ mph

11.

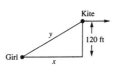

$\frac{dx}{dt} = 4$ mi/hr. Find $\frac{dy}{dt}$ when $y = 130$ ft.

$x^2 + 120^2 = y^2$. $2x\frac{dx}{dt} = 2y\frac{dy}{dt}$. $\frac{dy}{dt} = \frac{x\,dx}{y\,dt} = \frac{50}{130}(4) = 1.54$ mi/hr.

13. $V = (4/3)\pi r^3$. $\frac{dV}{dt} = -3$ ft^3/min. Find $\frac{dr}{dt}$ when $r = 20$ ft. $\frac{dV}{dt} = 4\pi r^2\frac{dr}{dt}$.

When $r = 20$, $-3 = 4\pi(400)\frac{dr}{dt}$. $\frac{dr}{dt} = \frac{-3}{1600\pi} = -.0005968$ ft/min

15. Volume of water $= (1/3)\pi(5)^2(15) - (1/3)\pi r^2(15 - h)$

$\frac{15}{5} = \frac{15 - h}{r}$, $r = (1/3)(15 - h)$

Volume of water $V = 125\pi - \frac{1}{27}\pi(15 - h)^3$

$\frac{dV}{dt} = -\frac{\pi}{9}(15 - h)^2(-1)\frac{dh}{dt}$. When $h = 10$, $1.5 = \frac{\pi}{9}(5)^2\frac{dh}{dt}$. $\frac{dh}{dt} = \frac{1.5(9)}{25\pi} = .172$ ft/min

17. $y = x^4 + 2x^3 - 5x^2 - 85$, $\frac{dx}{dt} = 2$. Find $\frac{dy}{dt}$ when $x = 3$.

$\frac{dy}{dt} = (4x^3 + 6x^2 - 10x)\frac{dx}{dt}$, $\frac{dy}{dt} = (108 + 54 - 30)(2) = 264$

y is increasing at the rate of 264 units/second.

19. $C(q) = 4 + 21q - 2q^2$, $0 \le q \le 5$. When $q = 3$, $dq/dt = \frac{50}{1000} = .05$ thousand radios/month.

$dC/dt = (21 - 4q)\frac{dq}{dt} = (21 - 12)(.05) = .45$.

The cost is increasing at the rate of \$450/month.

21. $C(q) = 14000 + 13428\sqrt{q}$, $R(q) = -4q^2 + 2400q$, $0 \le q \le 300$. When $q = 225$, $\frac{dq}{dt} = -10$

(a) $\frac{dC}{dt} = \frac{6714}{\sqrt{q}}\frac{dq}{dt} = \frac{6714(-10)}{15} = -\$4476/month$.

(b) $\frac{dR}{dt} = (-8q + 2400)\frac{dq}{dt} = 600(-10) = -\$6000/month$.

(c) Profit = Revenue − cost. $\frac{dP}{dt} = \frac{dR}{dt} - \frac{dC}{dt} = -6000 + 4476 = -\$1524/month$.

23. $C(q) = -.25q^2 + 25q + 600$, $R(q) = 100qe^{-.2q}$, $0 \le q \le 450$. When $q = 30$, $\frac{dq}{dt} = 3$.

(a) $dC/dq = (-.5q + 25)\frac{dq}{dt} = (-15 + 25)(3) = \$30/day$

(b) $dR/dq = 100(e^{-.2q} - .2qe^{-.2q})dq/dt = 100e^{-.2q}(1 - .2q)\frac{dq}{dt} = -\$3.718/day$

(c) $dP/dt = dR/dt - \frac{dC}{dt} = -3.718 - 30 = -\$33.718/day$

25. $P(q) = -.00005q^2 - .3q + 8000$. $C(q) = 500q + 4,500,000$

$\frac{dq}{dt} = 100$ computers/year when $q = 6000$

$R(q) = q(-.00005q^2 - .3q + 8000)$

$\quad = -.00005q^3 - .3q^2 + 8000q$

$\frac{dR}{dt} = (-.00015q^2 - .6q + 8000)dq/dt$

$\quad = -\$100,000/year$

EXERCISE SET 3.12 CHAPTER REVIEW

1. (a) $C'(q) = 30/\sqrt{q + 50}$

(b) $C'(175) = 30/\sqrt{225} = 2$.   The cost of the 175th unit is approximately \$2.

3. (a) $D(50) = 1140$, $D(200) = 960$.

The slope of the demand function is $\frac{1140 - 960}{-150} = -1.2$.

The demand function is

$$D(q) = -1.2(q-50) + 1140 = -1.2q + 1200$$

(b) $R(q) = qD(q) = -1.2q^2 + 1200q$

(c) $P(q) = R(q) - C(q) = -1.2q^2 + 1200q + q^2 - 980q - 20{,}000$

$$= -.2q^2 + 220q - 20{,}000$$

(d) $P(q) = 0$ if $.2q^2 - 220q + 20{,}000 = 0$, $0 \le q \le 490$

$$q = \frac{220 \pm \sqrt{32400}}{.4} = \frac{220 \pm 180}{.4} \cdot \quad q = 100$$

The profit becomes positive when $q = 100$

(e) $P'(q) = -.4q + 220$

(f) $P'(300) = 100$. The approximate profit from the 300th item is \$100.

(g) $P(q)$ is a maximum when $q = \frac{220}{.4} = 550$

(h) $A(q) = \frac{C(q)}{q} = -q + 980 + 20000/q$

(i) $A'(q) = -1 - 20000/q^2$

5. $\frac{dq}{dp} = -2p - 25.$

(a) When $p = 20$, $\frac{dp}{dq} = -65$ and $q = 6600$.

$$\eta = \frac{20}{6600}(-65) = -.197. \quad \text{The demand is inelastic.}$$

(b) When $p = 50$, $\frac{dp}{dq} = -125$ and $q = 3750$.

$$\eta = \frac{50}{3750}(-125) = -1.\overline{6} \quad \text{The demand is elastic.}$$

7. (a) $f(x) = 2x^3 + 9x^2 - 60x + 13$

$$f'(x) = 6x^2 + 18x - 60$$

$$f'(x) = 12x + 18 > 0 \text{ if } x > -3/2.$$

The graph is concave upward if $x > -1.5$, concave downward if $x < -1.5$. There is a point of inflection at $(-1.5, 116.5)$.

(b) $g(x) = (x^2 + 5x + 6)e^{-x}$

$g'(x) = (2x + 5)e^{-x} - (x^2 + 5x + 6)e^{-x} = e^{-x}(-x^2 - 3x - 1)$

$\quad = -e^{-x}(x^2 + 3x + 1)$

$g''(x) = e^{-x}(x^2 + 3x + 1) - e^{-x}(2x + 3) = e^{-x}(x^2 + x - 2)$

$\quad = e^{-x}(x + 2)(x - 1)$

x				-2			1			
$x+2$	$-$	$-$	$-$	0	$+$	$+$	$+$	$+$	$+$	$+$
$x-1$	$-$	$-$	$-$	$-$	$-$	$-$	0	$+$	$+$	$+$
$g''(x)$	$+$	$+$	$+$	0	$-$	$-$	0	$+$	$+$	$+$

The graph is concave upward on $(-\infty, -2)$ and $(1, +\infty)$. It is concave downward on $(-2,1)$.

The points of inflection are $(-2,0)$ and $(1,\frac{12}{e})$

(c) $h'(x) = 2x^{-2/3}$, $h''(x) = -\frac{4}{3}x^{-5/3}$. $h''(x) > 0$ if $x < 0$.

$h''(x) < 0$ if $x > 0$. The graph is concave upward on $(-\infty,0)$, concave downward on $(0,\infty)$ and has a point of inflection at $(0,0)$.

9. $f'(x) = 3x^2 - 6x - 45 = 3(x^2 - 2x - 15) = 3(x-5)(x+3)$.

5 and -3 are the critical values.

$f''(x) = 6x - 6$. $f''(5) > 0$, $f''(-3) < 0$. There is a local minimum when $x = 5$ and a local maximum when $x = -3$.

11. $f'(x) = 3x^2 + 18x - 21 = 3(x^2 + 6x - 7) = 3(x+7)(x-1)$.

The critical values are 1 and -7, but -7 is not in the interval. $f''(x) = 6x + 18$.
$f''(1) > 0$. There is a local minimum when $x = 1$.
$f(1) = -6$, $f(0) = 5$, $f(3) = 50$. The absolute maximum is 50 and the absolute minimum is -6.

13. (a) f is increasing when f' is positive, which occurs on $(-\infty, -3)$ and $(1,5)$.

(b) f is decreasing when $f'(x)$ is negative, which occurs on $(-3,1)$ and $(5, +\infty)$.

(c) The graph is concave down when f' is decreasing, which occurs on $(-\infty, -1)$ and $(3, +\infty)$.

(d) The graph is concave up when f' is increasing, which occurs on $(-1,3)$.

(e) $f'(x) = 0$ when $x = -3$, $x = 1$ and $x = 5$.

(f) A point of inflection occurs when the direction of concavity changes. From parts c and d, this occurs when $x = -1$ and $x = 3$.

15. (a) $dy = f'(x)dx = (2x+5)dx$

(b) $dy/dt = 2te^{3t} + 3t^2e^{3t} = e^{3t}(2t + 3t^2)$.
$dy = \frac{dy}{dt} \cdot dt = e^{3t}t(2 + 3t)dt$

(c) $dy/dx = D_x(\ln(x^4 + 3))(x^4 + 3)^{-1} = \frac{4x^3}{(x^4 + 3)^2} - \frac{4x^3\ln(x^4 + 3)}{(x^4 + 3)^2} =$

$\frac{4x^3(1 - \ln(x^4 + 3))}{(x^4 + 3)^2}$. $dy = \frac{4x^3(1 - \ln(x^4 + 3))}{(x^4 + 3)^2}dx$

17. Let $f(x) = \sqrt{x}$. $f'(x) = 1/(2\sqrt{x})$

$$f(623) \doteq f(625) + f'(625)(-2)$$
$$= 25 - .04 = 24.96$$

19.

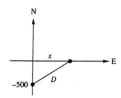

$$\frac{dD}{dt} = 96 \text{ ft/sec}$$

We wish to find $\frac{dx}{dt}$ when

$D = 1300$ feet.

$x^2 + 500^2 = D^2$. $x^2 + 25000 = D^2$. Differentiating both sides of the equation with respect to t,

$$2x\frac{dx}{dt} = 2D\frac{dD}{dt} \quad \cdot \quad \frac{dx}{dt} = \frac{D}{x}\frac{dD}{dt},$$

when $x = 1300$, $x = \sqrt{1300^2 - 500^2} = 1200$ and

$$\frac{dx}{dt} = \frac{1300}{1200}(96) = 104 \text{ ft/sec} = 70.9 \text{ mph}$$

21. $f(x) = (4x^2 - 2x + 2)e^{-x}$

(a) $f'(x) = (8x - 2)e^{-x} - e^{-x}(4x^2 - 2x + 2) = e^{-x}(8x - 2 - 4x^2 + 2x - 2)$

$\quad = -e^{-x}(4x^2 - 10x + 4) = -2e^{-x}(2x^2 - 5x + 2) = -2e^{-x}(2x - 1)(x - 2)$

The critical values are $\frac{1}{2}$ and 2.

x			1/2			2			
$2x - 1$	$-$	$-$	0	$+$	$+$	$+$	$+$	$+$	f is decreasing on $(-\infty, 1/2)$ and $(2, +\infty)$.
$x - 2$	$-$	$-$	$-$	$-$	$-$	0	$+$	$+$	f is increasing on $(1/2, 2)$
$f'(x)$	$-$	$-$	0	$+$	$+$	0	$-$	$-$	
$f(x)$	↘			↗			↘		

(b) $f''(x) = 2e^{-x}(2x^2 - 5x + 2) - 2e^{-x}(4x - 5)$

$\quad = 2e^{-x}(2x^2 - 5x + 2 - 4x + 5) = 2e^{-x}(2x^2 - 9x + 7)$

$\quad = 2e^{-x}(2x - 7)(x - 1) = 0$ if $x = 1$ or $x = 3.5$

f'' is concave upward on $(-\infty, 1)$ and $(3.5, +\infty)$.

f' is concave downward on $(1, 3.5)$.

(c)   Points of inflection are $(1, 4/e)$ and $(3.5, 44e^{-3.5})$.

(d) $f''(1/2) > 0$ and $f''(2) < 0$.   There is a local minimum when $x = 1/2$ and a local maximum when $x = 2$.

(e)

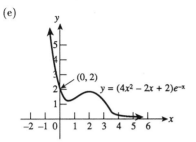

23. $\lim\limits_{x \to +\infty} y = 1 = \lim\limits_{x \to -\infty} y.$

$\lim\limits_{x \to 5^+} y = \lim\limits_{x \to \frac{13}{3}^-} y = -\infty$

$\lim\limits_{x \to 5^-} y = \infty = \lim\limits_{x \to \frac{13}{3}^+} y$

The line $y = 1$ is a horizontal asymptote to the left and right.   The lines $x = 5$ and $x = 13/3$ are vertical asymptotes.

The x-intercepts are 4 and 7.   The y-intercept is 84/65.

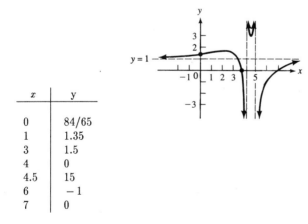

x	y
0	84/65
1	1.35
3	1.5
4	0
4.5	15
6	-1
7	0

25. $\lim\limits_{x \to +\infty} y = \lim\limits_{x \to -\infty} y = 4$

$\lim\limits_{x \to 2^+} y = \lim\limits_{x \to -2^-} y = +\infty$

$\lim\limits_{x \to 2^-} y = \lim\limits_{x \to -2^+} y = -\infty.$

The line $y = 4$ is a horizontal asymptote to the left and right.   The lines $x = 2$ and $x = -2$ are vertical asymptotes.

When $x = 0$, $y = -1/2$. y is never zero.

Thus, $-1/2$ is the y-intercept and there are no x-intercepts.

$$\frac{4x^2 + 2}{x^2 - 4} = 4 + \frac{18}{x^2 - 4} = 4 + 18(x^2 - 4)^{-1}$$

$$y' = -18(x^2 - 4)^{-2}2x = \frac{-36x}{(x^2 - 4)^2}.$$

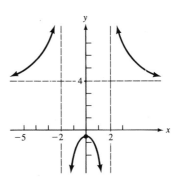

y'	$+$	$+$	$-$	$-$

| y | | | | |

-2 $\quad$ 0 $\quad$ 2

x	y
0	$-1/2$
± 1	-2
± 3	7.6
± 5	4.8

27. $y = (x + 4)\ln(x + 4)$

$y' = 1 + \ln(x + 4) = 0$ if $x = \frac{1}{e} - 4$.

$y'' = \frac{1}{x + 4}$.   When $x = \frac{1}{e} - 4$, $y'' > 0$ and the graph has a local minimum.

$y'' > 0$ if $x > -4$.

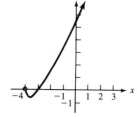

x	y
$-3.6 \sim \frac{1}{e} - 4$	$-1/e \sim -.37$
0	$4\ln4 \sim 5.5$
-3	0

$$\lim_{x \to +\infty} y = +\infty$$

$$\lim_{x \to -4^+} y = 0$$

29. $R(q) = (q^2 - 36q + 432)q = q^3 - 36q^2 + 432q$, $0 \le q \le 18$

$R'(q) = 3q^2 - 72q + 432 = 3(q^2 - 24q + 144) = 3(q - 12)^2$.

$R''(q) = 6q - 72$. $R''(12) = 0$.

$R(0) = 0$

$R(12) = 1728$

$R(18) = 1944$

The maximum revenue occurs when 18 thousand cameras are produced and sold.

31. Let x be the tuition increase.

$$R(x) = (150 + x)(225{,}000 - 98x^2 - 4794) = (150 + x)(220206 - 98x^2)$$
$$= 150(220{,}206) + 220206x - 14700x^2 - 98x^3.$$

$$R'(x) = 220206 - 29400x - 294x^2 = 0 \text{ if } x = 7.$$

$R''(x) = -29400 - 588x.$ If $x = 7$, $R'' < 0$ and the maximum revenue is reached when the tuition is \$157 per credit hour.

CHAPTER 4

INTEGRATION

EXERCISE SET 4.1 ANTIDIFFERENTIATION

1. $g'(x) = 2(3x^2) + 2(2x) - 3(1) + 0 = 6x^2 + 4x - 3 = f(x)$

3. $g'(x) = \frac{5}{32}(16)(x^2+1)^{15}D_x(x^2+1) + 0 = \frac{5}{2}(x^2+1)^{15}(2x) = 5(x^2+1)^{15}x = f(x)$

5. $g'(x) = \frac{3}{5}D_x(5x) \cdot e^{5x} + 2x + 3(1) = \frac{3 \cdot 5}{5}e^{5x} + 2x + 3 = f(x)$

7. $g'(x) = \frac{1}{3}\frac{D_x(x^3+1)}{x^3+1} = \frac{1}{3}\frac{(3x^2)}{x^3+1} = \frac{x^2}{x^3+1} = f(x)$

9. $g'(x) = \frac{xD_x(e^x) - e^x(D_x(x))}{x^2} + 0 = \frac{xe^x - e^x}{x^2} = \frac{e^x(x-1)}{x^2} = f(x)$

11. $g'(x) = xD_x(\ln|x| - 1) + (\ln|x| - 1)D_x(x) - 0 = x(\frac{1}{x}) + \ln|x| - 1 = \ln|x| = f(x)$

13. $\int (2x^3 + 5)\,dx = Kx^4 + 5x + c$

    $D_x(Kx^4 + 5x + c) = 4Kx^3 + 5 = 2x^3 + 5.$

    $4Kx^3 = 2x^3,\ 4K = 2,\ K = \frac{1}{2}$ and $\int (2x^3+5)\,dx = \frac{1}{2}x^4 + 5x + c$

15. $\int \frac{t^3}{t^4+1}\,dt = K\ln|t^4+1| + c$

    $D_x(K\ln|t^4+1| + c) = \frac{K(4t^3)}{t^4+1} = \frac{t^3}{t^4+1} \cdot 4K = 1,\ K = \frac{1}{4}$, and $\int \frac{t^3}{t^4+1}\,dt = \frac{1}{4}\ln|t^4+1| + c$

17. $\int 3(x^4+1)^{50}x^3\,dx = K(x^4+1)^{51} + c.\quad D_x(K(x^4+1)^{51}+c)$

    $= 51K(x^4+1)^{50}(4x^3) + 0 = 3(x^4+1)^{50}x^3.\quad 204K = 3,\ K = 3/204$

    $\int 3(x^4+1)^{50}x^3\,dx = \frac{3}{204}(x^4+1)^{51} + c.$

19. $\int 7s^3 e^{s^4}\, ds = K e^{s^4} + c.$ $D_s(K e^{s^4}) = 4K s^3 e^{s^4}$

$= 7 s^3 e^{s^4}$ if $4K = 7$ and $K = 7/4.$ Thus,

$$\int 7 s^3 e^{s^4}\, ds = \frac{7}{4} e^{s^4} + c$$

21. $\int \frac{x^2 + 1}{x^3 + 3x + 1}\, dx = K\ln|x^3 + 3x + 1| + c.$

$D_x(K\ln|x^3 + 3x + 1| + c) = \frac{K(3x^2 + 3x)}{x^3 + 3x + 1} = \frac{x^2 + 1}{x^3 + 3x + 1}$ if $3K = 1$ or $K = 1/3.$

$$\int \frac{x^2 + 1}{x^3 + 3x + 1}\, dx = \frac{1}{3}\ln|x^3 + 3x + 1| + c.$$

23. $D_t(t^{7/2} + t^2 + t) = \frac{7}{2} t^{5/2} + 2t + 1.$ Hence,

$$\int (2 t^{5/2} + 6t + 1)\, dt = t^{7/2} \cdot \frac{2}{7} \cdot 2 + 6 t^2 \cdot \frac{1}{2} + t + c$$

$$= \frac{4}{7} t^{7/2} + 3 t^2 + t + c.$$

25. $D_x(x^2 + 1)^{10/3} = \frac{10}{3}(x^2 + 1)^{7/3}(2x).$ Hence,

$$\int (x^2 + 1)^{7/3} x\, dx = (x^2 + 1)^{10/3} \cdot \frac{3}{20} + c = \frac{3}{20}(x^2 + 1)^{10/3} + c.$$

27. $D_x ln|x^4 + 1| = 4x^3/(x^4 + 1).$ Hence,

$$\int \frac{5x^3}{x^4 + 1}\, dx = 5\ln|x^4 + 1| \cdot \frac{1}{4} + c = \frac{5}{4}\ln|x^4 + 1| + c.$$

29. $D_t(e^{t^2 + 2t + 5}) = (2t + 2) e^{t^2 + 2t + 5} = 2(t + 1) e^{t^2 + 2t + 5}.$ Hence,

$$\int e^{t^2 + 2t + 5}(t + 1)\, dt = e^{t^2 + 2t + 5} \cdot \frac{1}{2} + c = .5 e^{t^2 + 2t + 5} + c.$$

31. $s(t) = \int (32t + 15)\, dt = 32 \cdot \frac{1}{2} t^2 + 15t + c = 16 t^2 + 15t + c.$

$s(0) = c = 20.$ $s(t) = 16 t^2 + 15t + 20.$

33. $s(t) = \int (6t^2 + 4t + 1)\, dt = 6 \cdot \frac{1}{3} t^3 + 4 \cdot \frac{1}{2} \cdot t^2 + t + c = 2t^3 + 2t^2 + t + c.$

$s(1) = 2 + 2 + 1 + c = 5 + c = -5.$ $c = -10.$

$s(t) = 2t^3 + 2t^2 + t - 10$

35. $D_t \ln|t^3 + 3t + 1| = (3t^2 + 3)/(t^3 + 3t + 1)$

$\qquad = 3(t^2 + 1)/(t^3 + 3t + 1)$. Hence $s(t) = \displaystyle\int \frac{t^2 + 1}{t^3 + 3t + 1} dt$

$\qquad = \frac{1}{3}\ln|t^3 + 3t + 1| + c. \quad s(0) = \frac{1}{3}\ln 1 + c = c = 50.$

$\qquad s(t) = \frac{1}{3}\ln|t^3 + 3t + 1| + 50.$

37. $v(t) = -6t^2 + 3t + c. \quad v(3) = -54 + 9 + c = -45 + c = 15. \ c = 60$

$\qquad v(t) = -6t^2 + 3t + 60.$

$\qquad s(t) = -2t^3 + \frac{3}{2}t^2 + 60t + K.$

$\qquad s(3) = -54 + \frac{27}{2} + 180 + K = 139.5 + K = 25. \quad K = -114.5.$

$\qquad s(t) = -2t^3 + 1.5t^2 + 60t - 114.5.$

39. $v(t) = \frac{1}{.03}e^{.03t} + c. \quad v(0) = \frac{1}{.03} + c = 5. \quad c = 5 - \frac{1}{.03} = -\frac{85}{3}$

$\qquad v(t) = \frac{100}{3}e^{.03t} - \frac{85}{3}.$

$\qquad s(t) = \frac{10000}{3}e^{.03t} - \frac{85}{3}t + K. \quad s(0) = \frac{10000}{9} + K = 100$

$\qquad s(t) = \frac{10000}{9}e^{.03t} - \frac{85}{3}t - \frac{9100}{9}.$

41. $C(q) = \frac{.01}{3}q^3 - 3q^2 + 15q + + c. \quad C(0) = 45000.$

$\qquad C(q) = \frac{.01}{3}q^3 - 3q^2 + 15q + 45000.$

43. $C(q) = \frac{-3}{.002}e^{-.002q} + c. \quad C(0) = 35,000 = -1500 + K. \quad K = 36,500$

$\qquad C(q) = -1500e^{-.002q} + 36,500.$

EXERCISE SET 4.2 THE U-SUBSTITUTION

1. $\displaystyle\int (x^3 + 5)^{10}x^2\, dx = \int \frac{1}{3}u^{10}\, du \qquad\qquad$ let $u = x^3 + 5 \quad du = 3x^2\, dx$

$\qquad = \frac{1}{3} \cdot \frac{1}{11}u^{11} + c = \frac{1}{33}(x^3 + 5)^{11} + c$

$\qquad$ Checking $D_x(\frac{1}{33}(x^3 + 5)^{11} + c) = \frac{11}{33}(x^3 + 5)^{10}(3x^2) = (x^3 + 5)^{10}x^2$

3. $\displaystyle\int \frac{x^3}{(x^4 + 3)^5} dx = \int u^{-5} \cdot \frac{1}{4}du \qquad\qquad$ let $u = x^4 + 3 \quad du = 4x^3\, dx$

$\qquad = u^{-4} \cdot \frac{1}{4}(-\frac{1}{4}) + c = \dfrac{-(x^4 + 3)^{-4}}{16} + c.$

$\qquad$ Checking, $D_x\dfrac{-(x^4 + 3)^{-4}}{16} + c = \dfrac{+4(x^4 + 3)^{-5}}{16}(4x^3) = \dfrac{x^3}{(x^4 + 3)^5}$

5. $\displaystyle\int (x^2+2x+5)^{10}(x+1)\,dx = \int u^{10}\cdot \tfrac{1}{2}u\,du$ $\qquad u=x^2+2x+5 \quad du=(2x+2)\,dx$

$\displaystyle = \frac{u^{11}}{2\cdot 11}+c = \frac{(x^2+2x+5)^{11}}{22}+c.$

Checking, $\displaystyle D_x\frac{(x^2+2x+5)^{11}}{22} = \frac{11}{22}(x^2+2x+5)^{10}(2x+2)$

$\displaystyle = (x^2+2x+5)^{10}(x+1)$

7. $\displaystyle\int (x^2+6x+1)^{1/3}(5x+15)\,dx$ $\qquad\qquad u=x^2+6x+1$
$\qquad\qquad\qquad\qquad\qquad\qquad\qquad\qquad\qquad\qquad du=(2x+6)\,dx \ =2(x+3)\,dx$

$\displaystyle = \int u^{1/3}\cdot 5(x+3)\,dx$

$\displaystyle = \int u^{1/3}\cdot \tfrac{5}{2}du = \tfrac{5}{2}u^{4/3}\cdot \tfrac{3}{4}+c = \tfrac{15}{8}(x^2+6x+1)^{4/3}+c$

Checking, $\displaystyle D_x(\tfrac{15}{8}(x^2+6x+1)^{4/3}) = \tfrac{15}{8}\cdot \tfrac{4}{3}(x^2+6x+1)^{1/3}(2x+6) = 5(x+3)\sqrt[3]{x^2+6x+1}$

9. $\displaystyle\int \frac{7x^4}{x^5+4}\,dx = \int \frac{7\,du}{5u}$ $\qquad\qquad\qquad u=x^5+4 \quad du=5x^4\,du$

$\displaystyle = \tfrac{7}{5}\ln|u|+c = \tfrac{7}{5}\ln|5x+4|+c$

Checking, $\displaystyle D_x\tfrac{7}{5}\ln|x^5+4| = \tfrac{7}{5}\cdot \frac{5x^4}{(x^5+4)} = \frac{7x^4}{x^5+4}$

11. $\displaystyle\int \frac{e^{2x}dx}{e^{2x}+1} = \int \frac{du}{2u}$ $\qquad\qquad\qquad u=e^{2x}+1 \quad du=2e^{2x}dx$

$\displaystyle = \tfrac{1}{2}\ln|u|+c = \tfrac{1}{2}\ln|e^{2x}+1|+c.$

Checking, $\displaystyle D_x\tfrac{1}{2}\ln|e^{2x}+1| = \tfrac{1}{2}\cdot \frac{2e^{2x}}{e^{2x}+1} = \frac{e^{2x}}{e^{2x}+1}$

13. $\displaystyle\int (e^{5x}+1)^{12}\cdot 2e^{5x}dx$ $\qquad\qquad\qquad u=e^{5x}+1 \quad du=5e^{5x}dx$

$\displaystyle = \int u^{12}\cdot \tfrac{2}{5}du = = \tfrac{2}{5}\cdot \tfrac{1}{13}u^{13}+c$

$\displaystyle = \tfrac{2}{65}(e^{5x}+1)^{13}+c.$

Checking, $\displaystyle D_x(\tfrac{2}{65}(e^{5x}+1)^{13}) = \tfrac{2}{65}(5e^{5x})\cdot 13(e^{5x}+1)^{12}$

$\displaystyle = 2e^{5x}(e^{5x}+1)^{12}$

15. $\displaystyle\int e^{6x}dx = \frac{1}{6}\int e^u du = \frac{1}{6}e^u + c$ $\qquad\qquad u = 6x \qquad du = 6dx$

$\qquad = \frac{1}{6}e^{6x} + c.$

Checking, $D_x(\frac{1}{6}e^{6x}) = \frac{1}{6} \cdot 6e^{6x} = e^{6x}$

17. $\displaystyle\int e^{x^3+6x+10}(x^2+2)dx = \int \frac{1}{3}e^u du$ $\qquad u = x^3 + 6x + 10$
$\qquad\qquad\qquad\qquad\qquad\qquad\qquad\qquad du = (3x^2 + 6)dx$
$\qquad = \frac{1}{3}e^u + c = \frac{1}{3}e^{x^3+6x+10} + c$ $\qquad\qquad = 3(x^2 + 2)dx$

Checking, $D_x(\frac{1}{3}e^{x^3+6x+10}) = \frac{1}{3}(3x^2+6)e^{x^3+6x+10} = (x^2+2)e^{x^3+6x+10}$

19. $\displaystyle\int \frac{e^{\sqrt{x}}}{\sqrt{x}}dx = \int 2e^u du = 2e^u + c = 2e^{\sqrt{x}} + c.$ $\qquad u = \sqrt{x}$

$\qquad\qquad\qquad\qquad\qquad\qquad\qquad\qquad\qquad\qquad du = \frac{1}{2\sqrt{x}}dx$

Checking, $D_x 2e^{\sqrt{x}} = 2 \cdot \frac{1}{2\sqrt{x}}e^{\sqrt{x}} = \frac{e^{\sqrt{x}}}{\sqrt{x}}$

21. $\displaystyle\int 2^{x^2} x\, dx = \int (\frac{1}{2})2^u du = \frac{1}{2\ln 2}2^u + c$ $\qquad u = x^2$

$\qquad\qquad\qquad\qquad\qquad\qquad\qquad\qquad\qquad\qquad du = 2x\, dx$

$\qquad = \frac{1}{2\ln 2}2^{x^2} + c.$

Checking, $D_x\frac{1}{2\ln 2}2^{x^2} = \frac{1}{2\ln 2} \cdot (2x\ln 2)(2^{x^2}) = x \cdot 2^{x^2}.$

23. If $\displaystyle\int 5xe^{3x}dx = (ax+b)e^{3x} + c$, differentiating $5xe^{3x} = ae^{3x} + 3(ax+b)e^{3x}$

$\qquad = e^{3x}(a + 3ax + 3b).$

Letting $x = 0$, $0 = a + 3b$, and letting $x = 1$, $5 = 4a + 3b$.

Solving the system $\begin{cases} a + 3b = 0 \\ 4a + 3b = 5 \end{cases}$, $3a = 5$, $a = \frac{5}{3}$ and $b = -\frac{5}{9}.$

$\displaystyle\int 5xe^{3x}dx = (\frac{5}{3}x - \frac{5}{9})e^{3x} + c$

Checking, $D_x(\frac{5}{3}x - \frac{5}{9})e^{3x} = \frac{5}{3}e^{3x} + (\frac{5}{3}x - \frac{5}{9})3e^{3x} = 5xe^{3x}$

25. $s(t) = \displaystyle\int (t^2+1)^{1/2}3t\, dt = \int \frac{3}{2}u^{1/2}du$ $\qquad u = t^2 + 1$
$\qquad\qquad\qquad\qquad\qquad\qquad\qquad\qquad\qquad du = 2t\, dt$

$\qquad = \frac{3}{2} \cdot \frac{2}{3}u^{3/2} + c = (t^2+1)^{3/2} + c.$

$s(0) = 1^{3/2} + c = 10.$ $\quad s(t) = (t^2+1)^{3/2} + 9.$

27. $s(t) = \int e^{-t^2} 3t\,dt = \int -\frac{3}{2}e^u\,du = -\frac{3}{2}e^u + c \qquad\qquad u = -t^2$

$\qquad\qquad\qquad\qquad\qquad\qquad\qquad\qquad\qquad du = -2t\,dt$

$\qquad = \frac{-3}{2}e^{-t^2} + c. \quad s(0) = \frac{-3}{2} + c = 4. \quad s(t) = \frac{-3}{2}e^{-t^2} + \frac{11}{2}.$

29. $s(t) = \int 5^{t^2+1}(4t)\,dt = \int (2)5^u\,du \qquad\qquad u = t^2 + 1$

$\qquad\qquad\qquad\qquad\qquad\qquad\qquad\qquad du = 2t\,dt$

$\qquad = \frac{2}{\ln 5}5^u + c = \frac{2}{\ln 5}5^{t^2+1} + c.$

$\qquad s(0) = \frac{2}{\ln 5}(5) + c = 25. \quad c = 25 - \frac{10}{\ln 5}$

$\qquad s(t) = \frac{2}{\ln 5}5^{t^2+1} + 25 - \frac{10}{\ln 5}.$

31. $C(q) = \int \frac{q+3}{(q^2+6q+1)^{2/3}}\,dq = \int \frac{1}{2}u^{-2/3}\,du \qquad\qquad u = q^2 + 6q + 1$

$\qquad\qquad\qquad\qquad\qquad\qquad\qquad\qquad\qquad\qquad du = (2q+6)\,dq$

$\qquad\qquad\qquad\qquad\qquad\qquad\qquad\qquad\qquad\qquad\quad = 2(q+3)\,dq$

$\qquad = \frac{3}{2}u^{1/3} + c = \frac{3}{2}\sqrt[3]{q^2+6q+1} + c.$

$\qquad C(0) = 3/2 + c = 30000. \quad C(q) = \frac{3}{2}\sqrt[3]{q^2+6q+1} + 29998.5$

33. $C(q) = \int \frac{2q+1}{q^2+q+5}\,dq = \int \frac{1}{u}\,du = \ln|u| + c \qquad\qquad u = q^2 + q + 5$

$\qquad\qquad\qquad\qquad\qquad\qquad\qquad\qquad\qquad\qquad du = (2q+1)\,dq$

$\qquad = \ln|q^2+q+5| + c.$

$\qquad C(0) = \ln 5 + c = 40000$

$\qquad C(q) = \ln|q^2+q+5| + 40000 - \ln 5$

35. $D_u\left[\int f(u)\,du - \int g(u)\,du\right] = D_u\left[\int f(u)\,du\right] - D_u\left[\int g(u)\,du\right] = f(u) - g(u)$

37. $D_u\left[\frac{u^{n+1}}{n+1} + c\right] = \frac{(n+1)u^n}{n+1} + 0 = u^n, \; n \neq -1$

39. $D_u\left[e^u + c\right] = e^u + 0 = e^u$

41. $v(t) = \int e^{4t}\,dt = \int \frac{1}{4}e^u\,du = \frac{1}{4}e^u + c$

$\qquad\qquad\qquad\qquad\qquad\qquad\qquad\qquad\qquad u = e^{4t}$

$\qquad\qquad\qquad\qquad\qquad\qquad\qquad\qquad\qquad du = 4e^{4t}\,dt$

$\qquad v(t) = \frac{e^{4t}}{4} + c. \quad v(0) = \frac{1}{4} + c = 12$

$\qquad v(t) = \frac{1}{4}e^{4t} + \frac{47}{4}.$

$\qquad s(t) = \int (\frac{1}{4}e^{4t} + \frac{47}{4})\,dt = \int \frac{1}{16}e^u\,du + \frac{47}{4}t = \frac{1}{16}e^{4t} + \frac{47}{4}t + K.$

$\qquad\qquad s(0) = \frac{1}{16} + K = 25$

$\qquad s(t) = \frac{1}{16}e^{4t} + \frac{47}{4}t + \frac{399}{16}.$

43. $v(t) = \int te^{3t}dt = (at + b)e^{3t} + c$

By Exercise 23,

$v(t) = (\frac{1}{3}t - \frac{1}{9})e^{3t} + c. \quad v(0) = -\frac{1}{9} + c = 8.$

$v(t) = (\frac{1}{3}t - \frac{1}{9})e^{3t} + \frac{73}{9}.$

$s(t) = \frac{1}{3}\int te^{3t}dt - \frac{1}{9}\int e^{3t}dt + \frac{73t}{9} = \frac{1}{3}(\frac{1}{3}t - \frac{1}{9})e^{3t} - \frac{1}{27}e^{3t} + \frac{73t}{9} + K$

$s(0) = -\frac{1}{27} - \frac{1}{27} + K = 40. \quad K = 40 + \frac{2}{27} = \frac{1082}{27}$ and

$s(t) = \frac{1}{9}te^{3t} - \frac{2}{27}e^{3t} + 73t/9 + 1082/27$

45. $D_x(\frac{2x}{5}e^{5x} + c) = e^{5x}D_x(\frac{2x}{5}) + \frac{2x}{5}D_xe^{5x} + 0$

$= e^{5x}(\frac{2}{5}) + 2xe^{5x} \neq 2xe^{5x}$

EXERCISE SET 4.3 INTEGRATION BY PARTS

1. $\int 3xe^{8x}dx = \frac{3}{8}xe^{8x} - \int \frac{1}{8}e^{8x} \cdot 3dx$

$\qquad\qquad\qquad\qquad u = 3x \qquad v = \frac{1}{8}e^{8x}$

$= \frac{3}{8}xe^{8x} - \frac{3}{64}e^{8x} + c$

$\qquad\qquad\qquad\qquad du = 3dx \quad dv = e^{8x}dx$

3. $\int (5x + 6)e^{2x}dx=$

$\qquad\qquad\qquad\qquad u = 5x + 6 \qquad v = \frac{1}{2}e^{2x}$

$\frac{1}{2}(5x + 6)e^{2x} - \int \frac{1}{2}e^{2x} \cdot 5dx$

$\qquad\qquad\qquad\qquad du = 5dx \qquad dv = e^{2x}dx$

$\frac{1}{2}(5x + 6)e^{2x} - \frac{5}{4}e^{2x} + c.$

5. $\int (3x + 1)e^{4x}dx$

$\qquad\qquad\qquad\qquad u = 3x + 1 \qquad v = (1/4)e^{4x}$

$\qquad\qquad\qquad\qquad du = 3dx \qquad dv = e^{4x}dx$

$= \frac{1}{4}(3x + 1)e^{4x} - \int \frac{1}{4}e^{4x} \cdot 3dx$

$= \frac{1}{4}(3x + 1)e^{4x} - \frac{3}{16}e^{4x} + c$

7. $\int x^5 e^{x^3} dx$

$\qquad\qquad\qquad\qquad u = x^3 \qquad v = (1/3)e^{x^3}$

$= \frac{1}{3}x^3 e^{x^3} - \int x^2 e^{x^3} dx = \frac{1}{3}x^3 e^{x^3} - \frac{1}{3}e^{x^3} + c$

$\qquad\qquad\qquad\qquad du = 3x^2 dx$

$\qquad\qquad\qquad\qquad\qquad\qquad dv = x^2 e^{x^3} dx$

9. $\int \ln|x^3|dx$

$\qquad\qquad\qquad\qquad u = \ln|x^3|dx \quad v = x$

$= x\ln|x^3| - \int 3dx$

$\qquad\qquad\qquad\qquad du = \frac{3}{x}dx \qquad dv = dx$

$= x\ln|x^3| - 3x + c.$

11. $\displaystyle\int 2x\ln|x+3|\,dx =$ $\qquad\qquad\qquad\qquad u = \ln|x+3|\quad v = x^2$

$= x^2\ln|x+3| - \displaystyle\int \frac{x^2}{x+3}\,dx$ $\qquad\qquad\qquad du = \dfrac{1}{x+3}dx\;\; dv = 2x\,dx$

$= x^2\ln|x+3| - \displaystyle\int\left(x - 3 + \frac{9}{x+3}\right)dx = x^2\ln|x+3| - \dfrac{x^2}{2} + 3x - 9\ln|x+3| + c$

13. $\displaystyle\int \frac{\ln|x|}{x^3}\,dx$ $\qquad\qquad\qquad\qquad\qquad u = \ln|x| \qquad v = -\tfrac{1}{2}x^{-2}$

$= \dfrac{-1}{2x^2}\ln|x| + \displaystyle\int \frac{1}{2}\cdot\frac{1}{x^2}\cdot\frac{1}{x}\,dx$ $\qquad\qquad du = \tfrac{1}{x}dx \qquad dv = x^{-3}\,dx$

$= \dfrac{-1}{2x^2}\ln|x| + \displaystyle\int \frac{1}{2}x^{-3}\,dx = \dfrac{-1}{2x^2}\ln|x| - \dfrac{1}{4}x^{-2} + c$

$= \dfrac{-1}{2x^2}\ln|x| - \dfrac{1}{4x^2} + c$

15. $\displaystyle\int \ln|x|\cdot x^{-1/2}\,dx$ $\qquad\qquad\qquad\quad u = \ln|x| \qquad v = 2x^{1/2}$

$= 2\sqrt{x}\ln|x| - \displaystyle\int \frac{2\sqrt{x}}{x}\,dx$ $\qquad\qquad\qquad du = \tfrac{1}{x}dx \qquad dv = x^{-1/2}\,dx$

$= 2\sqrt{x}\ln|x| - \displaystyle\int 2x^{-1/2}\,dx = 2\sqrt{x}\ln|x| - 4x^{1/2} + c$

$= 2\sqrt{x}\ln|x| - 4\sqrt{x} + c$

17. $\displaystyle\int x\cdot 3^x\,dx$ $\qquad\qquad\qquad\qquad\qquad u = x \qquad v = \dfrac{1}{\ln 3}\cdot 3^x$

$= \dfrac{x}{\ln 3}(3^x) - \displaystyle\int \frac{1}{\ln 3}(3^x)\,dx$ $\qquad\qquad du = dx \qquad dv = 3^x\,dx$

$= \dfrac{x}{\ln 3}(3^x) - \dfrac{1}{(\ln 3)^2}\cdot 3^x + c$

19. $\displaystyle\int x^3\ln|x+4|\,dx$ $\qquad\qquad\qquad\qquad u = \ln|x+4|\quad v = (1/4)x^4$

$= \tfrac{1}{4}x^4\ln|x+4| - \displaystyle\int (\tfrac{1}{4})\frac{x^4}{x+4}\,dx$ $\qquad\qquad du = \dfrac{1}{x+4}dx\;\; dv = x^3\,dx$

$= \tfrac{1}{4}x^4\ln|x+4| - \displaystyle\int \frac{1}{4}\left(x^3 - 4x^2 + 16x - 64 + \frac{256}{x+4}\right)dx$

$= \tfrac{1}{4}x^4\ln|x+4| - \dfrac{1}{16}x^4 + \dfrac{1}{3}x^3 - 2x^2 + 16x - 64\,\ln|x+4| + c$

21. $\displaystyle\int x\,\sqrt[3]{x+2}\,dx = \frac{3x}{4}(x+2)^{4/3} - \int \frac{3}{4}(x+2)^{4/3}\,dx$ $\qquad u = x \qquad\qquad v = \tfrac{3}{4}(x+2)^{4/3}$

$= \dfrac{3x}{4}(x+2)^{4/3} - \dfrac{9}{28}(x+2)^{7/3} + c$ $\qquad\qquad\qquad du = dx \qquad\qquad dv = \sqrt[3]{x+2}\,dx$

23. $\displaystyle\int \frac{x}{\sqrt{x+4}}\cdot\, dx =$

$\qquad u = x \qquad\qquad v = 2(x+4)^{1/2}$

$\qquad du = dx \qquad\quad dv = (x+4)^{-1/2}\,dx$

$2x\sqrt{x+4} - \displaystyle\int 2(x+4)^{1/2}\,dx$

$= 2x\sqrt{x+4} - \tfrac{4}{3}(x+4)^{3/2} + c$

25. $\displaystyle\int x^2(x+6)^{20}\,dx$

$\qquad u = x^2 \qquad\qquad v = \tfrac{1}{21}(x+6)^{21}$

$\qquad du = 2x\,dx \qquad dv = (x+6)^{20}\,dx$

$= \tfrac{x^2}{21}(x+6)^{21} - \displaystyle\int \tfrac{2}{21}(x+6)^{21}x\,dx$

$\qquad\qquad$ Now let $u = x \quad v = \dfrac{2}{21(22)}(x+6)^{22}$

$\qquad\qquad\qquad du = dx \qquad\quad dv = \tfrac{2}{21}(x+6)^{21}\,dx$

$= \tfrac{x^2}{21}(x+6)^{21} - \Big(\tfrac{x}{21(11)}(x+6)^{22} - \displaystyle\int \tfrac{1}{21(11)}(x+6)^{22}\,dx\Big)$

$= \tfrac{x^2}{21}(x+6)^{21} - \tfrac{x}{21(11)}(x+6)^{22} + \dfrac{(x+6)^{23}}{21(11)(23)} + c$

27. Letting $n = 2$ and $a = 10$, $\displaystyle\int x^2 e^{10x}\,dx = \tfrac{1}{10}x^2 e^{10x} - \tfrac{1}{5}\displaystyle\int x e^{10x}\,dx$

Next, letting $n = 1$ and $a = 10$,

$\tfrac{1}{10}x^2 e^{10x} - \tfrac{1}{5}\displaystyle\int x e^{10x}\,dx = \tfrac{1}{10}x^2 e^{10x} - \tfrac{1}{5}\Big[\tfrac{1}{10}x e^{10x} - \tfrac{1}{10}\displaystyle\int e^{10x}\,dx\Big]$

$= \tfrac{1}{10}x^2 e^{10x} - \tfrac{1}{50}x e^{10x} + \tfrac{1}{500}e^{10x} + c$

29. Letting $n = 4$ and $a = -2$

$\displaystyle\int x^4 e^{-2x}\,dx = \tfrac{-1}{2}x^4 e^{-2x} + 2\displaystyle\int x^3 e^{-2x}\,dx.$

Next, letting $n = 3$ and $a = -2$,

$-\tfrac{1}{2}x^4 e^{-2x} + 2\Big[-\tfrac{1}{2}x^3 e^{-2x} + \tfrac{3}{2}\displaystyle\int x^2 e^{-2x}\,dx\Big]$

$= -\tfrac{1}{2}x^4 e^{-2x} - x^3 e^{-2x} + 3\displaystyle\int x^2 e^{-2x}\,dx.\quad$ Letting $n = 2$ and $a = -2$,

$\displaystyle\int x^4 e^{-2x}\,dx = \tfrac{-1}{2}x^4 e^{-2x} - x^3 e^{-2x} + 3\Big[\tfrac{-1}{2}x^2 e^{-2x} + \displaystyle\int x e^{-2x}\,dx\Big]$

$= \tfrac{-1}{2}x^4 e^{-2x} - x^3 e^{-2x} - \tfrac{3}{2}x^2 e^{-2x} + 3\displaystyle\int x e^{-2x}\,dx.$

Finally, letting $n = 1$ and $a = -2$,

$\displaystyle\int x^4 e^{-2x}\,dx = \tfrac{-1}{2}x^4 e^{-2x} - x^3 e^{-2x} - \tfrac{3}{2}x^2 e^{-2x} + 3\Big[\tfrac{-1}{2}x e^{-2x} + \tfrac{1}{2}\displaystyle\int e^{-2x}\,dx\Big]$

$= \tfrac{-1}{2}x^4 e^{-2x} - x^3 e^{-2x} - \tfrac{3}{2}x^2 e^{-2x} - \tfrac{3}{2}x e^{-2x} - \tfrac{3}{4}e^{-2x} + c.$

31. $C(q) = \int (-q^2 + 10000)e^{-.1q}dq$ $\qquad\qquad$ $u = -q^2 + 10000 \quad v = -10e^{-.1q}$

$\qquad = -(-q^2 + 10000)10e^{-.1q} - \int 20qe^{-.1q}dq$ $\qquad du = -2qdq$ $\qquad dv = e^{-.1q}dq$

$\qquad\qquad\qquad\qquad\qquad$ Next, let $\qquad\qquad$ $u = 20q$ $\qquad\qquad$ $v = -10e^{-.1q}$

$\qquad\qquad\qquad\qquad\qquad\qquad\qquad\qquad\qquad\qquad$ $du = 20dq$ $\qquad\qquad$ $dv = e^{-.1q}dq$

$\qquad = 10(q^2 - 10000)e^{-.1q} - (-200qe^{-.1q} + \int 200e^{-.1q}dq)$

$\qquad = 10(q^2 - 10000)e^{-.1q} + 200qe^{-.1q} + 2000e^{-.1q} + K$

$\qquad C(q) = 10e^{-.1q}(q^2 + 20q - 9800) + K.$

$\qquad C(0) = -98,000 + K = 15000.$

$\qquad K = 113,000$

$\qquad C(q) = 10e^{-.1q}(q^2 + 20q - 9800) + 113,000.$

33. (a) $\;\; v(t) = \int te^{-2t}dt$ $\qquad\qquad\qquad\qquad\qquad$ $u = t$ $\qquad\qquad$ $y = -\frac{1}{2}e^{-2t}$

$\qquad\qquad\qquad\qquad\qquad\qquad\qquad\qquad\qquad\qquad\qquad$ $du = dt$ $\qquad\qquad$ $dy = e^{-2t}dt$

$\qquad\qquad = -\frac{1}{2}te^{-2t} + \int \frac{1}{2}e^{-2t}dt$

$\qquad\qquad = -\frac{1}{2}te^{-2t} - \frac{1}{4}e^{-2t} + C.$

$\qquad\qquad v(0) = -\frac{1}{4} + C = 25. \quad C = 25.25$

$\qquad\qquad v(t) = -\frac{1}{2}te^{-2t} - \frac{1}{4}e^{-2t} + 25.25$

$\qquad$ (b) $\;\; s(t) = \int (-\frac{1}{2}te^{-2t} - \frac{1}{4}e^{-2t} + 25.25)dt$

$\qquad\qquad = -\frac{1}{2}\int te^{-2t}dt + \frac{1}{8}e^{-2t} + 25.25t + K$

$\qquad\qquad = -\frac{1}{2}\left[-\frac{1}{2}te^{-2t} - \frac{1}{4}e^{-2t}\right] + \frac{1}{8}e^{-2t} + 25.25t + K$

$\qquad\qquad = \frac{1}{4}te^{-2t} + \frac{1}{4}e^{-2t} + 25.25t + K$

$\qquad\qquad s(5) = \frac{5}{4}e^{-10} + \frac{1}{4}e^{-10} + 126.25 + K = 25$

$\qquad\qquad K = -101.25 - \frac{3}{2}e^{-10}$

$\qquad\qquad$ and

$\qquad\qquad s(t) = \frac{1}{4}te^{-2t} + \frac{1}{4}e^{-2t} + 25.25t - 101.25 - 1.5e^{-10}$

35. $u = e^{5x}$ $\qquad v = x^2$ $\;\;$ and

$\qquad du = 5e^{5x}$ $\qquad dv = 2xdx$

$\qquad \int 2xe^{5x}dx = x^2e^{5x} - \int 5x^2e^{5x}dx$ and the latter integral is more complicated than $\int 2xe^{5x}dx$

1. Using $\displaystyle\int \frac{du}{\sqrt{u^2 \pm a^2}} = \ln|u + \sqrt{u^2 \pm a^2}|$ with $x = u$ and $a = 5$,

    $\displaystyle\int \frac{1}{\sqrt{x^2 - 25}} dx = \ln|x + \sqrt{x^2 - 25}| + C$

3. Using $\displaystyle\int \sqrt{u^2 \pm a^2}\, du = \frac{u}{2}\sqrt{u^2 \pm a^2} \pm \frac{a^2}{2}\ln|u + \sqrt{u^2 \pm a^2}|$

    with $a = 4$ and $u = x$, $\displaystyle\int \sqrt{x^2 + 16}\, dx = \frac{1}{2}\left[x\sqrt{x^2 + 16} + 16\ln|x + \sqrt{x^2 + 16}|\right] + C$

5. Let $u = e^x$, $du = e^x dx$, and using formula #14,

    $\displaystyle\int e^x\sqrt{e^{2x} + 16}\, dx = \int \sqrt{u^2 + 16}\, du = \frac{1}{2}\left[u\sqrt{u^2 + 16} + 16\ln|u + \sqrt{u^2 + 16}|\right] + C$

    $= \frac{1}{2}\left[e^x\sqrt{e^{2x} + 16} + 16\ln|e^x + \sqrt{e^{2x} + 16}|\right] + C$

7. Using $\displaystyle\int u\sqrt{au + b}\, du = \frac{2(3au - 2b)}{15a^2}(au + b)^{3/2}$

    with $a = 5$, $b = 8$, and $u = x$

    $\displaystyle\int x\sqrt{5x + 8}\, dx = \frac{2}{375}(15x - 16)(5x + 8)^{3/2} + C$

9. Using $\displaystyle\int \frac{1}{a^2 - u^2} du = \frac{1}{2a}\ln\left|\frac{a + u}{a - u}\right|$ with $a = 7$ and $u = x$

    $\displaystyle\int \frac{1}{49 - x^2} dx = \frac{1}{14}\ln\left|\frac{7 + x}{7 - x}\right| + C.$

11. Using $\displaystyle\int \frac{du}{a + be^{mu}} = \frac{u}{a} - \frac{1}{am}\ln|a + be^{mu}|$ with $a = 5$, $b = 3$, $m = 1$ and $u = x$,

    $\displaystyle\int \frac{1}{5 + 3e^x} dx = \frac{x}{5} - \frac{1}{5}\ln|5 + 3e^x| + C$

13. Using $\displaystyle\int u^n e^{au}\, du = \frac{u^n}{a}e^{au} - \frac{n}{a}\int u^{n-1} e^{au}\, du$ and then

    $\displaystyle\int ue^{au}\, du = \frac{1}{a^2}(au - 1)e^{au}$ with $a = 10$ and $n = 2$ and $u = x$

    $\displaystyle\int x^2 e^{10x}\, dx = \frac{1}{10}x^2 e^{10x} - \frac{1}{5}\int xe^{10x}\, dx = \frac{1}{10}x^2 e^{10x} - \frac{1}{500}(10x - 1)e^{10x} + C.$

15. Let $u = \ln|x|$, $du = \frac{1}{x}dx$ and use formula 1 with $n = 5$.

$$\int \frac{\ln^5|x|}{x}dx = \int u^5\,du = \frac{u^6}{6} + C = \frac{\ln^6|x|}{6} + C$$

17. $\int \frac{3\,dx}{25 - 4x^2}$ obtain $= \int \frac{3}{4} \cdot \frac{1}{\frac{25}{4} - x^2}dx.$ Next use formula #10 with $a = 5/2 = 2.5$ and $u = x$ to

$$\frac{.75}{2(2.5)}\ln\left|\frac{5/2 + x}{5/2 - x}\right| + C = \frac{3}{20}\ln\left|\frac{5 + 2x}{5 - 2x}\right| + C.$$

19. Letting $u = x^2$ and $du = 2x\,dx$, $x^5\,dx = \frac{1}{2}x^4\,du.$

We obtain $\int x^5 e^{x^2}dx = \int \frac{1}{2}u^2 e^u\,du.$ By formula #27 with $a = 1$ and $n = 2$, we have

$$\frac{1}{2}\left[u^2 e^u - 2\int ue^u\,du\right].$$ Using formula #26 we have $\frac{1}{2}u^2 e^u - (u - 1)e^u + C.$

$$= \frac{1}{2}x^4 e^{x^2} - (x^2 - 1)e^{x^2} + C.$$

21. $\int x^2\sqrt{4x^2 - 25}\,dx = \int 2x^2\sqrt{x^2 - \frac{25}{4}}\,dx.$ Using formula #15 with $a = \frac{5}{2} = 2.5$ and $u = x$

we obtain $2 \cdot \frac{x}{8}(2x^2 - 2.5^2)\sqrt{x^2 - 2.5^2} - \frac{2(2.5)^4}{8}\ln|x + \sqrt{x^2 - 2.5^2}| + C$

$$= \frac{x}{4}(2x^2 - \frac{25}{4})\frac{1}{2}\sqrt{4x^2 - 25} - \frac{625}{64}\ln|x + \frac{1}{2}\sqrt{4x^2 - 25}| + C$$

$$= \frac{x}{32}(8x^2 - 25)\sqrt{4x^2 - 25} - \frac{625}{64}\ln|2x + \sqrt{4x^2 - 25}| + C_1$$

23. Letting $u = e^x$, $du = e^x dx$ and using formula #21 with $a = 5$,

$$\int \frac{e^x}{(e^{2x} - 25)^{3/2}} = \int \frac{1}{(u^2 - 25)^{3/2}}du = -\frac{u}{25\sqrt{u^2 - 25}} + C = \frac{-e^x}{25\sqrt{e^{2x} - 25}} + C$$

25. $\int x^3(9x^2 + 64)^{3/2}dx = \int 27x^3(x^2 + \frac{64}{9})^{3/2}dx.$ Using formula #23 with $a = \frac{8}{3}$ and $u = x$

we obtain $\frac{27}{7}(x^2 + \frac{64}{9})^{7/2} - \frac{64}{45}(27)(x^2 + \frac{64}{9})^{5/2} + C$

$$= \frac{1}{567}(9x^2 + 64)^{7/2} - \frac{64}{405}(9x^2 + 64)^{5/2} + C$$

27. Letting $u = 5x$, $du = 5\,dx$ we have $\int x^3\sqrt{25x^2 - 121}\,dx = \int \frac{u^3}{625}\sqrt{u^2 - 121}\,du.$

Using formula #16 with $a = 11$ we obtain

$$\frac{1}{625}\left[\frac{1}{5}(u^2 - 121)^{5/2}) + \frac{121}{3}(u^2 - 121)^{3/2}\right] + C$$

$$= \frac{1}{3125}(25x^2 - 121)^{5/2} + \frac{121}{1875}(25x^2 - 121)^{3/2} + C$$

29. Letting $u = x^2$, $du = 2xdx$, and $\int 8x^7\sqrt{x^4 + 16}\,dx = \int 4u^3\sqrt{u^2 + 16}\,du$.

Using formula #17 with $a = 4$ we obtain $4(\frac{u^2}{5} - \frac{32}{15})(u^2 + 16)^{3/2} + C$

$= (\frac{4}{5}x^4 - \frac{128}{15})(x^4 + 16)^{3/2} + C$

31. $\int x(x^4 + 4x^2 + 4 + 9)^{3/2}\,dx = \int x[(x^2 + 2)^2 + 9]^{3/2}\,dx$.

Letting $u = x^2 + 2$ and $du = 2xdx$ we obtain

$\int \frac{1}{2}(u^2 + 9)^{3/2}\,du = \frac{u}{2(4)}(u^2 + 9)^{3/2} + \frac{3}{8} \cdot \frac{9}{2}u\sqrt{u^2 + 9} + \frac{3}{16}(81)\ln|u + \sqrt{u^2 + 9}| + C$

from formula #22 with $a = 3$. Replacing u by $x^2 + 2$ we have

$\frac{x^2 + 2}{8}(x^4 + 4x^2 + 13)^{3/2} + \frac{27}{16}(x^2 + 2)\sqrt{x^4 + 4x^2 + 13} + \frac{243}{16}\ln|x^2 + 2 + \sqrt{x^4 + 4x^2 + 13}| + C$

33. $\int \frac{1}{36 + 5x - x^2}\,dx = \int \frac{-1}{x^2 - 5x - 36}\,dx$

$= \int \frac{-1}{x^2 - 2(\frac{5}{2})x + \frac{25}{4} - \frac{169}{4}}\,dx = \int \frac{-1}{(x - \frac{5}{2})^2 - \frac{169}{4}}\,dx = \int \frac{1}{\frac{169}{4} - (x - \frac{5}{2})^2}\,dx$.

Letting $u = x - \frac{5}{2}$ and using formula #10 with $a = \frac{13}{2}$ we obtain $\int \frac{1}{\frac{169}{4} - u^2}\,du$

$= \frac{1}{13}\ln|\frac{6.5 + u}{6.5 - u}| + C = \frac{1}{13}\ln|\frac{6.5 + x - 2.5}{6.5 - x + 2.5}| + C = \frac{1}{13}\ln|\frac{x + 4}{9 - x}| + C$

35. $\int \frac{x(x^4 + 6x^2 + 9 - 25)^{1/2}}{(x^2 + 3)^2}\,dx = \int \frac{x[(x^2 + 3)^2 - 25]^{1/2}}{(x^2 + 3)^2}\,dx$. Letting $u = x^2 + 3$,

$du = 2xdx$ and using formula #20 with $a = 5$ we obtain

$\int \frac{1}{2}\frac{(u^2 - 25)^{1/2}}{u^2} = \frac{-\sqrt{u^2 - 25}}{2u} + \frac{1}{2}\ln/u + \sqrt{u^2 - 25}| + C$

$= \frac{-\sqrt{x^4 + 6x^2 - 16}}{2(x^2 + 3)} + \frac{1}{2}\ln|x^2 + 3 + \sqrt{x^4 + 6x^2 - 16}| + C$

EXERCISE SET 4.5 GUESSING AGAIN

1. $\int (x^4 + 3)^{25} 6x^3\,dx = K(x^4 + 3)^{26} + C$

$D_x K(x^4 + 3)^{26} = 26K(x^4 + 3)^{25}(4x^3)$

$= 104Kx^3(x^4 + 3)^{25}. \quad 104K = 6, \quad K = \frac{6}{104} = \frac{3}{52}$

$\int (x^4 + 3)^{25} 6x^3\,dx = \frac{3}{52}(x^4 + 3)^{26} + C$

3. $\int \sqrt[3]{x^5+4}\,(3x^4)\,dx = (x^5+4)^{4/3}K + C.$ $D_x(x^5+4)^{4/3}K = \frac{4K}{3}(x^5+4)^{1/3}(5x^4)$

$= \frac{20K}{3}(x^5+4)^{1/3}x^4.$ $\frac{20K}{3} = 3.$ $K = 9/20$

$\int \sqrt[3]{x^5+4}\,(3x^4)\,dx = \frac{9}{20}(x^5+4)^{4/3} + C$

5. $\int \dfrac{2x^2+4}{\sqrt[3]{(x^3+6x+1)^2}}\,dx = \int 2(x^2+2)(x^3+6x+1)^{-2/3}\,dx = K(x^3+6x+1)^{1/3} + C.$

$D_x K(x^3+6x+1)^{1/3} = K(3x^2+6)(1/3)(x^3+6x+1)^{-2/3}.$

$K(x^2+2) = 2x^2+4.$ $K(x^2+2) = 2(x^2+2).$ $K = 2$

$\int \dfrac{2x^2+4}{\sqrt[3]{(x^3+6x+1)^2}}\,dx = 2(x^3+6x+1)^{1/3} + C$

7. $\int (e^{2x}+1)^{50}4e^{2x}\,dx = K(e^{2x}+1)^{51} + C.$ $D_x K(e^{2x}+1)^{51} = 51K(e^{2x}+1)^{50}(2e^{2x})$

$102(e^{2x}+1)^{50}e^{2x}K = (e^{2x}+1)^{50}\cdot 4e^{2x}.$ $102K = 4.$ $K = 2/51$

$\int (e^{2x}+1)^{50}4e^{2x}\,dx = \frac{2}{51}(e^{2x}+1)^{51} + C$

9. Guess that $\int xe^{3x}\,dx = (ax+b)e^{3x} + C.$

$D_x((ax+b)e^{3x}) = ae^{3x} + (3ax+3b)e^{3x} = (3ax+a+3b)e^{3x} = xe^{3x},$

$3a = 1,\ a+3b = 0,\ a = 1/3,\ b = -1/9$ and

$\int xe^{3x}\,dx = (\frac{x}{3} - \frac{1}{9})e^{3x} + C = \frac{x}{3}e^{3x} - \frac{x}{9}e^{3x} + C$

11. Guess that $\int x^3 e^{-3x}\,dx = (ax^3 + bx^2 + cx + d)e^{-3x} + K.$

$D_x(ax^3 + bx^2 + cx + d)e^{-3x} = (3ax^2 + 2bx + c)e^{-3x} - 3e^{-3x}(ax^3 + bx^2 + cx + d),$

$x^3 = 3ax^2 + 2bx + c - 3ax^3 - 3bx^2 - 3cx - 3d,$

$-3a = 1$	$a = -1/3$
$3a - 3b = 0$	$b = -1/3$
$2b - 3c = 0$	$c = -2/9$
$c - 3d = 0$	$d = -2/27$

$\int x^3 e^{-3x}\,dx = (-\frac{1}{3}x^3 - \frac{1}{3}x^2 - \frac{2}{9}x - \frac{2}{27})e^{-3x} + K$

13. Guess that $\int x^2 5^x dx = (ax^2 + bx + c)5^x + K.$

$D_x((ax^2 + bx + c)5^x) = (2ax + b)5^x + (ax^2 + bx + c)(\ln5)5^x,$

$2ax + b + a\ln5(x^2) + \ln5(bx) + (\ln5)c = x^2,$

$a\ln5 = 1, \quad 2a + b\ln5 = 0 \quad b + c\ln5 = 0$

$a = \dfrac{1}{\ln5}, \quad b = \dfrac{-2}{(\ln5)^2}, \quad c = \dfrac{2}{(\ln5)^3}.$

$\int x^2 5^x dx = \left(\dfrac{x^2}{\ln5} - \dfrac{2}{(\ln5)^2}x + \dfrac{2}{(\ln5)^3}\right)5^x + K$

15. Guess that $\int (4x^2 + 22x + 9)e^{4x}dx = (ax^2 + bx + c)e^{4x} + K.$

$D_x(ax^2 + bx + c)e^{4x} = (2ax + b)e^{4x} + 4(ax^2 + bx + c)e^{4x} = (4x^2 + 22x + 9)e^{4x},$

$2ax + b + 4ax^2 + 4bx + 4c = 4x^2 + 22x + 9,$

$4a = 4, \quad 2a + 4b = 22, \quad b + 4c = 9,$

$a = 1, \quad 4b = 20, \quad 4c = 9 - b,$

$a = 1, \quad b = 5, \quad 4c = 9 - 5. \quad c = 1$

$\int (4x^2 + 22x + 9)e^{4x}dx = (x^2 + 5x + 1)e^{4x} + K$

17. Guess that $\int (3x^3 + 9x^2 + 19x + 8)e^{3x}dx = (ax^3 + bx^2 + cx + d)e^{3x} + K.$

$D_x(ax^3 + bx^2 + cx + d)e^{3x} = (3ax^2 + 2bx + c)e^{3x} + 3(ax^3 + bx^2 + cx + d)e^{3x},$

$3x^3 + 9x^2 + 19x + 8 = 3ax^2 + 2bx + c + 3ax^3 + 3bx^2 + 3cx + 3d,$

$3a = 3, \quad 3a + 3b = 9, \quad 2b + 3c = 19, \quad c + 3d = 8$

$a = 1, \quad b = 2, \quad c = 5, \quad d = 1.$

$\int (3x^3 + 9x^2 + 19x + 8)e^{3x}dx = (x^3 + 2x^2 + 5x + 1)e^{3x} + K$

19. Guess that $\int x^3(25 - x^2)^{5/2} dx = (ax^4 + bx^2 + c)(25 - x^2)^{5/2} + K.$

$D_x(ax^4 + bx^2 + c)(25 - x^2)^{5/2} = (4ax^3 + 2bx)(25 - x^2)^{5/2} - 2x(25 - x^2)^{3/2} \cdot \frac{5}{2}(ax^4 + bx^2 + c)$

$= (25 - x^2)^{3/2}\left[(4ax^3 + 2bx)(25 - x^2) - 5x(ax^4 + bx^2 + c)\right]$

$= (25 - x^2)^{3/2}\left[100ax^3 + 50bx - 4ax^5 - 2bx^3 - 5ax^5 - 5bx^3 - 5cx\right] = x^3(25 - x^2)^{5/2}.$

$100ax^3 - 9ax^5 - 7bx^3 + 50bx - 5cx = x^3(25 - x^2) = 25x^3 - x^5.$

$$9a = 1 \qquad\qquad\qquad a = 1/9$$

$$100a - 7b = 25 \qquad 7b = \frac{100}{9} - 25 \qquad b = \frac{-125}{63}$$

$$50b - 5c = 0 \qquad\quad 5c = 50b \qquad\qquad c = \frac{-1250}{63}$$

$\int x^3(25 - x^2)^{5/2} dx = (\frac{1}{9}x^4 - \frac{125}{63}x^2 - \frac{1250}{63})(25 - x^2)^{5/2} + K$

21. Guess that $\int x^5(16-x^2)^{3/4}\,dx = (ax^6 + bx^4 + cx^2 + d)(16-x^2)^{3/4} + K.$

$D_x(ax^6 + bx^4 + cx^2 + d)(16-x^2)^{3/4}$

$= (6ax^5 + 4bx^3 + 2cx)(16-x^2)^{3/4} - \frac{3}{2}x(16-x^2)^{-1/4}(ax^6 + bx^4 + cx^2 + d),$

$(6ax^5 + 4bx^3 + 2cx)(16-x^2) - \frac{3}{2}(ax^7 + bx^5 + cx^3 + dx) = x^5(16-x^2),$

$96ax^5 + 64bx^3 + 32cx - 6ax^7 - 4bx^5 - 2cx^3 - \frac{3}{2}ax^7 - \frac{3}{2}bx^5 - \frac{3}{2}cx^3 - \frac{3}{2}dx = 16x^5 - x^7.$

$$-6a - \frac{3}{2}a = -\frac{15}{2}a = -1 \qquad\qquad a = 2/15$$

$$96a - 4b - \frac{3}{2}b = 96a - \frac{11}{2}b = 16 \qquad \frac{11}{2}b = \frac{192}{15} - 16 = \frac{64}{5} - 16 = -\frac{16}{5}$$

$$64b - 2c - \frac{3}{2}c = 64b - \frac{7}{2}c = 0 \qquad\qquad \frac{7}{2}c = 64b$$

$$32c - \frac{3}{2}d = 0 \qquad\qquad\qquad \frac{3}{2}d = 32c$$

$a = \frac{2}{15} \quad b = -\frac{32}{55} \quad c = 64(\frac{2}{7})(-\frac{32}{55}) = -4096/385 \quad d = \frac{64}{3} \cdot 64(\frac{2}{7})(-\frac{32}{55}) = -\frac{262{,}144}{1155}$

$\int x^5(16-x^2)^{3/4}\,dx = \left(\frac{2}{15}x^6 - \frac{32}{55}x^4 - \frac{4096}{385}x^2 - \frac{262{,}144}{1155}\right)(16-x^2)^{3/4} + K$

23. Guess that $\int (x^3 - 3x)(16-x^2)^{5/2}\,dx = (ax^4 + bx^2 + c)(16-x^2)^{5/2} + K.$

$D_x(ax^4 + bx^2 + c)(16-x^2)^{5/2} = (4ax^3 + 2bx)(16-x^2)^{5/2} - 5x(16-x^2)^{3/2}(ax^4 + bx^2 + c),$

$(x^3 - 3x)(16-x^2) = (4ax^3 + 2bx)(16-x^2) - 5x(ax^4 + bx^2 + c),$

$16x^3 - 48x - x^5 + 3x^3 = 64ax^3 + 32bx - 4ax^5 - 2bx^3 - 5ax^5 - 5bx^3 - 5cx,$

$-x^5 + 19x^3 - 48x = -9ax^5 + (64a - 7b)x^3 + (32b - 5c)x.$

$$9a = 1 \qquad\qquad a = 1/9$$

$$64a - 7b = 19 \qquad 7b = 64a - 19 = \frac{64}{9} - 19 = -\frac{107}{9}, \quad b = -\frac{107}{63}$$

$$32b - 5c = -48 \qquad 5c = 32b + 48 = -\frac{107(32)}{63} + 48, \quad c = -80/63$$

$\int (x^3 - 3x)(16-x^2)^{5/2}\,dx = (\frac{1}{9}x^4 - \frac{107}{63}x^2 - \frac{80}{63})(16-x^2)^{5/2} + K$

25. Guess that $\int (36x^3 - 149x)(16-x^2)^{5/2}\,dx = (ax^4 + bx^2 + c)(16-x^2)^{5/2} + K.$

As in Exercise 23

$(36x^3 - 149x)(16-x^2) = -9ax^5 + (64a - 7b)x^3 + (32b - 5c)x,$

$576x^3 - 2384x - 36x^5 + 149x^3 = -9ax^5 + (64a - 7b)x^3 + (32b - 5c)x,$

$-36x^5 + 725x^3 - 2384x = -9ax^5 + (64a - 7b)x^3 + (32b - 5c)x.$

$$9a = 36 \qquad\qquad a = 4$$

$$64a - 7b = 725 \qquad 7b = 64(4) - 725 = -469 \quad b = -67$$

$$32b - 5c = -2384 \qquad 5c = 32b + 2384. \quad c = 48$$

$\int (36x^3 - 149x)(16-x^2)^{5/2}\,dx = (4x^4 - 67x^2 + 48)(16-x^2)^{5/2} + K$

27. Guess that $\int (10x^3 + 17x)(4 - x^2)^{1/2}\, dx = (ax^4 + bx^2 + c)(4 - x^2)^{1/2} + K.$

$D_x(ax^4 + bx^2 + c)(4 - x^2)^{1/2} = (4ax^3 + 2bx)(4 - x^2)^{1/2} - x(ax^4 + bx^2 + c)(4 - x^2)^{-1/2},$

$(10x^3 + 17x)(4 - x^2) = (4ax^3 + 2bx)(4 - x^2) - ax^5 - bx^3 - cx,$

$40x^3 + 68x - 10x^5 - 17x^3 = 16ax^3 + 8bx - 4ax^5 - 2bx^3 - ax^5 - bx^3 - cx,$

$-10x^5 + 2x^3 + 68x = -5ax^5 + (16a - 3b)x^3 + (8b - c)x,$

$$-5a = -10 \qquad\qquad a = 2$$
$$16a - 3b = 23 \qquad\qquad 3b = 32 - 23 = 9, \quad b = 3$$
$$8b - c = 68 \qquad\qquad c = 8b - 68 = 24 - 68 = -44$$

$\int (10x^3 + 17x)(4 - x^2)^{1/2}\, dx = (2x^4 + 3x^2 - 44)(4 - x^2)^{1/2} + K$

EXERCISE SET 4.6 TABULAR INTEGRATION

1.

k	$f^{(k)}(x)$		$g^{(4-k)}(x)$
0	x^3	$+$	e^{5x}
1	$3x^2$	$-$	$\frac{1}{5}e^{5x}$
2	$6x$	$+$	$\frac{1}{25}e^{5x}$
3	6	$-$	$\frac{1}{125}e^{5x}$
4	0		$\frac{1}{625}5e^{5x}$

$\int x^3 e^{5x}\, dx = \frac{1}{5}x^3 e^{5x} - \frac{3}{25}x^2 e^{5x} + \frac{6}{125}x e^{5x} - \frac{6}{625}e^{5x} + C.$

3.

k	$f^{(k)}(x)$		$g^{(6-k)}(x)$
0	x^5	$+$	e^{6x}
1	$5x^4$	$-$	$\frac{1}{6}e^{6x}$
2	$20x^3$	$+$	$1/36\, e^{6x}$
3	$60x^2$	$-$	$\frac{1}{216}e^{6x}$
4	$120x$	$+$	$(1/6^4)e^{6x}$
5	120	$-$	$(1/6^5)e^{6x}$
6	0		$\frac{1}{6^6}e^{6x}$

$\int x^5 e^{6x}\, dx = \frac{x^5}{6}e^{6x} - \frac{5x^4}{36}e^{6x} + \frac{20x^3}{216}e^{6x} - \frac{60x^2}{6^4}e^{6x} + \frac{120x}{6^5}e^{6x}$
$- \frac{120}{6^6}e^{6x} + C$

5.

k	$f^{(k)}(x)$		$g^{(4-k)}(x)$
0	$x^3 - 7x^2 + 3x + 3$	$+$	e^{-3x}
1	$3x^2 - 14x + 3$	$-$	$-\frac{1}{3}e^{-3x}$
2	$6x - 14$	$+$	$\frac{1}{9}e^{-3x}$
3	6	$-$	$-\frac{1}{27}e^{-3x}$
4	0		$\frac{1}{81}e^{-3x}$

$$\int (x^3 - 7x^2 + 3x + 3)e^{-3x}\,dx =$$
$$-(x^3 - 7x^2 + 3x + 3)\tfrac{1}{3}e^{-3x} - (3x^2 - 14x + 3)\tfrac{1}{9}e^{-3x} - (6x - 14)\tfrac{1}{27}e^{-3x} - \tfrac{6}{81}e^{-3x} + C$$

7.

k	$f^{(k)}(x)$		$g^{(5-k)}(x)$
0	$2x^4 + 5x^3 - 7x^2 + 6x - 13$		e^{2x}
1	$8x^3 + 15x^2 - 14x + 6$	$+$	$\frac{1}{2}e^{2x}$
2	$24x^2 + 30x - 14$	$-$	$\frac{1}{4}e^{2x}$
3	$48x + 30$	$+$	$\frac{1}{8}e^{2x}$
4	48	$-$	$\frac{1}{16}e^{2x}$
5	0	$+$	$\frac{1}{32}e^{2x}$

$$\int (2x^4 + 5x^3 - 7x^2 + 6x - 13)e^{2x}\,dx$$
$$= (2x^4 + 5x^3 - 7x^2 + 6x - 13)\tfrac{1}{2}e^{2x} - (8x^3 + 15x^2 - 14x + 6)\tfrac{1}{4}e^{2x} +$$
$$(24x^2 + 30x - 14)(\tfrac{1}{8}e^{2x}) - (48x + 30)(\tfrac{1}{16}e^{2x}) + \tfrac{48}{32}e^{2x} + C$$

9.

k	$f^{(k)}(x)$		$g^{(7-k)}(x)$
0	$x^6 + 2$	$+$	3^{4x}
1	$6x^5$	$-$	$(4\ln 3)^{-1}3^{4x}$
2	$30x^4$	$+$	$(4\ln 3)^{-2}3^{4x}$
3	$120x^3$	$-$	$(4\ln 3)^{-3}3^{4x}$
4	$360x^2$	$+$	$(4\ln 3)^{-4}3^{4x}$
5	$720x$	$-$	$(4\ln 3)^{-5}3^{4x}$
6	720	$+$	$(4\ln 3)^{-6}3^{4x}$
7	0		$(4\ln 3)^{-7}3^{4x}$

$$\int (x^6 + 2)3^{4x} dx = (x^6 + 2)(4\ln 3)^{-1}3^{4x} - 6x^5(4\ln 3)^{-2}3^{4x}$$

$$+ 30x^4(4\ln 3)^{-3}3^{4x} - 120x^3(4\ln 3)^{-4}3^{4x} + 360x^2(4\ln 3)^{-5}3^{4x}$$

$$- 720x(4\ln 3)^{-6}3^{4x} + 720(4\ln 3)^{-7}3^{4x} + C$$

11.

(1) $\ln^5|x|$ 1

                  $+$

(2) $5\ln^4|x|\frac{1}{x}$ x

(3) $5\ln^4|x|$ 1

                  $-$

(4) $20\ln^3|x| \cdot 1/x$ x

(5) $20\ln^3|x|$ 1

                  $+$

(6) $60\ln^2|x| \cdot \frac{1}{x}$ x

(7) $60\ln^2|x|$ 1

                  $-$

(8) $120\ln|x| \cdot \frac{1}{x}$ x

(9) $120\ln|x|$ 1

                  $+$

(10) $120/x$ x

(11) 120 1

                  $-$

(12) 0 x

$$\int \ln^5|x|\, dx = x\ln^5|x| - 5x\ln^4|x| + 20x\ln^3|x| - 60x\ln^2|x| + 120x\ln|x| - 120x + C$$

13. $\displaystyle\int x\ln^2|x+3|\, dx = \int (u-3)\ln^2|u|\, du$ $u = x+3$

                                                      $du = dx$

$\ln^2|u|$ $u - 3$

              $+$

$2\ln|u|(1/u)$ $\dfrac{u^2}{2} - 3u$

       $\downarrow$

       $\ln|u|$ $u - 6$

               $-$

       $\dfrac{1}{u}$ $\dfrac{u^2}{2} - 6u$

       $\downarrow$

       1 $\dfrac{u}{2} - 6$

              $+$

       0 $\dfrac{u^2}{4} - 6u$

$$\int x\ln^2|x+3|\,dx = \int (u-3)\ln^2|u|\,du =$$

$$\ln^2|u|(\frac{u^2}{2}-3u) - \ln|u|(\frac{u^2}{2}-6u) + \frac{u^2}{4} - 6u + C$$

$$= \ln^2|x+3|\left(\frac{(x+3)^2}{2} - 3(x+3)\right) - \ln|x+3|\left(\frac{(x+3)^2}{2} - 6(x+3)\right) + \frac{(x+3)^2}{4} - 6(x+3) + C$$

15. $\displaystyle\int x^3\ln^2|x+3|\,dx = \int (u-3)^3\ln^2|u|\,du$ $\qquad\qquad u = x+3$
$\qquad\qquad\qquad\qquad\qquad\qquad\qquad\qquad\qquad\qquad\qquad\qquad\qquad du = dx$

$\ln^2|u|$ $\qquad\qquad\qquad (u-3)^3 = u^3 - 9u^2 + 27u - 27$

$\dfrac{2\ln|u|}{u}$ $\qquad$ $+$ $\qquad$ $\dfrac{1}{4}u^4 - 3u^3 + \dfrac{27}{2}u^2 - 27u = \dfrac{1}{4}(u-3)^4 - \dfrac{81}{4}$

$\downarrow$

$\ln|u|$ $\qquad\qquad$ $-$ $\qquad\qquad$ $\dfrac{1}{2}u^3 - 6u^2 + 27u - 54$

$\dfrac{1}{u}$ $\qquad\qquad\qquad\qquad\qquad$ $\dfrac{1}{8}u^4 - 2u^3 + \dfrac{27}{2}u^2 - 54u$

$\downarrow$

1 $\qquad\qquad\qquad\qquad\qquad$ $\dfrac{1}{8}u^3 - 2u^2 + \dfrac{27}{2}u - 54$

$\downarrow$ $\qquad$ $+$

0 $\longrightarrow$ $\qquad\qquad$ $\dfrac{1}{32}u^4 - \dfrac{2}{3}u^3 + \dfrac{27}{4}u^2 - 54u$

$$\int x^3\ln^2|x+3|\,dx = \ln^2|x+3|(\frac{1}{4})(x^4-81) - \ln|x+3|(\frac{1}{8}(x+3)^4 - 2(x+3)^2 + \frac{27}{2}(x+3)^2$$

$$- 54(x+3)) + \frac{1}{32}(x+3)^4 - \frac{2}{3}(x+3)^3 + \frac{27}{4}(x+3)^2 - 54(x+3) + C$$

17. $\qquad\qquad$ $+$
$\displaystyle\int (x^2 - 3x + 1)\ln^2|x+1|\,dx =$ $\qquad\qquad\qquad\qquad u = x+1$
$\qquad\qquad\qquad\qquad\qquad\qquad\qquad\qquad\qquad\qquad\qquad\qquad du = dx$
$\displaystyle\int \left((u-1)^2 - 3(u-1) + 1\right)\ln^2|u|\,du$

$\ln^2|u|$ $\qquad\qquad\qquad (u-1)^2 - 3(u-1) + 1 = u^2 - 5u + 5$

$\qquad\qquad\qquad +$

$2\ln|u|/u$ $\qquad\qquad\qquad$ $\dfrac{u^3}{3} - \dfrac{5}{2}u^2 + 5u$

$\downarrow$

$\ln|u|$ $\qquad\qquad\qquad\qquad$ $\dfrac{2u^2}{3} - 5u + 10$

$\downarrow$

$\dfrac{1}{u}$ $\qquad\qquad\qquad\qquad$ $\dfrac{2u^3}{9} - \dfrac{5}{2}u^2 + 10u$

$\downarrow$ $\qquad\qquad$ $-$

1 $\qquad\qquad$ $+$ $\qquad\qquad$ $2u^2/9 - \dfrac{5}{2}u + 10$

$\downarrow$

0 $\longrightarrow$ $\qquad\qquad$ $\dfrac{2u^3}{27} - \dfrac{5}{4}u^2 + 10u$

$$\int (x^2 - 3x + 1)\ln^2|x+1|\,dx =$$

$$\ln^2|x+1|\left(\frac{(x+1)^3}{3} - \frac{5}{2}(x+1)^2 + 5(x+1)\right) - \ln|x+1|\left(\frac{2(x+1)^3}{9} - \frac{5}{2}(x+1)^2 + 10(x+1)\right)$$

$$+ \frac{2(x+1)^3}{27} - \frac{5}{4}(x+1)^2 + 10(x+1) + C$$

19. Find $\int x^3(25-x^2)^{3/2}\,dx$.

$$
\begin{array}{ll}
x^2 \quad + & x(25-x^2)^{3/2} \\
2x \quad \downarrow & \frac{-1}{5}(25-x^2)^{5/2} = \frac{-1}{5}\cdot\frac{x}{x}(25-x^2)^{5/2} \\
-\frac{2}{5} & x(25-x^2)^{5/2} \\
0 & \quad\;\; - \\
& -\frac{1}{7}(25-x^2)^{7/2}
\end{array}
$$

$$\int x^3(25-x^2)^{3/2}\,dx = -\frac{x^2}{5}(25-x^2)^{5/2} - \frac{2}{35}(25-x^2)^{7/2} + C$$

which equals the solution obtained in sections 9.4 and 9.5.

21.

$$
\begin{array}{ll}
x^3 + 2x^2 - 5x + 2 & e^{3x} \\
\quad\quad + & \\
3x^2 + 4x - 5 & \frac{1}{3}e^{3x} + 6 \\
\quad\quad - & \\
6x + 4 & \frac{1}{9}e^{3x} + 6x + 2 \\
\quad\quad + & \\
6 & \frac{1}{27}e^{3x} + 3x^2 + 2x + 1 \\
\quad\quad - & \\
0 & \frac{1}{81}e^{3x} + x^3 + x^2 + x
\end{array}
$$

$$\int (x^3 + 2x^2 - 5x + 2)e^{3x}\,dx = (x^3 + 2x^2 - 5x + 2)(\tfrac{1}{3}e^{3x} + 6)$$

$$- (3x^2 + 4x - 5)(\tfrac{1}{9}e^{3x} + 6x + 2) + (6x+4)(\tfrac{1}{27}e^{3x} + 3x^2 + 2x + 1)$$

$$- 6(\tfrac{1}{81}e^{3x} + x^3 + x^2 + x) + C$$

$$= \frac{e^{3x}}{3}(x^3 + 2x^2 - 5x + 2) - \tfrac{1}{9}e^{3x}(3x^2 + 4x - 5) + (6x+4)\frac{e^{3x}}{27}$$

$$- \frac{6}{81}e^{3x} + 6(x^3 + 2x^2 - 5x + 2) - (6x+2)(3x^2 + 4x - 5)$$

$$+ (6x+4)(3x^2 + 2x + 1) - 6(x^3 + x^2 + x) + C$$

$$= \frac{e^{3x}}{27}(9x^3 + 9x^2 - 51x + 35) + 6x^3 - 18x^3 + 18x^3 - 6x^3$$

$$+ 12x^2 - 6x^2 - 24x^2 + 12x^2 + 12x^2 - 6x^2$$

$$- 30x - 8x + 30x + 8x + 6x - 6x + 12 + 10 + 4 + C$$

$$= \frac{e^{3x}}{27}(9x^3 + 9x^2 - 51x + 35) + K$$

1. $\lim\limits_{n \to \infty} \frac{2}{n} = 0$

3. $\lim\limits_{n \to \infty} \frac{2n+1}{n+3} = \lim\limits_{n \to \infty} \frac{2n+1}{n+3} \cdot \frac{1/n}{1/n} = \lim\limits_{n \to \infty} \frac{2+1/n}{1+3/n} = \frac{2+0}{1+0} = 2$

5. $\frac{-1}{n+1} \le \frac{(-1)^n}{n+1} \le \frac{1}{n+1}$.   Since $\lim\limits_{n \to \infty} \frac{1}{n+1} = \lim\limits_{n \to \infty} \frac{-1}{n+1} = 0$, $\lim\limits_{n \to \infty} \frac{(-1)^n}{n+1} = 0$
    by the squeezing theorem.

7. $\lim\limits_{n \to \infty} (4 - (\frac{1}{2})^n) = \lim\limits_{n \to \infty} 4 - \lim\limits_{n \to \infty} (\frac{1}{2})^n = 4 - 0 = 4$

9. $\lim\limits_{n \to \infty} \frac{3}{2 - (\frac{4}{5})^n} = \frac{3}{2 - 0} = 1.5$

11. $\lim\limits_{n \to \infty} \frac{1}{n+1} = 0 = \lim\limits_{n \to \infty} \frac{1}{n}$.   By the squeezing theorem, $\lim\limits_{n \to \infty} s_n = 0$.

13. Let $n = 5k$. $\lim\limits_{n \to \infty} (1 + \frac{5}{n})^n = \lim\limits_{k \to \infty} (1 + \frac{5}{5k})^{5k} = \lim\limits_{k \to \infty} ((1 + \frac{1}{k})^k)^5 = e^5$.

15. Let $n = k/4$. $\lim\limits_{n \to \infty} (1 + \frac{1}{4n})^n = \lim\limits_{k \to \infty} (1 + \frac{1}{k})^{k/4} = \lim\limits_{k \to \infty} ((1 + \frac{1}{k})^k)^{1/4} = e^{1/4} = \sqrt[4]{e}$

17. (a) $(1.005)^{1000} = 146.5756256$

     (b) $(1.0005)^{10,000} = 148.2278203$

     (c) $(1.00005)^{100,000} = 148.3946092$ $\qquad\qquad e^5 \sim 148.4131591$

19. (a) $(1 + \frac{1}{4000})^{1000} = 1.283985298$

     (b) $(1 + \frac{1}{40000})^{10,000} = 1.284021404$

     (c) $(1 + \frac{1}{400,000})^{100,000} = 1.284025015$ $\qquad \sqrt[4]{e} \sim 1.284025417$

21. $a = 18$, $r = -\frac{1}{3}$, $\frac{a}{1-r} = \frac{18}{4/3} = \frac{54}{4} = 13.5$

23. $a = 6$, $r = -\frac{1}{2}$, $\frac{a}{1-r} = \frac{6}{3/2} = \frac{12}{3} = 4$

25. $a = 2$, $2r = -\frac{4}{3}$, $r = -\frac{2}{3}$, $\frac{a}{1-r} = \frac{2}{5/3} = 1.2$

27. $a = 2.197$, $2.197r = -1.69$, $r = -\frac{1.69}{2.197} = -\frac{1690}{2197} = -\frac{10}{13}$

     $\frac{a}{1-r} = \frac{2.197}{23/13} = \frac{(2.197)13}{23} = \frac{28,561}{23,000} \sim 1.242$

29. $a = 7$, $7r = -\sqrt{7}$, $r = -\sqrt{7}/7 = -1/\sqrt{7}$. $\frac{a}{1-r} = \frac{7}{1 + \frac{1}{\sqrt{7}}} = \frac{7\sqrt{7}}{\sqrt{7}+1} \sim 5.08$

31. $a = 5$, $\frac{a}{1-r} = \frac{5}{1-r} = 15$. $1 - r = \frac{1}{3}$, $r = \frac{2}{3}$. $s_3 = 5(\frac{2}{3})^2 = \frac{20}{9}$, $s_5 = 5(\frac{2}{3})^4 = \frac{80}{81}$

33. $a + ar^2 + ar^4 + ... = \frac{a}{1-r^2} = \frac{729}{4}$. $ar + ar^3 + ar^5 + ... = \frac{ar}{1-r^2} = \frac{243}{4}$.

$\frac{a}{1-r^2} = \frac{243}{4r} = \frac{729}{4}$. $\frac{243}{729} = r = \frac{3^5}{3^6} = \frac{1}{3}$ and $a = \frac{8}{9}(\frac{729}{4}) = 162$.

$s_2 = 162/3 = 54$, $s_1 = 162$, $s_3 = \frac{54}{3} = 18$.

35. $3.\overline{27} = 3 + .27 + .27(.01) + .27(.01)^2 + ... = 3 + \frac{.27}{1-.01} = 3 + \frac{.27}{.99} = 3 + \frac{27}{99} = \frac{324}{99}$

37. $6.0121212... = 6 + .012 + .012(.01) + .012(.01)^2 + ... = 6 + \frac{.012}{1-.01} = 6 + \frac{.012}{.99} = 6 + \frac{12}{990}$

$= 6 + \frac{2}{165} = \frac{992}{165}$

39. $24.625013013.... = 24.625 + .000013 + .000013(.001) + .000013(.001)^2 + ...$

$= \frac{24625}{1000} + \frac{.000013}{1-.001} = \frac{24,600,388}{999,000}$

41. $3.6\overline{784} = 3.6 + .0784 + (.0784)(.001) + (.0784)(.001)^2 + ... = \frac{36}{10} + \frac{.0784}{.999} = \frac{18}{5} + \frac{784}{9990}$

$= \frac{183,740}{49950}$

43. $13.\overline{7653} = 13 + .7653 + .7653(.0001) + .7653(.0001)^2 + ... = 13 + \frac{.7653}{.9999} = 13 + \frac{7653}{9999}$

$= 13 + \frac{2551}{3333} = \frac{45,880}{3333}$

45. The ball travels (in feet) $30 + 2(20) + 2(20)\frac{2}{3} + 2(20)(\frac{2}{3})^2 + ... = 30 + \frac{40}{1-\frac{2}{3}}$

$= 30 + 120 = 150$ feet.

47. The ball travels (in feet) $15 + 2(9) + 2(9)\frac{3}{5} + 2(9)(\frac{3}{5})^2 + ... = 15 + \frac{18}{1-\frac{3}{5}}$

$= 15 + 18(\frac{5}{2}) = 60$ feet.

49. The pendulum travels (in inches) $25 + 25(.85) + 25(.85)^2 + ... = \frac{25}{1-.85} = \frac{25}{.15} = 166\frac{2}{3}$ inches.

51. Since $\lim_{n \to \infty} \frac{9}{2}(1 - \frac{1}{n})(2 - \frac{1}{n}) = \frac{9}{2}(1)(2) = 9$ and $\lim_{n \to \infty} \frac{9}{2}(1 + \frac{1}{n})(2 + \frac{1}{n}) = \frac{9}{2}(1)(2) = 9$,

by the squeezing theorem $A = 9$ square units.

53. Let P_i be the amount which must be deposited now in order to have \$300 in i years.

$P_i = 300(1.08)^{-i}$. The total amount needed is $P_0 + P_1 + P_2 + ... = 300 + 300(1.08)^{-1}$

$+ 300(1.08)^{-2} + ... = \frac{300}{1-1.08^{-1}} = \$4050$. Note that the interest earned on $\$4050 - 300$ or

\$3750 in one year is \$300.

1. $P = (0,1,2,3,4,5)$

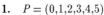

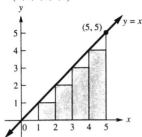

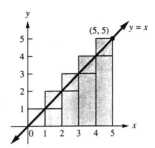

Inscribed rectangles Circumscribed rectangles

$$\underline{A} = 1\big(f(0) + f(1) + f(2) + f(3) + f(4)\big)$$
$$= (0 + 1 + 2 + 3 + 4) = 10$$

$$\overline{A} = 1\big((f(1) + f(2) + f(3) + f(4) + f(5)\big)$$
$$= 1 + 2 + 3 + 4 + 5 = 15$$

3. $P = (0,.5,1,1.5,2,2.5,3,3.5,4)$

$f(x)$ $f(x)$

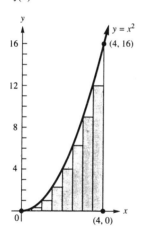

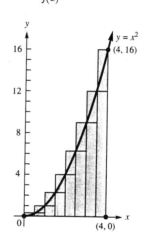

Inscribed rectangles Circumscribed rectangles

$$\underline{A} = \tfrac{1}{2}\big(f(0) + f(\tfrac{1}{2}) + f(1) + \dots + f(3.5)\big)$$
$$= \tfrac{1}{2}\big(0^2 + (\tfrac{1}{2})^2 + 1^2 + (\tfrac{3}{2})^2 + 2^2 + (\tfrac{5}{2})^2 + 3^2 + (\tfrac{7}{2})^2\big)$$
$$= \tfrac{1}{2}\big(\tfrac{1}{4} + 1 + \tfrac{9}{4} + 4 + \tfrac{25}{4} + 9 + \tfrac{49}{4}\big)$$
$$= \tfrac{1}{2}\big(\tfrac{84}{4} + 14\big) = \tfrac{1}{2}(21 + 14)$$
$$= \tfrac{35}{2}$$

$$\overline{A} = \frac{1}{2}\Big(f(.5) + f(1) + f(1.5) + f..... + f(4)\Big)$$
$$= \frac{1}{2}(\frac{1}{4} + 1 + \frac{9}{4} + 4 + \frac{25}{4} + 9 + \frac{49}{4} + 16)$$
$$= \frac{1}{2}(\frac{84}{4} + 30) = \frac{1}{2}(21 + 30)$$
$$= \frac{51}{2}$$

5.

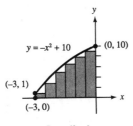

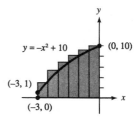

Inscribed
rectangles

Circumscribed
rectangles

$$\underline{A}_p = \frac{1}{2}\Big(f(-3) + f(-2.5) + f(-2) + f(-1.5) + f(-1) + f(-1/2)\Big)$$
$$= \frac{1}{2}(1 + 3.75 + 6 + 7.75 + 9 + 9.75) = 18.625$$
$$\overline{A}_p = \underline{A}_p - \frac{1}{2}(1) + \frac{1}{2}(10) = 18.625 + 4.5 = 23.125$$

7. $f(x)$

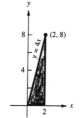

area $= \frac{1}{2}(2)(8) = 8$

(a) $\quad \underline{A}_p = \frac{2}{3}f(0) + \frac{1}{3}f(\frac{2}{3}) + \frac{1}{4}f(1) + \frac{3}{4}f(\frac{5}{4})$

$\qquad = \frac{2}{3}(0) + \frac{1}{3}(\frac{8}{3}) + \frac{1}{4}(4) + \frac{3}{4}(5)$

$\qquad = \frac{8}{9} + 1 + \frac{15}{4} = \frac{32}{36} + \frac{36}{36} + \frac{135}{36} = \frac{203}{36} = 5.63\overline{8}$

$\quad \overline{A}_p = \frac{2}{3}f(2/3) + \frac{1}{3}f(1) + \frac{1}{4}f(5/4) + \frac{3}{4}f(2)$

$\qquad = \frac{2}{3}(\frac{8}{3}) + \frac{1}{3}(4) + \frac{1}{4}(\frac{20}{4}) + \frac{3}{4}(8) = \frac{16}{9} + \frac{4}{3} + \frac{5}{4} + 6 = \frac{373}{36} = 10.36$

(b) $\quad q = (0, \frac{1}{4}, \frac{2}{4}, \frac{3}{4}, \frac{4}{4}, \frac{5}{4}, \frac{6}{4}, \frac{7}{4}, \frac{8}{4})$

$\quad \underline{A}_q = \frac{1}{4}\Big(f(0) + f(\frac{1}{4}) + f(\frac{2}{4}) + ... + f(\frac{7}{4})\Big) = \frac{1}{4}(0 + 1 + 2 + 3 + 4 + 5 + 6 + 7) = \frac{28}{4} = 7$

$\quad \overline{A}_q = \frac{1}{4}\Big(f(\frac{1}{4}) + f(\frac{2}{4}) + ... + f(\frac{8}{4})\Big) = \frac{1}{4}(1 + 2 + 3 + ... + 8) = \frac{36}{4} = 9$

(c) $r_n = (0, \frac{2}{n}, 2 \cdot \frac{2}{n}, ..., n \cdot \frac{2}{n})$

$$\underline{A}_{r_n} = \frac{2}{n} \sum_{i=1}^{n} f((i-1) \cdot \frac{2}{n}) = \frac{2}{n} \cdot \frac{8}{n} \sum_{i=1}^{n} (i-1) = \frac{16(n-1)(n)}{n^2} \cdot \frac{}{2} = \frac{8(n-1)}{n}$$

$$\overline{A}_{r_n} = \frac{2}{n} \sum_{i=1}^{n} f(\frac{2i}{n}) = \frac{2}{n} \cdot 4(\frac{2}{n}) \sum_{i=1}^{n} i = \frac{16}{n^2} \cdot \frac{n(n+1)}{2} = \frac{8(n+1)}{n}$$

(d) $\lim_{n \to \infty} \underline{A}_{r_n} = \lim_{n \to \infty} (8 - \frac{8}{n}) = 8$

$\lim_{n \to \infty} \overline{A}_{r_n} = \lim_{n \to \infty} (8 + \frac{8}{n}) = 8$

(e) $g(x) = \int 4x\,dx = 2x^2 + C$

$g(2) - g(0) = 8 + C - 0 - C = 8$

9. (a) $f'(x) = 3x^2 > 0$ for $x \epsilon (0,3)$. Thus, f is increasing on $[0,3]$.

(b) $\underline{A}_p = \frac{3}{4}f(0) + \frac{5}{4}f(\frac{3}{4}) + \frac{1}{3}f(2) + \frac{2}{3}f(\frac{7}{3}) = \frac{5}{4}(\frac{27}{64}) + \frac{8}{3} + \frac{2(343)}{3\ 27} = 11.666$

$\overline{A}_p = \frac{3}{4}f(\frac{3}{4}) + \frac{5}{4}f(2) + \frac{1}{3}f(\frac{7}{3}) + \frac{2}{3}f(3)$

$= \frac{3}{4}(\frac{27}{64}) + 10 + \frac{343}{81} + 18 = 32.551$

(c) $q = (0, \frac{1}{2}, 1, 3/2, 2, 5/2, 3)$

$\underline{A}_q = \frac{1}{2}(0 + \frac{1}{8} + 1 + \frac{27}{8} + 8 + \frac{125}{8}) = 4.5 + \frac{153}{16} = 14.0625$

$\overline{A}_q = \frac{1}{2}(1/8 + 1 + \frac{27}{8} + 8 + \frac{125}{8} + 27) = 18 + \frac{153}{16} = 27.5625$

(d) $\underline{A}_{r_n} = = \frac{3}{n} \sum_{i=1}^{n-1} (\frac{3i}{n})^3 = \frac{81}{n^4} \cdot \frac{(n-1)^2 n^2}{4} = \frac{81(n-1)^2}{4n^2}$

$\overline{A}_{r_n} = = \frac{3}{n} \sum_{i=1}^{n} (\frac{3i}{n})^3 = \frac{81}{n^4} \cdot \frac{(n+1)^2 n^2}{4} = \frac{81(n+1)^2}{4n^2}$

(e) $\lim_{n \to \infty} \frac{81(n-1)^2}{4\ n^2} = \lim_{n \to \infty} \frac{81}{4}(1 - \frac{1}{n})^2 = 81/4$

$\lim_{n \to \infty} \frac{81(n+1)^2}{4\ n^2} = \lim_{n \to \infty} \frac{81}{4}(1 + \frac{1}{n})^2 = 81/4$

(f) $81/4$

(g) $h(x) = \int x^3\,dx = \frac{x^4}{4} + C.$ $h(3) - h(0) = \frac{81}{4} - 0 = \frac{81}{4}$

11. (a) $f'(x) = -3x^2 < 0$ if $x \epsilon (-2,0)$. The function is decreasing on $[-2,0]$.

(b) $\underline{A}_p = \frac{2}{3}f(-\frac{4}{3}) + \frac{1}{3}f(-1) + \frac{2}{3}f(-\frac{1}{3}) + \frac{1}{3}f(0)$

$\qquad = \frac{2}{3}(2 + \frac{64}{27}) + \frac{1}{3}(3) + \frac{2}{3}(2 + \frac{1}{27}) + \frac{1}{3}(2)$

$\qquad = \frac{4}{3} + \frac{128}{81} + 1 + \frac{4}{3} + \frac{2}{81} + \frac{2}{3} = \frac{13}{3} + \frac{130}{81} = \frac{481}{81} = 5.938$

$\overline{A}_p = \frac{2}{3}f(-2) + \frac{1}{3}f(-\frac{4}{3}) + \frac{2}{3}f(-1) + \frac{1}{3}f(-\frac{1}{3})$

$\qquad = \frac{2}{3}(10) + \frac{1}{3}(2 + \frac{64}{27}) + \frac{2}{3}(3) + \frac{1}{3}(2 + \frac{1}{27})$

$\qquad = \frac{20}{3} + \frac{2}{3} + \frac{64}{81} + \frac{6}{3} + \frac{2}{3} + \frac{1}{81} = 10 + \frac{65}{81} = \frac{875}{81} = 10.802$

(c) $(-2, -1.75. -1.5, -1.25, -1, -.75, -.5, -.25, 0)$

$\underline{A}_q = \frac{2}{8}\Big(2(8) + (1.75^3 + 1.5^3 + 1.25^3 + 1^3 + .75^3 + .5^3 + .25^3)\Big)$

$\qquad = \frac{2}{8}(16 + 12.25) = 7.0625$

$\overline{A}_q = \frac{2}{8}\Big(2(8) + (2^3 + 1.75^3 + 1.5^3 + 1.25^3 + 1^3 + .75^3 + .5^3 + .25^3)\Big)$

$\qquad = 9.0625$

(d) $\underline{A}_{r_n} = \frac{2}{n}\sum_{i=1}^{n}\Big(2 - (-2 + \frac{2i}{n})^3\Big) = \frac{2}{n}\Big(2n + \sum_{i=1}^{n}(2 - \frac{2i}{n})^3\Big)$

$\qquad = \frac{2}{n}\Big(2n + \sum_{i=1}^{n}8(1 - \frac{i}{n})^3\Big) = 4 + \frac{16}{n}\sum_{i=1}^{n}(1 - \frac{3i}{n} + \frac{3i^2}{n^2} - \frac{i^3}{n^3})$

$\qquad = 4 + \frac{16}{n}(n - \frac{3(n)(n+1)}{2n} + \frac{3n(n+1)(2n+1)}{6n^2} - \frac{n^2(n+1)^2}{4n^3})$

$\qquad = 4 + 16(1 - \frac{3(n+1)}{2n} + \frac{(n+1)(2n+1)}{2n^2} - \frac{(n+1)^2}{4n^2})$

$\overline{A}_{r_n} = \frac{2}{n}\sum_{i=0}^{n-1}\Big(2 - (-2 + \frac{2i}{n})^3\Big) = \frac{2}{n}\Big(2n + \sum_{i=0}^{n-1}8(1 - \frac{i}{n})^3\Big)$

$\qquad = \frac{2}{n}\Big(2n + 8(\sum_{i=0}^{n-1}(1 - \frac{3i}{n} + \frac{3i^2}{n^2} - \frac{i^3}{n^3}))\Big)$

$\qquad = 4 + \frac{16}{n}(n - \frac{3(n-1)(n)}{n} + \frac{3(n-1)(n)(2n-1)}{6n^2} - \frac{(n-1)^2 n^2}{4n^3})$

$\qquad = 4 + 16\Big(1 - \frac{3}{2}(1 - \frac{1}{n}) + \frac{1}{2}(1 - \frac{1}{n})(2 - \frac{1}{n}) - \frac{(1 - \frac{1}{n})^2}{4}\Big).$

(e) $\lim_{n\to\infty} A_{r_n} = \lim_{n\to\infty}\left(4 + 16(1 - \frac{3}{2}(1+\frac{1}{n}) + \frac{1}{2}(1+\frac{1}{n})(2+\frac{1}{n}) - \frac{1}{4}(1+\frac{1}{n})^2\right)$

$= 4 + 16(1 - \frac{3}{2} + 1 - \frac{1}{4}) = 4 + 16(\frac{1}{4}) = 8$

$\lim_{n\to\infty} \overline{A}_{r_n} = \lim_{n\to\infty}\left(4 + 16(1 - \frac{3}{2} + 1 - 1/4)\right)$

$= 4 + 16(1/4) = 8$

(f) 8

(g) $h(x) = \int(2 - x^3)\,dx = 2x - \frac{x^4}{4} + C$

$h(0) - h(-2) = 0 - (-4 - 4) = 8.$

13. (a) $P_n = (0, \frac{5}{n}, 2 \cdot \frac{5}{n}, 3 \cdot \frac{5}{n}, \ldots\ldots n \cdot \frac{5}{n} = 5)$

(b) $R_{P_n} = \sum_{i=1}^{n} \frac{5}{n} f(\frac{5i}{n}) = \frac{5}{n}\sum_{i=1}^{n} 7(\frac{5i}{n}) = \frac{175}{n^2}\sum_{i=1}^{n} i$

$= \frac{175}{n^2} \cdot \frac{n(n+1)}{2} = \frac{175(n+1)}{2n}$

(c) $\lim_{n\to\infty} R_{P_n} = \lim_{n\to\infty} \frac{175}{2}(1+\frac{1}{n}) = \frac{175}{2} = 87.5$

(d) $g(x) = \frac{7}{2}x^2$

$g(5) - g(0) = \frac{7(5^2)}{2} - 0 = \frac{175}{2} = 87.5$ which is the same as the answer for part (c).

15. (a) $P = (0, \frac{3}{n}, 2(\frac{3}{n}), 3(\frac{3}{n}), \ldots n(\frac{3}{n}) = 3)$

(b) $R_{P_n} = \sum_{i=1}^{n} \frac{3}{n}(\frac{3i}{n})^2 = \frac{27}{n^3} \cdot \frac{n(n+1)(2n+1)}{6} = \frac{9(n+1)(2n+1)}{2n^2}$

(c) $\lim_{n\to\infty} \frac{9}{2}(1+\frac{1}{n})(2+\frac{1}{n}) = \frac{9}{2}(1)(2) = 9$

(d) $g(x) = \frac{x^3}{3}$, $g(3) - g(0) = \frac{27}{3} - 0 = 9$

17. (a) $P = (0, \frac{9}{n}, 2(\frac{9}{n}), 3(\frac{9}{n}), \ldots, n(\frac{9}{n}) = 9)$

(b) $R_{P_n} = \sum_{i=1}^{n} \frac{9}{n} f(\frac{9i}{n}) = \frac{9}{n}\sum_{i=1}^{n}(\frac{81i^2}{n^2} + 1) = \frac{9}{n}(\frac{81(n)(n+1)(2n+1)}{6n^2} + n)$

$= \frac{243(n+1)(2n+1)}{2n^2} + 9$

(c) $\lim_{n\to\infty} R_{P_n} = \lim_{n\to\infty}\left(9 + \frac{243}{2}(1+\frac{1}{n})(2+\frac{1}{n})\right) = 9 + 243 = 252.$

(d) $g(x) = \frac{1}{3}x^3 + x.$ $g(9) - g(0) = \frac{9^2 \cdot 9}{3} + 9 = 243 + 9 = 252.$

19. (a) $P_n = (0, \frac{6}{n}, 2(\frac{6}{n}), 3(\frac{6}{n}), \ldots n(\frac{6}{n}) = 6)$

(b) $R_{P_n} = \frac{6}{n} \sum\limits_{i=1}^{n} (\frac{6i}{n})^3 = \frac{6(216)}{n^4} \sum\limits_{i=1}^{n} i^3 = \frac{6(216)}{n^4} \cdot \frac{n^2(n+1)^2}{4}$

$= 6\frac{(54)(n+1)^2}{n^2} = \frac{324(n+1)^2}{n^2}$

(c) $\lim\limits_{n\to\infty} R_{P_n} = \lim\limits_{n\to\infty} 324(1+\frac{1}{n})^2 = 324$

(d) $g(x) = \frac{x^4}{4}.$ $\quad g(6) - g(0) = \frac{6^4}{4} = 3^2 \cdot 36 = 324$

21. (a) $P_n = (0, \frac{8}{n}, 2(\frac{8}{n}), 3(\frac{8}{n}), \ldots n(\frac{8}{n}) = 8)$

(b) $R_{P_n} = \frac{8}{n} \sum\limits_{i=1}^{n} \left[(\frac{8i}{n})^3 + 2(\frac{8i}{n})^2 - 5(\frac{8i}{n}) + 3 \right]$

$= \frac{8}{n} \sum\limits_{i=1}^{n} \left[\frac{512 i^3}{n^3} + \frac{128}{n^2} i^2 - \frac{40}{n} i + 3 \right]$

$= \frac{8}{n} \left[\frac{512 n^2(n+1)^2}{4n^3} + \frac{128 n(n+1)(2n+1)}{6n^2} - \frac{40 n(n+1)}{2n} + 3n \right]$

$= 8 \left[128(1+\frac{1}{n})^2 + \frac{64}{3}(1+\frac{1}{n})(2+\frac{1}{n}) - 20(1+\frac{1}{n}) + 3 \right]$

(c) $\lim\limits_{n\to\infty} R_{P_n} = 8\left[128 + \frac{64}{3}(2) - 20 + 3 \right] = 8\frac{[461]}{3} = \frac{3688}{3}$

(d) $g(x) = \frac{1}{4}x^4 + \frac{2}{3}x^3 - \frac{5}{2}x^2 + 3x$

$g(8) - g(0) = 8^3(2) + \frac{2}{3}(8^3) - 160 + 24$

$= 1024 + \frac{1024}{3} - 136 = \frac{3688}{3}$

23. (a) $P_n = (1, 1+\frac{3}{n}, 1+2(\frac{3}{n}), \ldots, 1 + n(\frac{3}{n}) = 4)$

(b) $R_{P_n} = \frac{3}{n} \sum\limits_{i=1}^{n} (1+\frac{3i}{n})^2 = \frac{3}{n} \sum\limits_{i=1}^{n} (1 + \frac{6i}{n} + \frac{9i^2}{n^2})$

$= \frac{3}{n}\left(n + 3(n+1) + \frac{3}{2} \frac{(n+1)(2n+1)}{n} \right)$

(c) $\lim\limits_{n\to+\infty} R_{P_n} = \lim\limits_{n\to+\infty} \left(3 + 9(1+\frac{1}{n}) + \frac{9}{2}(1+\frac{1}{n})(2+\frac{1}{n}) \right)$

$= 3 + 9 + 9 = 21$

(d) $g(x) = \frac{x^3}{3}.$ $\quad g(4) - g(1) = \frac{64}{3} - 1/3 = 21$

25. (a) $P_n = (-1, -1 + \frac{4}{n}, -1 + 2(\frac{4}{n}), ..., -1 + n(\frac{4}{n}) = 3)$

(b) $R_{P_n} = \frac{4}{n} \sum\limits_{i=1}^{n} \left((-1 + \frac{4i}{n})^2 + 3(-1 + \frac{4i}{n}) - 1 \right)$

$$= \frac{4}{n} \sum\limits_{i=1}^{n} (1 - \frac{8i}{n} + \frac{16i^2}{n^2} - 3 + \frac{12i}{n} - 1) = \frac{4}{n} \sum\limits_{i=1}^{n} (\frac{4i}{n} + \frac{16i^2}{n^2} - 3)$$

$$= \frac{4}{n}(\frac{4(n+1)}{2} + \frac{16(n+1)(2n+1)}{6} \cdot \frac{1}{n} - 3n) = \frac{4}{n}(-n + 2 + \frac{8(n+1)(2n+1)}{3} \cdot \frac{1}{n})$$

(c) $\lim\limits_{n \to \infty} R_{P_n} = \lim\limits_{n \to \infty} \left(\frac{8}{n} - 4 + \frac{32}{3}(1 + \frac{1}{n})(2 + \frac{1}{n}) \right) = -4 + \frac{64}{3} = \frac{52}{3}$

(d) $g(x) = \frac{1}{3}x^3 + \frac{3x^2}{2} - x$

$g(3) - g(-1) = (9 + \frac{27}{2} - 3) - (-\frac{1}{3} + \frac{3}{2} + 1) = 52/3$

EXERCISE SET 4.9 THE DEFINITE INTEGRAL—ANOTHER APPROACH

1. $\sum\limits_{i=1}^{n} (i - 1) = 0 + 1 + 2 + 3 + ... + (n - 1) = (1 + 2 + 3 + ... + n) - n$

$$= \frac{n(n+1)}{2} - n = \frac{n^2 + n - 2n}{n} = \frac{(n-1)n}{2}$$

3. $\sum\limits_{i=1}^{n} (i - 1)^3 = 0^3 + 1^3 + ... + (n - 1)^3 = (1^3 + 2^3 + ... + n^3) - n^3$

$$= \frac{n^2(n+1)^2}{4} - n^3 = \frac{n^4 + 2n^3 + n^2 - 4n^3}{4}$$

$$= \frac{n^4 - 2n^3 + n^2}{4} = \frac{n^2(n^2 - 2n + 1)}{4} = \frac{(n-1)^2 n^2}{4}$$

5. $P = (0, 1/2, 1, 3/2, 2, 5/2, 3)$

(a) $0 = \frac{1}{2}v(0) < d_1 < \frac{1}{2}v(1/2) = 1/8$

$\frac{1}{8} = \frac{1}{2}v(\frac{1}{2}) < d_2 < \frac{1}{2}v(1) = 1/2$

$\frac{1}{2} = \frac{1}{2}v(1) < d_3 < \frac{1}{2}v(3/2) = 9/8$

$\frac{9}{8} = \frac{1}{2}v(3/2) < d_4 < \frac{1}{2}v(2) = 2$

$2 = \frac{1}{2}v(2) < d_5 < \frac{1}{2}v(5/2) = 25/8$

$\frac{25}{8} = \frac{1}{2}v(\frac{5}{2}) < d_6 < \frac{1}{2}v(3) = 9/2$

$0 + \frac{1}{8} + \frac{1}{2} + \frac{9}{8} + 2 + \frac{25}{8} <$ distance traveled $< \frac{1}{8} + \frac{1}{2} + \frac{9}{8} + 2 + \frac{25}{8} + \frac{9}{2}$

$55/8$ ft $<$ distance traveled $< 91/8$ ft.

(b) $P_n = (0, 3/n, 2 \cdot \frac{3}{n}, ..., n \cdot \frac{3}{n} = 3)$

$$\sum_{i=1}^{n} d_i \geq \ = \frac{3}{n} \sum_{i=1}^{n-1} (\frac{3i}{n})^2 = \frac{27}{n^3}\frac{(n-1)(2n-1)(n)}{6} = \frac{9(n-1)(2n-1)}{2n^2}$$

$$\sum_{i=1}^{n} d_i \leq \frac{3}{n} \sum_{i=1}^{n} (\frac{3i}{n})^2 = \frac{27}{n^3}\frac{(n+1)(2n+1)(n)}{6} = \frac{9(n+1)(2n+1)}{2n^2}$$

$$\frac{9(n-1)(2n-1)}{2n^2} \leq \text{distance traveled} \leq \frac{9(n+1)(2n+1)}{2n^2}$$

(c) $\lim_{n \to \infty} \frac{9}{2}(1-\frac{1}{n})(2-\frac{1}{n}) = \lim_{n \to \infty} \frac{9}{2}(1+\frac{1}{n})(2+\frac{1}{n}) = 9 \text{ ft}$

7. (a) $P = (1, 1.5, 2, 2.5, 3, 3.5, 4)$

$$\sum_{i=1}^{6} d_i \geq \frac{1}{2}\Big(v(1) + v(1.5) + v(2) + v(2.5) + v(3) + v(3.5)\Big)$$

$$= \frac{1}{2}(7 + 10.75 + 16 + 22.75 + 31 + 40.75) = 64.125$$

$$\sum_{i=1}^{6} d_i \leq \frac{1}{2}(10.75 + 16 + 22.75 + 31 + 40.75 + 52) = 86.625$$

$64.125 \leq \text{distance traveled} \leq 86.625$

(b) Distanced traveled $= \sum_{i=1}^{n} d_i \geq \frac{3}{n} \sum_{i=0}^{n-1} [3(1+\frac{3i}{n})^2 + 4] = \frac{3}{n}\sum_{i=0}^{n-1}[3(1 + \frac{6i}{n} + \frac{9i^2}{n^2}) + 4]$

$$= \frac{3}{n} \sum_{i=0}^{n-1} (7 + \frac{18i}{n} + \frac{27i^2}{6n^2}) = \frac{3}{n}\Big[7n + 9(n-1) + \frac{27(n-1)(2n-1)}{6n}\Big]$$

$$= 21 + 27(1 - \frac{1}{n}) + \frac{27}{2}(1 - \frac{1}{n})(2 - \frac{1}{n})$$

distanced traveled $= \sum_{i=1}^{n} d_i \leq \frac{3}{n} \sum_{i=1}^{n} [3(1+\frac{3i}{n})^2 + 4] = \frac{3}{n} \sum_{i=1}^{n} (7 + \frac{18i}{n} + \frac{27i^2}{n^2})$

$$= \frac{3}{n}\Big(7n + \frac{18(n+1)}{2} + \frac{27(n+1)(2n+1)}{6n}\Big)$$

$$= 21 + 27(1 + \frac{1}{n}) + \frac{27(1+\frac{1}{n})(2+\frac{1}{n})}{2}$$

(c) Distance traveled $= \lim_{n \to 0} 21 + 27(1 + \frac{1}{n}) + \frac{27 + (1+\frac{1}{n})(2+\frac{1}{n})}{2} = 21 + 27 + 27 = 75 \text{ feet.}$

9. (a) $P = (2, 2.5, 3, 3.5, 4, 4.5, 5)$

$$\sum_{i=1}^{6} d_i \geq \frac{1}{2}\Big(f(2) + f(2.5) + f(3) + f(3.5) + f(4) + f(4.5)\Big)$$

$$= \frac{1}{2}(11 + 14.75 + 19 + 23.75 + 29 + 34.75) = 66.125 \text{ feet}$$

$$\sum_{i=1}^{6} d_i \leq \frac{1}{2}\Big(f(2.5) + f(3) + f(3.5) + f(4) + f(4.5) + f(5)\Big)$$

$$= \frac{1}{2}(14.75 + 19 + 23.75 + 29 + 34.75 + 41) = 81.125 \text{ feet}$$

(b) $\sum_{i=1}^{n} d_i \geq \frac{3}{n} \sum_{i=0}^{n-1} \left((2 + \frac{3i}{n})^2 + 3(2 + \frac{3i}{n}) + 1 \right)$

$= \frac{3}{n} \sum_{i=0}^{n-1} (4 + \frac{12i}{n} + \frac{9i^2}{n^2} + 6 + \frac{9i}{n} + 1) = \frac{3}{n} \sum_{i=0}^{n-1} (11 + \frac{21i}{n} + \frac{9i^2}{n^2})$

$= \frac{3}{n}(11n + 21 \cdot \frac{(n-1)}{2} + \frac{9(n-1)(2n-1)}{6n}) = 33 + \frac{63}{2}(1 - \frac{1}{n}) + \frac{9}{2}(1 - \frac{1}{n})(2 - \frac{1}{n})$

$\sum_{i=1}^{n} d_i \leq \frac{3}{n}(11n + \frac{21(n+1)}{2} + \frac{9(n+1)(2n+1)}{6n}) = 33 + \frac{63}{2}(1 + \frac{1}{n}) + \frac{9}{2}(1 + \frac{1}{n})(2 + \frac{1}{n})$

(c) The distance traveled is $\lim_{n \to \infty} 33 + \frac{63}{2}(1 + \frac{1}{n}) + 9(1 + \frac{1}{n})(2 + \frac{1}{n}) = 33 + \frac{63}{2} + 9 = 73.5$

feet

11. (a) $D(q) = .01q^2 - q + 100$

$D'(q) = .02q - 1 < 0$ if $.02q < 1$ or $q < 50$.

D is decreasing on $[0,50]$

(b) The highest price is $D(0) = \$100$ per unit.

The lowest price is $D(50) = \$75$ per unit.

(c) Total spent $\doteq \frac{50}{n} \sum_{i=1}^{n} (.01(\frac{50i}{n})^2 - \frac{50i}{n} + 100)$

$= \frac{50}{n} \sum_{i=1}^{n} (\frac{25i^2}{n^2} - \frac{50i}{n} + 100) = \frac{50}{n}(\frac{25(n+1)(2n+1)}{6n} - \frac{50(n+1)}{2} + 100n)$

$= 50(\frac{25}{6}(1 + \frac{1}{n})(2 + \frac{1}{n}) - 25(1 + \frac{1}{n}) + 100)$

The exact amount spent is

$\lim_{n \to \infty} (50(\frac{25}{6})(1 + \frac{1}{n})(2 + \frac{1}{n}) - 25(1 + \frac{1}{n}) + 100)$

$= 50(\frac{50}{6} - 25 + 100) = 50(\frac{25}{3} + 75) = \$4166.67$

13. (a) $D'(q) = -.02q < 0$ if $q > 0$.   Hence, D is decreasing on $[0,150]$

(b) The highest price is $D(0) = \$400$

The lowest price is $D(150) = -.01(150)^2 + 400 = \$175$

(c) Total spent $\doteq \frac{150}{n} \sum_{i=1}^{n} (-.01(\frac{150i}{n})^2 + 400)$

$= \frac{150}{n} \sum_{i=1}^{n} (\frac{-225i^2}{n^2} + 400) = \frac{150}{n}(\frac{-225(n+1)(2n+1)}{6n} + 400n)$.

Total spent $= \lim_{n \to \infty} 150(\frac{-225}{6}(1 + \frac{1}{n})(2 + \frac{1}{n}) + 150(400)$

$= \$\frac{-150(225)(2)}{6} + 60000 = \$48,750$

15. (a) $D'(q) = .0003q^2 - .06q + 1.08 = .0003(q-180)(q-20)$

$D'(q) < 0$ if $20 < q < 180$. D is decreasing on $[20,180]$

(b) The highest price per unit is $D(20) = \$210.40$

The lowest price per unit is $D(180) = \$5.60$

(c) Total spent $\doteq \sum_{i=1}^{n} \frac{160}{n}(.0001(20 + \frac{160i}{n})^3 - .03(20 + \frac{160i}{n})^2 + 1.08(20 + \frac{160i}{n}) + 200)$

$= \frac{160}{n}\left[\sum_{i=1}^{n} .1(2 + \frac{16i}{n})^3 - 3(2 + \frac{16i}{n})^2 + 21.6 + \frac{172.8i}{n} + 200 \right]$

$= \left[\frac{128}{n} \sum_{i=1}^{n} (1 + \frac{8i}{n})^3 \right] - \left[\frac{160(12)}{n} \sum_{i=1}^{n} (1 + \frac{8i}{n})^2 \right] + \left[16(216) \right] + \left[\frac{160}{n} \sum_{i=1}^{n} \frac{172.8i}{n} \right] + 32,000$

$= \frac{128}{n} \sum_{i=1}^{n} (1 + \frac{24i}{n} + \frac{192i^2}{n^2} + \frac{512i^3}{n^3}) - \frac{160(12)}{n} \sum_{i=1}^{n} (1 + \frac{16i}{n} + \frac{64i^2}{n^2})$

$\quad + 3456 + \frac{27648}{n} \sum_{i=1}^{n} \frac{i}{n} + 32,000$

$= \frac{128}{n}\left(n + (24)\frac{n+1}{2} + \frac{192(n+1)(2n+1)}{6n} + \frac{512(n+1)^2}{4n}\right)$

$\quad - \frac{1920}{n}\left(n + \frac{16(n+1)}{2} + \frac{64(n+1)(2n+1)}{6n}\right) + 35456 + \frac{27648(n+1)}{2n}$

Total spent is the limit as $n \to \infty$ of the above expression which is

$128(1 + 12 + 64 + 128) - 1920(1 + 8 + \frac{64}{3}) + 35456 + 13824$

$= 26240 - 58240 + 49280 = \$17,280.$

EXERCISE SET 4.10 THE FUNDAMENTAL THEOREM OF CALCULUS

1. $\displaystyle\int_{1}^{2} (4x^3 + 6x^2 + 2)\,dx = x^4 + 2x^3 + 2x\Big|_{1}^{2} = (36) - (5) = 31$

3. $\displaystyle\int_{1}^{4} \sqrt{x}\,dx = \frac{2}{3}x^{3/2}\Big|_{1}^{4} = \frac{2}{3}(8) - \frac{2}{3} = 14/3$

5. $\displaystyle\int_{0}^{2} e^{3x}\,dx = \frac{1}{3}e^{3x}\Big|_{0}^{2} = (e^6 - 1)/3$

7. $\displaystyle\int_{0}^{4} \frac{6}{x+1}\,dx = 6\ln|x+1|\,\Big|_{0}^{4} = 6\ln 5$

9. $\displaystyle\int_{0}^{1} (x^2 + 1)^5 x\,dx = \frac{1}{12}(x^2 + 1)^6\Big|_{0}^{1} = \frac{63}{12} = \frac{21}{4}$

11. Let $u = x^3 + 6x + 1$, $du = (3x^2 + 6)dx = 3(x^2 + 2)dx$. When $x = 0$, $u = 1$. When $x = 2$, $u = 21$.

$$\int_0^2 (x^2 + 2)(x^3 + 6x + 1)^{1/2}\,dx = \int_1^{21} \frac{1}{3}u^{1/2}\,du = \frac{2}{9}u^{3/2}\Big|_1^{21} = \frac{2}{9}(21^{3/2} - 1)$$

13. $\displaystyle\int_{-1}^2 x^2 e^{x^3}\,dx = \frac{1}{3}e^{x^3}\Big|_{-1}^2 = \frac{1}{3}(e^8 - e^{-1})$

15. $\displaystyle\int_{-1}^2 \frac{3e^{2x}}{(e^{2x} + 1)^2}\,dx = \int_{-1}^2 3e^{2x}(e^{2x} + 1)^{-2}\,dx = \frac{-3}{2}(e^{2x} + 1)^{-1}\Big|_{-1}^2$

$$= \frac{-3}{2}(e^4 + 1)^{-1} + \frac{3}{2}(e^{-2} + 1)^{-1}$$

17. $\displaystyle\int_0^2 \frac{e^{5x}}{e^{5x} + 1}\,dx = \frac{1}{5}\ln(e^{5x} + 1)\Big|_0^2 = \frac{1}{5}\ln(e^{10} + 1) - \frac{1}{5}\ln(2)$

19. $\displaystyle\int 3xe^{4x}\,dx = \qquad\qquad\qquad\qquad u = 3x \quad v = 1/4\,e^{4x}$

$\qquad\qquad\qquad\qquad\qquad\qquad\qquad\qquad\qquad du = 3dx \quad dv = e^{4x}dx$

$$\frac{3xe^{4x}}{4} - \int \frac{3}{4}e^{4x}\,dx = \frac{3xe^{4x}}{4} - \frac{3}{16}e^{4x} + C$$

$$\int_1^3 3xe^{4x}\,dx = e^{4x}\left(\frac{3x}{4} - \frac{3}{16}\right)\Big|_1^3 = e^{12}\left(\frac{9}{4} - \frac{3}{16}\right) - e^4\left(\frac{3}{4} - \frac{3}{16}\right)$$

$$= e^{12}\left(\frac{33}{16}\right) - e^4\left(\frac{9}{16}\right) = \frac{33e^{12} - 9e^4}{16}$$

21. $\displaystyle\int (x^2 + 1)e^{3x}\,dx \qquad\qquad\qquad\qquad u = x^2 + 1 \quad v = (1/3)e^{3x}$

$\qquad\qquad\qquad\qquad\qquad\qquad\qquad\qquad\qquad du = 2x\,dx \qquad dv = e^{3x}dx$

$\displaystyle = \frac{1}{3}(x^2 + 1)e^{3x} - \int \frac{2}{3}xe^{3x}\,dx$

$\displaystyle = \frac{1}{3}(x^2 + 1)e^{3x} - \left[\frac{2}{9}xe^{3x} - \int \frac{2}{9}e^{3x}\,dx\right] \qquad u = \frac{2}{3}x \qquad v = (1/3)e^{3x}$

$\displaystyle = \frac{1}{3}(x^2 + 1)e^{3x} - \frac{2}{9}xe^{3x} + \frac{2}{27}e^{3x} + C. \qquad\qquad du = \frac{2}{3}dx \qquad dv = e^{3x}dx$

$\displaystyle \int_0^2 (x^2 + 1)e^{3x}\,dx = \left(\frac{1}{3}(x^2 + 1) - \frac{2}{9}x + \frac{2}{27}\right)e^{3x}\Big|_0^2$

$\displaystyle = e^6\left(\frac{5}{3} - \frac{4}{9} + \frac{2}{27}\right) - \left(\frac{1}{3} + \frac{2}{27}\right) = e^6\left(\frac{35}{27}\right) - \frac{11}{27}$

23. $\displaystyle\int_2^6 \ln(x + 1)\,dx = x\ln(x + 1)\Big|_2^6 - \int_2^6 \frac{x}{x + 1}\,dx \qquad u = \ln(x + 1) \quad v = x$

$\qquad\qquad\qquad\qquad\qquad\qquad\qquad\qquad\qquad\qquad\qquad du = \dfrac{dx}{x + 1} \qquad dv = dx$

$\displaystyle = x\ln(x + 1)\Big|_2^6 - \int_2^6 \left(1 - \frac{1}{x + 1}\right)dx$

$\displaystyle = x\ln(x + 1) - x + \ln(x + 1)\Big|_2^6 = 7\ln 7 - 3\ln 3 - 4$

25. $\displaystyle d = \int_3^5 (9.8t + 3)\,dt = 4.9t^2 + 3t\Big|_3^5 = 137.5 - 53.1 = 84.4$ meters

27. $d = \displaystyle\int_2^7 (3t^2 + 1)\, dt = t^3 + t \Big|_2^7 = 350 - 10 = 340$ meters

29. $d = \displaystyle\int_0^3 e^{-3t}\, dt = -\frac{1}{3}e^{-3t}\Big|_0^3 = \frac{-e^{-9}}{3} + \frac{1}{3} = \frac{1 - e^{-9}}{3}$ meters

31. $\displaystyle\int_0^{400} (.00002)q^3 - .015q^2 + 1250)\, dq$

$= .000005q^4 - .005q^3 + 1250q \Big|_0^{400}$

$= \$308{,}000$

33. $\displaystyle\int_{26}^{7999} (30 - (q+1)^{1/3})\, dq = 30q - \frac{3}{4}(q+1)^{4/3}\Big|_{26}^{7999}$

$= 30(7999) - \frac{3}{4}(160000) - 780 + 60.75 = \$119{,}250.79$

35. $\displaystyle\int_0^{1000} \frac{110}{1 + .001q}\, dq = \frac{110}{.001}\ln(1 + .001q)\Big|_0^{1000} = 110000(\ln 2) = \$76{,}246.19$

37. $\displaystyle\int_2^7 D_x\!\left(\frac{\ln(x+1)}{e^{x^3}}\right) dx = \frac{\ln(x+1)}{e^{x^3}}\Big|_2^7 = \frac{\ln 8}{e^{343}} - \frac{\ln 3}{e^8}$

39. $D_x\!\left[\displaystyle\int_3^9 (x^3 + x^2 - 1)e^{4x}\, dx\right] = D_x(\text{ constant}) = 0$

41. $\displaystyle\int_{-2}^2 (x^4 + 2x^3 + 3x + 1)\, dx = \int_{-2}^2 x^4\, dx + \int_{-2}^2 2x^3\, dx + \int_{-2}^2 3x\, dx + \int_{-2}^2 1\, dx$

$= 2\displaystyle\int_0^2 x^4\, dx + 2\int_0^2 1\, dx = \frac{2}{5}x^5 + 2x\Big|_0^2 = \frac{64}{5} + 4 = \frac{84}{5}$

43. $\displaystyle\int_{-5}^5 xe^{x^4}\, dx = 0$, since $f(x) = xe^{x^4}$ is an odd function.

45. $\displaystyle\int_{-3}^3 (4x^5 + 3x^3\sqrt{x^2 + 1} + 6x)\, dx + \int_{-3}^3 (5x^2 + 1)\, dx = 0 + \frac{5}{3}x^3 + x\Big|_{-3}^3 = 96$

47. $\displaystyle\int_0^{10} 5000 e^{-.06t}\, dt = \frac{-5000}{.06}e^{-.06t}\Big|_0^{10} = \$37{,}599$

49. $\displaystyle\int_0^8 3000 e^{-.0725t}\, dt = \frac{-3000}{.0725}e^{-.0725t}\Big|_0^8 = \$18{,}211$

EXERCISE SET 4.11 APPROXIMATE INTEGRATION

1. (a) $P = (1,1.5,2,2.5,3)$

$$\int_1^3 x^3\,dx \doteq \tfrac{2}{8}(1^3 + 2(1.5)^3 + 2(2^3) + 2(2.5)^3 + 3^3) = 20.5$$

(b) $\int_1^3 x^3\,dx = \left.\tfrac{x^4}{4}\right|_1^3 = 20$

The percentage error is $\dfrac{20.5 - 20}{20}(100) = 2.5\%$

3. (a) $(\tfrac{4}{4},\tfrac{5}{4},\tfrac{6}{4},\tfrac{7}{4},\tfrac{8}{4},\tfrac{9}{4},\tfrac{10}{4},\tfrac{11}{4},\tfrac{12}{4})$

$$\int_1^3 x^3\,dx \doteq \tfrac{1}{8}(1 + 2(1.25^2 + 1.5^3 + 1.75^3 + 2^3 + 2.25^3 + 2.5^3 + 2.75^3) + 27)$$
$$= 20.125$$

(b) $\dfrac{20.125 - 20}{20}(100) = .625\%$

5. $P = (2,2.5,3,3.5,4,4.5,5)$

(a) $\displaystyle\int_2^5 \tfrac{1}{x}\,dx \doteq \tfrac{3}{2(6)}(\tfrac{1}{2} + 2(\tfrac{1}{2.5} + \tfrac{1}{3} + \tfrac{1}{3.5} + \tfrac{1}{4} + \tfrac{1}{4.5}) + \tfrac{1}{5})$

$= .92063492$

(b) $\displaystyle\int_2^5 \tfrac{1}{x}\,dx = \ln(5) - \ln(2) = \ln(2.5) = .916290731$

The percentage error is $\dfrac{.92063492 - .916290731}{.916290731}(100) = .47\%$

7. (a) $\displaystyle\int_2^5 \tfrac{1}{x}\,dx = \tfrac{1}{12}(\tfrac{1}{2} + \tfrac{4}{2.25} + \tfrac{2}{2.5} + \tfrac{4}{2.75} + \tfrac{2}{3} + \tfrac{4}{3.25} + \tfrac{2}{3.5} + \tfrac{4}{3.75} + \tfrac{2}{4} + \tfrac{4}{4.25}$

$+ \tfrac{2}{4.5} + \tfrac{4}{4.75} + \tfrac{1}{5}) = .916298378$

(b) The percentage error is $\dfrac{.916298378 - .916290731}{.916290731}(100) \doteq .00083\%$

9. (a) $P_6 = (0,.5,1,1.5,2,2.5,3)$

$$\int_0^3 \tfrac{x}{x^2+1}\,dx \doteq \tfrac{1}{4}(0 + 2(\tfrac{.5}{1.25}) + 2(\tfrac{1}{2}) + 2(\tfrac{1.5}{3.25}) + 2(.4) + 2(\tfrac{2.5}{7.25}) + .3)$$
$$= 1.128183024$$

(b) $\displaystyle\int_0^3 \tfrac{x}{x^2+1}\,dx = \left.\tfrac{1}{2}\ln(x^2+1)\right|_0^3 = \tfrac{\ln 10}{2} = 1.151292547$

The percentage error is $\dfrac{1.128183024 - 1.15129547}{1.151292547}(100) \doteq -2.15\%$

11. $P_{12} = (0,.25,.5,.75,1,1.25,1.5,1.75,2,2.25,2.5,2.75,3)$

(a) $\displaystyle\int_0^3 \frac{x}{x^2+1}\,dx \doteq \frac{1}{8}(0 + 2(\frac{.25}{.0625}) + 2(\frac{.5}{1.25}) + 2(\frac{.75}{1.5625}) + 2(\frac{1}{2}) + 2(\frac{1.25}{2.5625}) + 2(\frac{1.5}{3.25})$

$+ 2(\frac{1.75}{4.0625}) + 2(\frac{2}{5}) + 2(\frac{2.25}{6.0625}) + 2(\frac{2.5}{7.25}) + 2(\frac{2.75}{8.5625}) + .3) = 1.145634045$

(b) The percentage error is approximately $-.63\%$

since $\dfrac{1.145634045 - \dfrac{\ln 10}{2}}{\dfrac{\ln 10}{2}} \doteq -.0063$

13. (a) $P_8 = (0,.25,.5,.75,1,1.25,1.5,1.75,2)$

$\displaystyle\int_0^2 e^{2x}\,dx \doteq \frac{1}{8}(e^0 + 2e^{.5} + 2e^1 + 2e^{1.5} + 2e^2 + 2e^{2.5} + 2e^3 + 2e^{3.5} + e^4)$

$= 27.35507653$

(b) $\dfrac{27.35507653 - 26.79907502}{26.79907502}(100) = 2.07\%$

15. $P_{12} = (0.25,.5,.75,1,1.25,1.5,1.75,2,2.25,2.5,2.75,3)$

$\displaystyle\int_0^3 \frac{1}{x^2+1}\,dx \doteq \frac{1}{8}(1 + \frac{2}{1.0625} + \frac{2}{1.25} + \frac{2}{1.5625} + \frac{2}{2} + \frac{2}{2.5625} + \frac{2}{3.25} + \frac{2}{4.0625} + \frac{2}{5}$

$+ \frac{2}{6.0625} + \frac{2}{7.25} + \frac{2}{8.5625} + .1) = 1.2487$

17. $P_6 = (0,.5,1,1.5,2,2.5,3)$

$\displaystyle\int_0^3 \sqrt{9-x^2}\,dx \doteq \frac{1}{4}(3 + 2\sqrt{8.75} + 2\sqrt{8} + 2\sqrt{6.75} + 2\sqrt{5} + 2\sqrt{2.75} + 0) = 6.8894618$

19. $P_{10}(0,.5,1,1.5,2,2.5,3,3.5,4,4.5,5)$

$\displaystyle\int_0^5 \ln(x^2+1)\,dx \doteq \frac{1}{4}(0 + 2\ln(1.25) + 2\ln 2 + 2\ln(3.25) + 2\ln(5)$

$+ 2\ln(7.25) + 2\ln(10) + 2\ln(13.25) + 2\ln(17)$

$+ 2\ln(21.25) + \ln(26)) \doteq 9.04529$

21. $\displaystyle\int_1^3 f(x)\,dx \doteq \frac{(3-1)}{3(4)}(2 + 4(3) + 2(3.1) + 4(2.9) + 2.1) = 5.65$

23. $\displaystyle\int_1^4 f(x)\,dx \doteq \frac{2}{3(6)}(3 + 4(2) + 2(2.5) + 4(3.1) + 2(4.2) + 4(5) + 4.8) = 10.2\overline{6}$

25. $\displaystyle\int_4^9 D(q)\,dq \doteq \frac{5}{2(10)}(9 + 2(8.7 + 8.5 + 8.1 + 7.9 + 7 + 6.5 + 5.9 + 5 + 3.9) + 2)$

$= 33.5$, using the trapezoidal rule. Approximately \$33,500 is spent by the consumers.

27. $\displaystyle\int_{5}^{10} D(q)\,dq \doteq \dfrac{5}{2(10)}(7 + 2(6.5 + 6.1 + 5.9 + 5.2 + 5 + 4.5 + 4.1 + 4 + 3.2) + 3)$

$= 24.75$, using the trapezoidal rule.   Approximately \$24,750 is spent by the consumers.

29. $\displaystyle\int_{3}^{8} v(t)\,dt \doteq \dfrac{5}{3(10)}(6 + 4(7.3) + 2(9.2) + 4(8.3) + 2(7.6) + 4(7) + 2(6.8)$

$+ 4(6.1) + 2(6) + 4(5.4) + 3) = 34.1$ meters

31. $\displaystyle\int_{5}^{10} v(t)\,dt \doteq \dfrac{1}{6}(4 + 4(3.4) + 2(3.2) + 4(2.8) + 2(2.3) + 4(2) + 2(1.6)$

$+ 4(1.2) + 2(1) + 4(.6) + 1) = 10.2$ meters

33. (a) $\displaystyle\int_{-h}^{h}(ax^2 + bx + c)\,dx = \dfrac{a}{3}x^3 + \dfrac{b}{2}x^2 + cx\Big|_{-h}^{h} = \dfrac{2ah^3}{3} + 2ch$

$= \dfrac{h}{3}(2ah^2 + 6c)$

(b) $y_0 = ah^2 - bh + c$
$y_1 = c$
$y_2 = ah^2 + bh + c$

$\dfrac{h}{3}(y_0 + 4y_1 + y_2) = \dfrac{h}{3}(ah^2 - bh + c + 4c + ah^2 + bh + c)$

$= \dfrac{h}{3}(2ah^2 + 6c) = \displaystyle\int_{-h}^{h}(ax^2 + bx + c)\,dx$

EXERCISE SET 4.12 CHAPTER REVIEW

1. (a) $D_x(3x^4 - 2x^3 + 7x - 9) = 12x^3 - 6x^2 + 7 = f(x)$

(b) $D_x(3e^{x^4} + \pi^2) = 3(4x^3)e^{x^4} + 0 = 12x^3 e^{x^4} = f(x)$

(c) $D_x(xe^{5x} + \ln 5) = 1(e^{5x}) + x(5e^{5x}) + 0 = (1 + 5x)e^{5x} = f(x)$

(d) $D_x\left((x^2 + 1)\ln(x^2 + 1) + \dfrac{3}{4}\right) = 2x\ln(x^2 + 1) + \dfrac{(x^2 + 1)2x}{x^2 + 1} + 0$

$= 2x\ln(x^2 + 1) + 2x = 2x(1 + \ln(x^2 + 1)) = f(x)$

(e) $D_x\left(\dfrac{e^{2x}}{x} + \ln(e + 3)\right) = \dfrac{x(2e^{2x}) - e^{2x}}{x^2} + 0 = \dfrac{(2x - 1)e^{2x}}{x^2} = f(x)$

3. (a) $s(t) = \displaystyle\int (32t + 10)\,dt = 16t^2 + 10t + c$

$s(0) = c = 15$
$s(t) = 16t^2 + 10t + 15$

(b) $\displaystyle\int (-12t + 4)\,dt = -6t^2 + 4t + c.$ $s(2) = -24 + 8 + c = 52.$

$c = 68,\quad s(t) = -6t^2 + 4t + 68$

(c) $\int (9t^2 + 6t - 4) dt = 3t^3 + 3t^2 - 4t + c$

$s(1) = 3 + 3 - 4 + c = 60. \quad c = 58$

$s(t) = 3t^3 + 3t^2 - 4t + 58$

(d) $\int (3e^{.002t} + 6t^2 + 2) dt = \frac{3}{.002} e^{.002t} + 2t^3 + 2t + c$

$s(0) = \frac{3}{.002} + c = 10. \quad c = 10 - \frac{3}{.002} = -1490$

$s(t) = 1500 e^{.002t} + 2t^3 + 2t - 1490$

5. (a) $C(q) = \int (.003q^2 - 6q + 1200) dq = .001q^3 - 3q^2 + 1200q + c$

$C(0) = c = 50,000$

$C(q) = .001q^3 - 3q^2 + 1200q + 50,000$

(b) $C(q) = \int 25q^{-1/2} dq = 50q^{1/2} + c$

$C(0) = c = 75,000. \quad C(q) = 50\sqrt{q} + 75,000$

(c) $C(q) = \int \frac{10q}{q^2 + 3} dq = 5\ln|q^2 + 3| + c$

$C(0) = 5\ln 3 + c = 62000, \quad C = 62000 - 5\ln 3$

$C(q) = 5\ln(q^2 + 3) + 62000 - 5\ln 3$

7. (a) $\int 5xe^{12x} dx$ \hspace{2cm} $u = 5x \qquad v = \frac{1}{12}e^{12x}$

$= \frac{5x}{12}e^{12x} - \int \frac{5}{12}e^{12x} dx$ \hspace{2cm} $du = 5dx \quad dv = e^{12x} dx$

$= \frac{5x}{12}e^{12x} - \frac{5}{144}e^{12x} + c$

(b) $\int 4x^2 e^{13x} dx$ \hspace{2cm} $u = 4x^2 \qquad v = \frac{1}{13}e^{13x}$

$= \frac{4x^2}{13}e^{13x} - \int \frac{8}{13}xe^{13x} dx.$ \hspace{1cm} $du = 8xdx \quad dv = e^{13x} dx$

$= \frac{4x^2}{13}e^{13x} - \left(\frac{8}{169}xe^{13x} - \int \frac{8}{169}e^{13x} dx \right)$ \hspace{0.5cm} Now let $u = \frac{8x}{13} \quad v = \frac{1}{13}e^{13x}$

$= \frac{4}{13}x^2 e^{13x} - \frac{8}{169}xe^{13x} + \frac{8}{169(13)}e^{13x} + c$ \hspace{0.5cm} $du = \frac{8}{13}dx \quad dv = e^{13x} dx$

(c) $\int \ln|x + 12| dx$ \hspace{2cm} $u = \ln|x + 12| \quad v = x$

\hspace{6cm} $du = \frac{dx}{x + 12} \qquad dv = dx$

$= x\ln|x + 12| - \int \frac{x}{x + 12} dx = x\ln|x + 12| - \int 1 - \frac{12}{x + 12} dx$

$= x\ln|x + 12| - x + 12\ln|x + 12| + C$

(d) $\displaystyle\int 5x\ln|x+6|\,dx =$ $\qquad\qquad\qquad u = \ln|x+6| \qquad v = \dfrac{5x^2}{2}$

$\dfrac{5}{2}x^2\ln|x+6| - \displaystyle\int \dfrac{5}{2}\dfrac{x^2}{x+6}\,dx \qquad\qquad du = \dfrac{1}{x+6}\,dx \qquad dv = 5x\,dx$

$= \dfrac{5}{2}x^2\ln|x+6| - \dfrac{5}{2}\displaystyle\int\left(x - 6 + \dfrac{36}{x+6}\right)dx$

$= \dfrac{5}{2}x^2\ln|x+6| - \dfrac{5}{2}\left(\dfrac{x^2}{2} - 6x + 36\ln|x+6|\right) + C$

9. (a) $\displaystyle\int \dfrac{3}{4x\sqrt{25+9x^2}}\,dx \qquad\qquad\qquad u = 3x \qquad x = u/3$

$\qquad\qquad\qquad\qquad\qquad\qquad\qquad\qquad\qquad du = 3\,dx \qquad 4x = 4u/3$

$= \displaystyle\int \dfrac{3\,du}{(3)\dfrac{4u}{3}\sqrt{25+u^2}} = \int \dfrac{3}{4}\dfrac{du}{u\sqrt{25+u^2}} \qquad \text{(let } a = 5)$

$= -\dfrac{1}{5}\dfrac{3}{4}\ln\left(\dfrac{5+\sqrt{25+u^2}}{u}\right) + C$

$= (-3/20)\ln\left(\dfrac{5+\sqrt{25+9x^2}}{3x}\right) + C$

(b) $\displaystyle\int \dfrac{2}{25-9x^2}\,dx = \int \dfrac{2}{3}\dfrac{du}{25-u^2} \qquad\qquad u = 3x \qquad du = 3\,dx$

$\qquad\qquad\qquad\qquad\qquad\qquad\qquad\qquad\qquad \text{(let } a = 5)$

$= \dfrac{2}{3}\cdot\dfrac{1}{10}\ln\left|\dfrac{5+u}{5-u}\right| + C = \dfrac{1}{15}\ln\left|\dfrac{5+3x}{5-3x}\right| + C$

(c) $\displaystyle\int x^3 e^{7x}\,dx = \dfrac{x^3 e^{7x}}{7} - \dfrac{3}{7}\int x^2 e^{7x}\,dx \qquad\qquad (n = 3,\ b = 7)$

$= \dfrac{x^3 e^{7x}}{7} - \dfrac{3}{7}\left[\dfrac{x^2 e^{7x}}{7} - \dfrac{2}{7}\int x e^{7x}\,dx\right] \qquad (n = 2,\ b = 7)$

$= \dfrac{x^3 e^{7x}}{7} - \dfrac{3x^2 e^{7x}}{49} + \dfrac{6}{49}\int x e^{7x}\,dx \qquad\qquad (n = 1,\ b = 7)$

$= \dfrac{x^3 e^{7x}}{7} - \dfrac{3x^2 e^{7x}}{49} + \dfrac{6}{49}\left(\dfrac{x e^{7x}}{7} - \dfrac{1}{7}\int e^{7x}\,dx\right)$

$= \dfrac{x^3 e^{7x}}{7} - \dfrac{3x^2 e^{7x}}{49} + \dfrac{6x e^{7x}}{343} - \dfrac{6}{343(7)}e^{7x} + C$

(d) $\displaystyle\int (x^2 + 4x + 4)\ln|x+2|\,dx$

$= \displaystyle\int (x+2)^2\ln|x+2|\,dx \qquad\qquad u = x + 2$

$\qquad\qquad\qquad\qquad\qquad\qquad\qquad du = dx$

$= \displaystyle\int u^2\ln|u|\,du = u^3\left(\dfrac{\ln|u|}{3} - \dfrac{1}{9}\right) + C$

$= (x+2)^3\left(\dfrac{\ln|x+2|}{3} - \dfrac{1}{9}\right) + C$

11. $P_6 = (-1, -1/2, 0, 1/2, 1, 3/2, 2)$

$$R = \frac{1}{2}\left[f\left(-\frac{3}{4}\right) + f\left(-\frac{1}{4}\right) + f\left(\frac{1}{4}\right) + f\left(\frac{3}{4}\right) + f\left(\frac{5}{4}\right) + f\left(\frac{7}{4}\right) \right]$$

$$= \frac{1}{2}\left[\frac{19}{16} + \frac{51}{16} + \frac{75}{16} + \frac{91}{16} + \frac{99}{64} + \frac{99}{64} \right]$$

$$= \frac{434}{32} = \frac{217}{16}$$

13. (a)

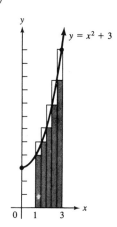

$y = x^2 + 3$

$P_5 = (1, 1.4, 1.8, 2.2, 2.6, 3)$

(b) $\underline{A}_{P_5} = .4\left(f(1) + f(1.4) + f(1.8) + f(2.2) + f(2.6) \right)$
$= .4(4 + 4.96 + 6.24 + 7.84 + 9.76) = 13.12$

(c) $\overline{A}_{P_5} = .4\left(f(1.4) + f(1.8) + f(2.2) + f(2.6) + f(3) \right)$
$= .4(4.96 + 6.24 + 7.84 + 9.76 + 12) = 16.32$

(d) $\displaystyle\int_1^3 (x^2 + 3)\,dx = \frac{x^3}{3} + 3x\Big|_1^3 = (9 + 9) - \left(\frac{1}{3} + 3\right) = 18 - \frac{8}{3} = \frac{46}{3}$

(e) $11.92 < \dfrac{46}{3} = 15.\overline{3} < 16.32$

15. $P = (0, \frac{5}{n}, 2\cdot\frac{5}{n}, \ \text{........} \ , n\cdot\frac{5}{n} = 5)$

(a) $f'(x) = 3x^2 > 0$ if $x > 0$. Thus, f is increasing on $[0, 5]$

(b) $\underline{S}_{P_n} = \frac{5}{n} \sum_{i=0}^{n-1} \left(\frac{5i}{n}\right)^3 = \frac{625}{n^4} \sum_{i=1}^{n-1} i^3 = \frac{625\,n^2(n-1)^2}{4\,n^4}$

$$= \frac{625(n-1)^2}{4\,n^2} = \frac{625}{4}\left(1 - \frac{1}{n}\right)^2$$

(c) $\overline{S}_{P_n} = \frac{5}{n}\sum_{i=1}^{n}(\frac{5i}{n})^3 = \frac{625}{n^4}\sum_{i=1}^{n}i^3 = \frac{625}{4n^4}\cdot n^2(n+1)^2$

$$= \frac{625(n+1)^2}{4n^2}$$

(d) $\int_0^5 x^3\,dx = \frac{x^4}{4}\Big|_0^5 = \frac{625}{4}$

(e) $\lim_{n\to\infty}\frac{625}{4}(1-\frac{1}{n})^2 = \frac{625}{4} = \lim_{n\to\infty}\frac{625}{4}(1+\frac{1}{n})^2$

17. (a) $\int_0^{300}(-.001q^2 - .15q + 800)\,dq = -\frac{.001}{3}q^3 - \frac{.15q^2}{2} + 800q\Big|_0^{300} = \$224{,}250$

(b) $\int_2^{618}(75 - \sqrt{q-7})\,dq = 75q - \frac{2}{3}(q+7)^{3/2}\Big|_2^{618} = \$35801.33$

19. $\int_3^{10}(6t^2 + 4t + 2)\,dt = 2t^3 + 2t^2 + 2t\Big|_3^{10} = 2220 - 78 = 2142$ feet

21. $P_5 = (1,2,3,4,5,6)$

$\int_1^6 x^2\,dx \doteq \frac{5}{10}(1^2 + 2\cdot 2^2 + 2\cdot 3^2 + 2\cdot 4^2 + 2\cdot 5^2 + 6^2) = 72.5$

$\int_1^6 x^2\,dx = \frac{x^3}{3}\Big|_1^6 = \frac{216}{3} - \frac{1}{3} = 71.\overline{6}$

The percentage error is $\dfrac{72.5 - 71.\overline{6}}{71.\overline{6}} \cdot 100 = 1.16\%$

23. $P = (1,1.25,1.5,1.75,2,2.25,2.5,2.75,3)$

$\int_1^3 \frac{1}{x^2+1}\,dx \doteq \frac{1}{8}(\frac{1}{2} + \frac{2}{2.5625} + \frac{2}{3.25} + \frac{2}{4.0625} + \frac{2}{5} + \frac{2}{6.0625} + \frac{2}{7.25} + \frac{2}{8.5625} + .1)$

$= .4659$

25. (a) $(1 + \frac{1}{20})^{10} = 1.6288946$

(b) $(1 + \frac{1}{200})^{100} = 1.6466685$

(c) $(1 + \frac{1}{2000})^{1000} = 1.6485153$

$\sqrt{e} \doteq 1.6487213.$ We conclude, $\lim_{n\to\infty}(1 + \frac{1}{2n})^n = \sqrt{e}$

27. $.\overline{23} = .23 + .23(.01) + .23(.01)^2 + \ldots\ldots = \frac{.23}{1-.01} = \frac{.23}{.99} = \frac{23}{99}$

29. $s_1 + s_3 + s_5 + \ldots\ldots = a + ar^2 + ar^4 + \ldots\ldots = \dfrac{a}{1 - r^2} = \dfrac{16}{3}$

$s_2 + s_4 + s_6 + \ldots\ldots = ar + ar^3 + ar^5 + \ldots\ldots = \dfrac{ar}{1 - r^2} = \dfrac{4}{3}$

$\dfrac{3a}{16} = 1 - r^2 = \dfrac{3ar}{4}. \quad r = \dfrac{4}{16} = \dfrac{1}{4}$ and $\dfrac{3a}{16} = 1 - r^2 = \dfrac{15}{16}. \quad a = 5.$

The first three terms are 5, 5/4, 5/16.

CHAPTER 5

APPLICATIONS OF INTEGRATION

EXERCISE SET 5.1 DIFFERENTIAL EQUATIONS

1. $y = f_1(x) = e^{2x}$, $y' = 2e^{2x}$, $y'' = 4e^{2x}$.

$4e^{2x} - 10e^{2x} + 6e^{2x} = 0$, so f_1 is a solution

$y = f_2(x) = e^{3x}$, $y' = 3e^{3x}$, $y'' = 9e^{3x}$.

$9e^{3x} - 15e^{3x} + 6e^{3x} = 0$, so f_2 is a solution.

If $y = f_3(x) = C_1 e^{2x} + C_2 e^{3x}$,

$y' = 2C_1 e^{2x} + 3C_2 e^{3x}$, and $y'' = 4C_1 e^{2x} + 9C_2 e^{3x}$

$4C_1 e^{2x} + 9C_2 e^{3x} - 5(2C_1 e^{2x} + 3C_2 e^{3x}) + 6(C_1 e^{2x} + C_2 e^{3x}) = 0$

so f_3 is a solution.

3. If $y = f_1(x) = e^{-2x}$, $y' = -2e^{-2x}$ and $y'' = 4e^{-2x}$.

$y'' - y' - 6y = 4e^{-2x} + 2e^{-2x} - 6e^{-2x} = 0$ so f_1 is a solution.

If $y = f_2(x) = e^{3x}$, $y' = 3e^{3x}$, $y'' = 9e^{3x}$,

$y'' - y' - 6y = 9e^{3x} - 3e^{3x} - 6e^{3x} = 0$ so f_2 is a solution.

If $y = f_3(x) = C_1 f_1(x) + C_2 f_2(x) = C_1 e^{-2x} + C_2 e^{3x}$

$y' = -2C_1 e^{-2x} + 3C_2 e^{3x}$, $y'' = 4C_1 e^{-2x} + 9C_2 e^{3x}$ and

$y'' - y' - 6y = 4C_1 e^{-2x} + 9C_2 e^{3x} + 2C_1 e^{-2x} - 3C_2 e^{3x} - 6C_1 e^{-2x} - 6C_2 e^{3x} = 0$

so f_3 is a solution, also.

5. $y = g(x) = xe^{2x}$, $y' = e^{2x} + 2xe^{2x} = e^{2x}(1 + 2x)$

$y'' = 2e^{2x} + 2e^{2x} + 4xe^{2x} = 4e^{2x} + 4xe^{2x} = 4e^{2x}(1 + x)$.

$y'' - 3y' + 2y = 4e^{2x}(1 + x) - 3e^{2x}(1 + 2x) + 2xe^{2x} =$

$e^{2x}(4 + 4x - 3 - 6x + 2x) = e^{2x}$ verifying that g is a solution.

7. $\int (y^2 + 3y + 5) dy = \int (\sqrt{x} + 4) dx$

$y^3/3 + \frac{3}{2}y^2 + 5y = \frac{2}{3}x^{3/2} + 4x + C$

9. $\int (2y+6)\,dy = \int e^{3x}\,dx$

$y^2 + 6y = \frac{1}{3}e^{3x} + C.$

11. $\int \frac{dy}{y} = \int 5\,dx, \quad \ln|y| = 5x + C,$

$e^{\ln|y|} = e^{5x+c}, \quad |y| = C_1 e^{5x}, \quad y = C_2 e^{5x}$

13. $\int \frac{dy}{y} = \int 4x\,dx, \quad \ln|y| = 2x^2 + C.$ If $y = 3$, $x = 0$ and $\ln 3 = C.$ $\ln|y| = 2x^2 + \ln 3,$

$e^{\ln|y|} = e^{2x^2 + \ln 3}, \quad |y| = 3e^{2x^2}, \quad y = 3e^{2x^2}$

15. $t\frac{dx}{dt} = x - xt^2 = x(1 - t^2).$

$\int \frac{dx}{x} = \int \frac{(1 - t^2)}{t}\,dt = \int (\frac{1}{t} - t)\,dt.$

$\ln|x| = \ln|t| - \frac{t^2}{2} + C.$ If $t = 1$, $x = e$, and $\ln|e| = 1 = -1/2 + C.$ $C = 3/2$ and

$\ln|x| = \ln|t| - \frac{t^2}{2} + 3/2.$ $e^{\ln|x|} = e^{\ln|t|}e^{-t^2/2 + 3/2}.$ $x = te^{(-t^2 + 3)/2}$

17. $\int e^{-y}\,dy = \int xe^{-x}\,dx$

$-e^{-y} = -xe^{-x} - e^{-x} + C$ when $x = 0$, $y = 0$ and $-1 = -1 + C.$ $C = 0$
and $e^{-y} = xe^{-x} + e^{-x} = e^{-x}(x + 1)$
$\ln e^{-y} = \ln(e^{-x})(x + 1), \quad -y = \ln(e^{-x}) + \ln(x + 1)$
$-y = -x + \ln(x + 1)$
$y = x - \ln(x + 1)$

19. $3y\sqrt{x^2 + 1}\frac{dy}{dx} = x + xy^2 = x(1 + y^2).$

$\int \frac{3y}{1 + y^2}\,dy = \int \frac{x}{\sqrt{x^2 + 1}}\,dx$

$\frac{3}{2}\ln(1 + y^2) = \sqrt{x^2 + 1} + C.$ When $x = \sqrt{3}$, $y = 0$ and $0 = 2 + C.$
$C = -2$ and $3\ln(1 + y^2) = 2\sqrt{x^2 + 1} - 4$

21. $P = P_0 e^{it}$
$P = 20{,}000\,e^{(.08)10} = 20{,}000\,e^{.8} = \$44{,}510.82$

23. $8000 = 5000\,e^{.09t}.$ $1.6 = e^{.09t}.$ $\ln 1.6 = .09t.$
$t = \frac{\ln(1.6)}{.09} = 5.22$ years.

25. Let V be the value after t years. $\frac{dV}{dt} = K\sqrt{V}.$ $\int V^{-1/2}\,dV = \int K\,dt.$ $2\sqrt{V} = Kt + C.$
When $t = 0$, $V = 90{,}000$ and $C = 600.$

$2\sqrt{V} = Kt + 600.$ When $t = 4$, $V = 90601$, and $K = .5.$
In 2 more years, t = 28 and $2\sqrt{V} = 614.$ $V = \$94{,}249$

27. $\dfrac{dx}{dt} = K(5 - \dfrac{24}{\sqrt{.03t + 36}}). \quad \displaystyle\int 1\,dx = \int K(5 - \dfrac{24}{\sqrt{.03t + 36}})\,dt$

$x = K(5t - 1600\sqrt{.03t + 36}) + C$

When t = 0, $x = 0 = K(-1600(6)) + C = -9600K + C.$

After 105 hours, $t = 6300$, $x = 342{,}000 = K(31{,}500 - 1600(15)) + 9600K$

$342{,}000 = K(17{,}100)$ and $K = 20$

After another 375 hours, $t = 480(60)$ minutes and

$x = 20(144{,}000 - 48{,}000) + 9600(20) = 2{,}112{,}000$

The number of words typed in the next 375 hours is $2{,}112{,}000 - 342{,}000 = 1{,}770{,}000.$

29. $\eta = \dfrac{p\,dq}{q\,dp} = \dfrac{12p^4 - 600p^3 + 12p^2 - 600p}{3p^4 - 200p^3 + 6p^2 - 600p + 6265000}, \quad 0 \le p \le 50$

$\displaystyle\int \dfrac{dq}{q} = \int \dfrac{12p^3 - 600p^2 + 12p - 600}{3p^4 - 200p^3 + 6p^2 - 600p + 6265000}\,dp$

$\ln q = \ln(3p^4 - 200p^3 + 6p^2 - 600p + 6265000) + \ln C$

$q = C(3p^4 - 200p^3 + 6p^2 - 600p + 6265000).$

If $p = 10$, $q = 6{,}089{,}600$ and $C = 1$

and $q = f(p) = 3p^4 - 200p^3 + 6p^2 - 600p + 6265000.$

31. $\eta = \dfrac{p\,dq}{q\,dp} = \dfrac{-p}{4(4096 - p)}, \quad 0 < p < 4000$

$\displaystyle\int \dfrac{dq}{q} = \int \dfrac{-1}{4(4096 - p)}\,dp. \quad \ln q = \tfrac{1}{4}\ln(4096 - p) + \ln C$

When $p = 3471$, $q = 250{,}000$ and $\ln(250{,}000) = \ln C(4096 - 3471)^{1/4}$

$C = 50{,}000.$

$q = 50{,}000\ \sqrt[4]{4096 - p},\ 0 < p < 4000.$

33. $\dfrac{dP}{dt} = KP(600000 - P) \quad 0 < P < 60000.$

$\displaystyle\int \dfrac{dP}{P(600000 - P)} = \int K\,dt.$

$\displaystyle\int \dfrac{dP}{600000}(\dfrac{1}{P} + \dfrac{1}{600000 - P}) = Kt + C.$

$\dfrac{1}{600000}\Big(\ln P - \ln(600000 - P)\Big) = Kt + C.$

When $t = 0$, $P = 100000$. When $t = 6$, $P = 150{,}000.$

$\dfrac{1}{600{,}000}\ln(\dfrac{100{,}000}{500{,}000}) = C$

$\ln\dfrac{P}{600{,}000 - P} = 600{,}000Kt - \ln 5.$

$\ln\dfrac{150{,}000}{450{,}000} = 3{,}600{,}000K - \ln 5$

$\ln 5 - \ln 3 = 3{,}600{,}000K$

$K = \ln(\tfrac{5}{3}) \cdot \dfrac{1}{3{,}600{,}000}$

$\ln\dfrac{P}{600{,}000 - P} = \tfrac{t}{6}\ln\tfrac{5}{3} - \ln 5$

$$\frac{P}{600,000 - P} = \left(e^{\ln\frac{5}{3}}\right)^{t/6} e^{-\ln 5} = \frac{1}{5}\left(\frac{5}{3}\right)^{t/6}$$

$$P = \frac{1}{5}\left(\frac{5}{3}\right)^{t/6}(600,000) - \frac{1}{5}\left(\frac{5}{3}\right)^{t/6}P$$

$$P = \frac{\frac{1}{5}\left(\frac{5}{3}\right)^{t/6}(600,000)}{1 + \frac{1}{5}\left(\frac{5}{3}\right)^{t/6}} = \frac{600,000}{5\left(\frac{5}{3}\right)^{-t/6} + 1}$$

35. Let x be the amount of substance present at time t and let x_0 be the original amount.

$$\frac{dx}{dt} = Kx. \quad \int \frac{dx}{x} = \int K\,dt. \quad \ln x = Kt + C.$$

$x = C_1 e^{Kt}$. When $t = 0$, $x = x_0$ and $C_1 = x_0$,

$x = x_0 e^{Kt}. \quad .75x_0 = x_0 e^{K(2324)}$

$\dfrac{\ln .75}{2324} = K$ and $x = x_0 e^{t\ln(.75)/2324}$

When $\dfrac{x_0}{2} = x_0 e^{t\ln(.75)/2324}$,

$\dfrac{(-\ln 2)(2324)}{\ln(.75)} = t \doteq 6000$ years

37. The solutions of $r^2 - 3r + 2 = (r - 2)(r - 1) = 0$ are $r = 2$ and $r = 1$. Hence, $y = e^{2x}$ and $y = e^x$ are solutions to $y'' - 3y' + 2y = 0$

39. The solutions of $r^2 + 3r - 4 = (r + 4)(r - 1) = 0$ are $r = -4$ and $r = 1$. Hence, $y = e^{-4x}$ and $y = e^x$ are solutions to $y'' + 3y' - 4 = 0$.

41. $2r^2 - 5r - 3 = (2r + 1)(r - 3) = 0$ if $r = -1/2$ or $r = 3$.
$y = e^{-x/2}$ and $y = e^{3x}$ are solutions

43. $r^3 + 2r^2 - r - 2 = r^2(r + 2) - (r + 2) = (r + 2)(r^2 - 1) = 0$
has solutions $r = -2$, $r = 1$ and $r = -1$.
Thus, $y = e^{-2x}$, $y = e^x$ and $y = e^{-x}$ are solutions.

45. $\displaystyle\int ye^{-y^2}\,dy = \int x^2 e^{x^3}\,dx$

$-\dfrac{1}{2}e^{-y^2} = \dfrac{1}{3}e^{x^3} + C$

$e^{-y^2} = -\dfrac{2}{3}e^{x^3} + C$

$-y^2 = \ln\left(C_1 - \dfrac{2}{3}e^{x^3}\right), \quad y^2 = -\ln\left(C_1 - \dfrac{2}{3}e^{x^3}\right)$

47. $(x + 1)y\ln|x + 1|\dfrac{dy}{dx} = (y^2 + 1)^2$

$\displaystyle\int \frac{y}{(y^2 + 1)^2}\,dy = \int \frac{dx}{(x + 1)\ln|x + 1|}$

$u = \ln|x + 1|$

$du = \dfrac{1}{x + 1}\,dx$

$$-\tfrac{1}{2}(y^2+1)^{-1} = \int \tfrac{1}{u}du = \ln|\ln|x+1|| + C$$

$$\frac{1}{y^2+1} = -2\ln|\ln|x+1|| + C_1$$

$$y^2+1 = \frac{1}{-2\ln|\ln|x+1|| + C_1}$$

EXERCISE SET 5.2 AREA

1.

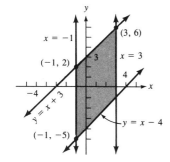

$$A = \int_{-1}^{3} (x+3) - (x-4)\,dx$$

$$= \int_{-1}^{3} 7\,dx = 7x\Big|_{-1}^{3} = 21+7 = 28$$

3.

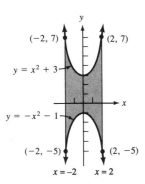

$$A = 2\int_{0}^{2} (x^2+3) - (-x^2-1)\,dx$$

$$= 2\int_{0}^{2} (2x^2+4)\,dx$$

$$= 4\int_{0}^{2} (x^2+2)\,dx$$

$$= 4(\tfrac{x^3}{3}+2x)\Big/_{0}^{2} = 4(\tfrac{8}{3}+4)$$

$$= \frac{80}{3}$$

5.

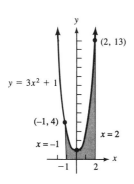

$$A = \int_{-1}^{2} ((3x^2+1) - 0)\,dx$$

$$= x^3+x\Big|_{-1}^{2}$$

$$= (10) - (-2) = 12$$

7.

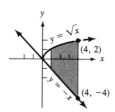

$$A = \int_1^4 (\sqrt{x} + x)\, dx$$

$$= \frac{2}{3}x^{3/2} + \frac{x^2}{2}\Big|_1^4$$

$$= (\frac{16}{3} + 8) - (\frac{2}{3} + \frac{1}{2})$$

$$= \frac{14}{3} + \frac{15}{2} = 73/6$$

9.

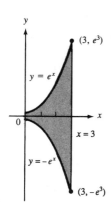

$$A = \int_0^3 (e^x - (-e^x))\, dx$$

$$= \int_0^3 2e^x dx$$

$$= 2e^x\Big|_0^3 = 2e^3 - 2 \doteq 38.17$$

11.

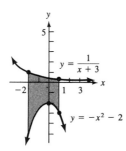

$$A = \int_{-2}^1 \Big(\frac{1}{x+3} - (-x^2 - 2)\Big)dx$$

$$= \int_{-2}^1 \Big(\frac{1}{x+3} + x^2 + 2\Big)dx$$

$$= \ln(x+3) + \frac{1}{3}x^3 + 2x\Big|_{-2}^1$$

$$= (\ln 4 + \frac{1}{3} + 2) - (\ln 1 - \frac{8}{3} - 4)$$

$$= \ln 4 + 3 + 2 + 4 = 9 + \ln 4$$

13.
$$\left.\begin{array}{c} y = x^2 + 3x - 1 \\ y = 2x + 1 \end{array}\right\} \longrightarrow x^2 + 3x - 1 = 2x + 1$$

$x^2 + x - 2 = 0$, $(x+2)(x-1) = 0$, $x = -2$ or $x = +1$ at the points of intersection of the graphs.

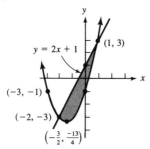

$$A = \int_{-2}^1 (2x + 1 - x^2 - 3x + 1)\, dx$$

$$= \int_{-2}^1 (2 - x - x^2)\, dx = 2x - \frac{x^2}{2} - \frac{x^3}{3}\Big|_{-2}^1$$

$$= (2 - \frac{1}{2} - \frac{1}{3}) - (-4 - 2 + \frac{8}{3}) = 4.5$$

15. $x^2 + x - 7 = -x + 1$ when $x^2 + 2x - 8 = (x+4)(x-2) = 0$
The graphs intersect at $(2, -1)$ and $(-4, 5)$

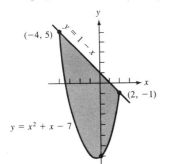

$$A = \int_{-4}^{2} 1 - x - (x^2 + x - 7)\,dx = \int_{-4}^{2} (-x^2 - 2x + 8)\,dx$$

$$= -\tfrac{1}{3}x^3 - x^2 + 8x\Big|_{-4}^{2}$$

$$= (-\tfrac{8}{3} - 4 + 16) - (\tfrac{64}{3} - 16 - 32)$$

$$= 36$$

17. $x^2 + 3x + 1 = 2x^2 + 4x - 1$ when $x^2 + x - 2 = (x+2)(x-1) = 0$
The graphs intersect at $(1, 5)$ and $(-2, -1)$

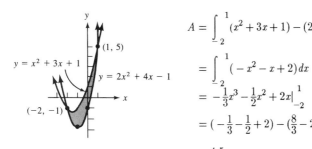

$$A = \int_{-2}^{1} (x^2 + 3x + 1) - (2x^2 + 4x - 1)\,dx$$

$$= \int_{-2}^{1} (-x^2 - x + 2)\,dx$$

$$= -\tfrac{1}{3}x^3 - \tfrac{1}{2}x^2 + 2x\Big|_{-2}^{1}$$

$$= (-\tfrac{1}{3} - \tfrac{1}{2} + 2) - (\tfrac{8}{3} - 2 - 4)$$

$$= 4.5$$

19. $-x^2 + x + 5 = x^2 - x - 7$ if $2x^2 - 2x - 12 = 2(x^2 - x - 6) = 2(x-3)(x+2) = 0$
The graphs intersect at $(3, -1)$ and $(-2, -1)$

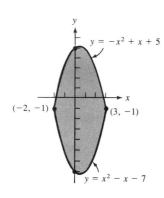

$$A = \int_{-2}^{3} (-x^2 + x + 5) - (x^2 - x - 7)\,dx$$

$$= \int_{-2}^{3} (-2x^2 + 2x + 12)\,dx$$

$$= -\tfrac{2}{3}x^3 + x^2 + 12x\Big|_{-2}^{3}$$

$$= 27 - (\tfrac{16}{3} - 20) = 125/3$$

21. $x^2 + 3x - 1 = 2x + 1$
$x^2 + x - 2 = 0$
$(x+2)(x-1) = 0$
$x = -2$ and $x = 1$.   The parabolas intersect at $(-2, -3)$ and $(1, 3)$

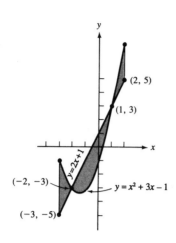

$$\text{Area} = \int_{-3}^{-2} (x^2 + 3x - 1 - 2x - 1)\,dx$$

$$+ \int_{-2}^{1} (2x + 1 - x^2 - 3x + 1)\,dx$$

$$+ \int_{1}^{3} (x^2 + 3x - 1 - 2x - 1)\,dx$$

$$= \int_{-3}^{-2} (x^2 + x - 2)\,dx + \int_{-2}^{1} (2 - x - x^2)\,dx + \int_{1}^{3} (x^2 + x - 2)\,dx$$

$$= \left(-2x + \frac{x^3}{3} + \frac{x^2}{2}\Big|_{-3}^{-2}\right) + \left(-\frac{x^3}{3} - \frac{x^2}{2} + 2x\Big|_{-2}^{1}\right) + \left(-2x + \frac{x^3}{3} + \frac{x^2}{2}\Big|_{1}^{3}\right)$$

$$= \frac{11}{6} + \frac{9}{2} + \frac{26}{3} = 15$$

23. $x^2 + x - 7 = -x + 1$ when $x = -4$ and $x = 2$
Points of intersection are $(2, -1)$ and $(-4, 5)$

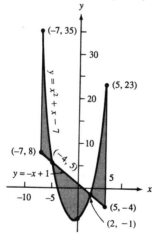

$$A_1 = \int_{2}^{-4} (x^2 + x - 7) - (-x + 1)\,dx$$

$$= \int_{-7}^{4} (x^2 + 2x - 8)\,dx = 36$$

$$A_2 = \int_{-4}^{2} (8 - 2x - x^2)\,dx = 36$$

$$A_3 = \int_{2}^{5} (x^2 + 2x - 8)\,dx = 36$$

The total area $= 36(3) = 108$

25. The parabolas intersect at $(1,5)$ and $(-2,-1)$.

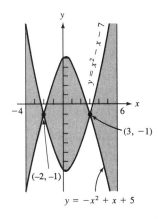

$y = 2x^2 + 4x - 1$

$(3, 29)$

$(3, 19)$

$y = x^2 + 3x + 1$

$(-3, 5)$ $(1, 5)$

$(-3, 1)$

$(-2, -1)$

$$A_1 = \int_{-3}^{-2} (2x^2 + 4x - 1) - (x^2 + 3x + 1)\,dx = 1.8\overline{3}$$

$$A_2 = \int_{-2}^{1} \left((x^2 + 3x + 1) - (2x^2 + 4x - 1)\right)dx = 4.5$$

$$A_3 = \int_{1}^{3} \left((2x^2 + 4x - 1) - (x^2 + 3x + 1)\right)dx = 8.\overline{6}$$

$$A = A_1 + A_2 + A_3 = 15$$

27. The parabolas intersect at $(3, -1)$ and $(-2, -1)$

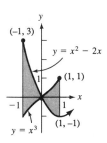

$y = x^2 - x - 7$

$(3, -1)$

$(-2, -1)$

$y = -x^2 + x + 5$

$$A_1 = \int_{-4}^{-2} (x^2 - x - 7) - (-x^2 + x + 5)\,dx = \frac{76}{3}$$

$$A_2 = \int_{-2}^{3} -x^2 + x + 5 - (x^2 - x - 7)\,dx = \frac{125}{3}$$

$$A_3 = \int_{3}^{6} x^2 - x - 7 - (-x^2 + x + 5)\,dx = 63$$

$$A = A_1 + A_2 + A_3 = 120$$

29.

$(-1, 3)$

$y = x^2 - 2x$

$(1, 1)$

$(1, -1)$

$y = x^3$

$$A = \int_{-1}^{0} (x^2 - 2x - x^3)\,dx + \int_{0}^{1} (x^3 - x^2 + 2x)\,dx$$

$$= \frac{x^3}{3} - x^2 - \frac{x^4}{4}\Big|_{-1}^{0} + \left(\frac{x^4}{4} - \frac{x^3}{3} + x^2\right)\Big|_{0}^{1}$$

$$= -\left(-\frac{1}{3} - 1 - \frac{1}{4}\right) + \left(\frac{1}{4} - \frac{1}{3} + 1\right) = 2.5$$

31. The graphs intersect at the points $(0,4)$, $(-2,1)$ and $(1,3)$
If $f(x) = (-5x^3 - 15x^2 + 8x + 48)/12$ and $g(x) = (-3x^3 - 13x^2 + 4x + 48)/12$.

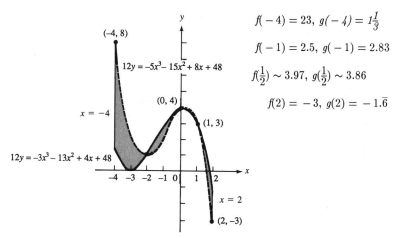

(-4, 8)

$12y = -5x^3 - 15x^2 + 8x + 48$

(0, 4)

$x = -4$

(1, 3)

$12y = -3x^3 - 13x^2 + 4x + 48$

$x = 2$

(2, -3)

$f(-4) = 23$, $g(-4) = 1\frac{1}{3}$

$f(-1) = 2.5$, $g(-1) = 2.83$

$f(\frac{1}{2}) \sim 3.97$, $g(\frac{1}{2}) \sim 3.86$

$f(2) = -3$, $g(2) = -1.\overline{6}$

The relative positions of the graphs are shown,

------------------------   for the graph of $f(x)$
———————————   for the graph of $g(x)$

The area is $\displaystyle\int_{-4}^{-2}\Big(f(x) - g(x)\Big)\,dx + \int_{-2}^{0}\Big(g(x) - f(x)\Big)dx + \int_{0}^{1}\Big(f(x) - g(x)\Big)dx$

$+ \displaystyle\int_{1}^{2}\Big(g(x) - f(x)\Big)dx = \frac{44}{9} + \frac{4}{9} + \frac{5}{72} + \frac{37}{72} = \frac{48}{9} + \frac{42}{72} = \frac{16}{3} + \frac{7}{12} = \frac{55}{12}$

33.

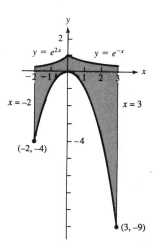

$y = e^{2x}$

$y = e^{-x}$

$x = -2$

$x = 3$

(-2, -4)

(3, -9)

$\text{Area} = \displaystyle\int_{-2}^{0} e^{2x}\,dx + \int_{0}^{3} e^{-x}\,dx - \int_{-2}^{3} -x^2\,dx$

$= \frac{1}{2}e^{2x}\Big|_{-2}^{0} + \Big(-e^{-x}\Big|_{0}^{3}\Big) + \Big(\frac{x^3}{3}\Big|_{-2}^{3}\Big)$

$= \frac{1}{2} - \frac{1}{2}e^{-4} - e^{-3} + 1 + 9 + \frac{8}{3}$

$= \frac{3}{2} - \frac{1}{2}e^{-4} - e^{-3} + \frac{35}{3} = \frac{79}{6} - \frac{1}{2}e^{-4} - e^{-3}$

35.

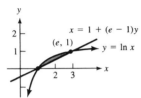

(−17, 2)   (9, 2)

$y = \sqrt[4]{-x-1}$ $y = \sqrt[3]{x-1}$

$$\text{Area} = \int_{y=0}^{2}\Big((y^3+1)-(-y^4-1)\Big)dy$$

$$= \int_{y=0}^{2}(y^4+y^3+2)\,dy$$

$$= \frac{y^5}{5}+\frac{y^4}{4}+2y\Big|_{y=0}^{2} = \frac{32}{5}+4+4 = \frac{72}{5}$$

$y = \sqrt[3]{x-1} \longleftrightarrow y^3 = x-1 \longleftrightarrow x = y^3+1$

If $x < -1$, $y = \sqrt[4]{-x-1} \longleftrightarrow y^4 = -x-1 \longleftrightarrow x = -y^4-1$

37.

$y = \ln x \longleftrightarrow x = e^y$

The graphs intersect at $(1,0)$ and $(e,1)$

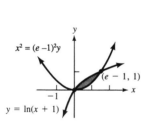

$x = 1 + (e-1)y$

$(e, 1)$

$y = \ln x$

$$\text{Area} = \int_{0}^{1}\Big((1+(e-1)y)-e^y\Big)dy$$

$$= y+\frac{(e-1)}{2}y^2-e^y\Big|_{0}^{1} = 1+\frac{e-1}{2}-e+1 = 1.5-\frac{e}{2}$$

39. The graphs intersect at $(0,0)$ and $(e-1,1)$

$x^2 = (e-1)^2 y$

$(e-1, 1)$

$y = \ln(x+1)$

$$\text{Area} = \int_{0}^{1}\Big((e-1)\sqrt{y}-e^y+1\Big)dy$$

$$= \frac{2}{3}(e-1)y^{3/2}-e^y+y\Big|_{0}^{1} = \frac{2}{3}(e-1)-e+1+1$$

$$= \frac{4}{3}-\frac{e}{3}$$

EXERCISE SET 5.3 CONSUMERS' AND PRODUCERS' SURPLUS

1. (a) $S(q) = D(q)$ if $2+q = 14-2q$. $3q = 12$, $q = 4$.
When $q = 4$, $p = 6$.
The equilibrium quantity is 4,000 units.
The equilibrium price is \$6/unit.

(b) $C.S. = \displaystyle\int_{0}^{4}(14-2q-6)\,dq = 8q-q^2\Big|_{0}^{4} = \$16,000$

(c) $P.S. = \displaystyle\int_{0}^{4}\Big(6-(2+q)\Big)dq = 4q-\frac{1}{2}q^2\Big|_{0}^{4} = \$8,000$

3. (a) $S(q) = D(q)$ if $q^2 + 6q = -q^2 - 4q + 132$, $0 \leq q \leq 9$.

$2q^2 + 10q - 132 = 0$

$q^2 + 5q - 66 = 0$

$(q - 6)(q + 11) = 0$.

$q = 6$. When $q = 6$, $p = 72$.

The equilibrium quantity is 6,000 units
The equilibrium price is $72/unit.

(b) $C.S. = \int_0^6 (132 - q^2 - 4q - 72) dq = 60q - \frac{1}{3}q^3 - 2q^2 \Big|_0^6 = \$216{,}000$

(c) $P.S. = \int_0^6 (72 - q^2 - 6q) dq = 72q - \frac{1}{3}q^3 - 3q^2 \Big|_0^6 = \$252{,}000$

5. (a) $S(q) = D(q)$ if $q^2 + 8q + 5 = -q^2 - 4q + 635$, $0 \leq q \leq 23$.

$2q^2 + 12q - 630 = 0$

$q^2 + 6q - 315 = (q + 21)(q - 15) = 0$

$q = 15$ and $p = 225 + 120 + 5 = 350$.

The equilibrium quantity is 15,000 units.
The equilibrium price is $350/unit.

(b) $C.S. = \int_0^{15} (635 - q^2 - 4q - 350) dq = 285q - \frac{1}{3}q^3 - 2q^2 \Big|_0^{15}$

$= \$2{,}700{,}000$

(c) $P.S. = \int_0^{15} (350 - q^2 - 8q - 5) dq = 345q - \frac{1}{3}q^3 - 4q^2 \Big|_0^{15}$

$= \$3{,}150{,}000$

7. (a) $D(q) = S(q)$ if $q^2 + 8q + 5 = -q^2 - 7q + 1408$, $0 \leq q \leq 32$,

$2q^2 + 15q - 1403 = 0$. $(2q + 61)(q - 23) = 0$. $q = 23$ and $p = 718$.

The equilibrium quantity is 23,000 units.
The equilibrium price is $718/unit.

(b) $C.S. = \int_0^{23} (-q^2 - 7q + 1408 - 718) dq = 690q - \frac{1}{3}q^3 - \frac{7}{2}q^2 \Big|_0^{23}$

$= \$9{,}962{,}833.$

(c) $P.S. = \int_0^{23} (718 - q^2 - 8q - 5) dq = 713q - \frac{1}{3}q^3 - 4q^2 \Big|_0^{23}$

$= \$10{,}227{,}333$

9. (a) $S(q) = D(q)$ if $q^3 + q^2 + 1775 = q^3 - 1200q + 20000$; $0 \le q \le 20$.

$q^2 + 1200q - 18225 = (q + 1215)(q - 15)$.

$q = 15$ and $p = 5375$.

The equilibrium quantity is 15,000 units.
The equilibrium price is \$5,375/unit.

(b) C.S. $= \displaystyle\int_0^{15} (q^3 - 1200q + 20,000 - 5375)\,dq$

$= \frac{1}{4}q^4 - 600q^2 + 14625q \Big|_0^{15}$

$= \$97,031,250$

(c) P.S. $= \displaystyle\int_0^{15} (5375 - q^3 - q^2 - 1775)\,dq = 3600q - \frac{1}{4}q^4 - \frac{1}{3}q^3 \Big|_0^{15}$

$= \$40,218,750$

11. (a) $S(q) = D(q)$ if $\frac{q}{3} + 6 = \sqrt{144 - 7q}$; $0 \le q \le 20$

$q + 18 = 3\sqrt{144 - 7q}$; $q^2 + 36q + 324 = 1296 - 63q$,

$q^2 + 99q - 972 = (q + 108)(q - 9)$. $q = 9$, $p = 9$.

The equilibrium quantity is 9,000 units.
The equilibrium price is \$9/unit.

(b) C.S. $= \displaystyle\int_0^9 (\sqrt{144 - 7q} - 9)\,dq = \frac{-2}{21}(144 - 7q)^{3/2} - 9q \Big|_0^9$

$= \$14,143$

(c) P.S. $= \displaystyle\int_0^9 (9 - \frac{q}{3} - 6)\,dq = 3q - \frac{q^2}{6}\Big|_0^9 = \$13,500$

13. (a) $S(q) = D(q)$ if $\frac{q}{3} + 3 = \sqrt{225 - 3q}$; $0 \le q \le 50$.

$q + 9 = 3\sqrt{225 - 3q}$. $q^2 + 18q + 81 = 2025 - 27q$.

$q^2 + 45q - 1944 = 0 = (q + 72)(q - 27)$

$q = 27$ and $p = 12$. The equilibrium quantity is 27,000 units.
The equilibrium price is \$12/unit.

(b) C.S. $= \displaystyle\int_0^{27} (\sqrt{225 - 3q} - 12)\,dq = -\frac{2}{9}(225 - 3q)^{3/2} - 12q \Big|_0^{27} = \$42,000$

(c) P.S. $= \displaystyle\int_0^{27} (12 - \frac{q}{3} - 3)\,dq = 9q - q^2/6\Big|_0^{27} = \$121,500$

15. (a) $S(q)=D(q)$ if $\sqrt{12q+25} = \sqrt{1225 - 12q}$; $0 \le q \le 100$.

$12q + 25 = 1225 - 12q$. $24q = 1200$. $q = 50, p = 25$.

The equilibrium quantity is 50,000 units.
The equilibrium price is \$25/unit.

(b) $C.S. = \displaystyle\int_0^{50} (\sqrt{1225 - 12q} - 25)\,dq = -\frac{1}{18}(1225 - 12q)^{3/2} - 25q\Big|_0^{50} = \$263,889$

(c) $P.S. = \displaystyle\int_0^{50} (25 - \sqrt{12q + 25}\,dq = 25q - \frac{1}{18}(12q + 25)^{3/2}\Big|_0^{50} = \$388,889$

17. (a) $S(q) = D(q)$ if $q + 20 = \dfrac{300}{.1q + 3}$; $0 \le q \le 70$. $(q + 20)(.1q + 3) = 300$.

$.1q^2 + 5q - 240 = 0 = (q - 30)(.1q + 8)$. $q = 30, \ p = 50$.

The equilibrium quantity is 30,000 units.
The equilibrium price is \$50/unit.

(b) $C.S. = \displaystyle\int_0^{30} (\frac{300}{.1q + 3} - 50)\,dq = 3000 \ln(.1q + 3) - 50q\Big|_0^{30} = \$579,442$

(c) $P.S. = \displaystyle\int_0^{30} (50 - q - 20)\,dq = 30q - q^2/2\Big|_0^{30} = \$450,000$

19. (a) $D(q) = S(q)$ if $e^{5-.1q} = e^{2+.2q}$; $0 \le q \le 30$.

$5 - .1q = 2 + .2q$, $.3q = 3$ $q = 10, p = e^4$

The equilibrium quantity is 10,000 units.
The equilibrium price is \$54,598.15/unit.

(b) $C.S. = \displaystyle\int_0^{10} (e^{5-.1q} - e^4)\,dq = -10e^5\,e^{-.1q} - e^4 q\Big|_0^{10} = -10e^4 - 10e^4 + 10e^5$

$= 10e^5 - 20e^4 = \$392,169.$

(c) $P.S. = \displaystyle\int_0^{10} (e^4 - e^2 e^{.2q})\,dq = e^4 q - \frac{e^2 e^{.2q}}{.2}\Big|_0^{10}$

$= 10e^4 - 5e^2 e^2 + 5e^2 = 5e^2 + 5e^4 \doteq \$309,936$

21. (a) $P(q) = R(q) - C(q) = qD(q) - C(q) = 150q - 2q^2 - .5q^2 - 1000 = -2.5q^2 + 150q - 1000$

$P'(q) = -5q + 150 = 0$ if $q = 30$.
$P''(q) = -5$. The profit is a maximum when the number of units is 30,000 and the price per unit is $150 - 60 = \$90$.

(b) $C.S. = \displaystyle\int_0^{30} (150 - 2q - 90)\,dq = 60q - q^2\Big|_0^{30} = \$900,000.$

23. (a) $P(q) = qD(q) - C(q) = 60q - .9q^2 - .1q^2 - 100 = -q^2 + 60q - 100.$

$P'(q) = -2q + 60 = 0$ if $q = 30$

$P''(q) = -2.$ the profit is a maximum when the number of units is 30,000 and the price per unit is \$33.

(b) $C.S. = \displaystyle\int_0^{30} (60 - .9P - 33)\,dp = 27p - .45p^2\Big|_0^{30} = \$405,000.$

25. (a) $P(q) = qD(q) - C(q) = -.2q^3 - 2q^2 + 48q - q^2 - 3q - 150; \ 0 \le q \le 10.$

$P(q) = -.2q^3 - 3q^2 + 45q - 150.$

$P'(q) = -.6q^2 - 6q + 45 = 0$ if $6q^2 + 60q - 450 = 0$

or $q^2 + 10q - 75 = (q + 15)(q - 5) = 0.$

$P''(q) = -.12q - 6. \ \ P''(5) < 0.$

The profit is a maximum for 5,000 units and a unit price of \$33.

(b) $C.S. = \displaystyle\int_0^5 (-.2q^2 - 2q + 48 - 33)\,dq$

$= \dfrac{-1q^3}{15} - q^2 + 15q\Big|_0^5 = \$41,666.67$

EXERCISE SET 5.4 MORE APPLICATIONS

1. (a)

(b) $I(.2) = (.2)^3 = .008$

(c) $2\displaystyle\int_0^1 (x - x^3)\,dx = 2(\dfrac{x^2}{2} - \dfrac{x^4}{4}\Big|_0^1) = 2(\dfrac{1}{4}) = \dfrac{1}{2}$

3. (a)

(b) $I(.15) = .133$

(c) $2\displaystyle\int_0^1 (x - \dfrac{2x^2}{15} - \dfrac{13}{15}x)\,dx = 2\displaystyle\int_0^1 (\dfrac{2}{15}x - \dfrac{2x^2}{15})\,dx = \dfrac{2}{15}(x^2 - \dfrac{2}{3}x^3)\Big|_0^1 = \dfrac{2}{45} = .0\overline{4}$

5. (a)

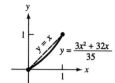

(b) $1 - I(.35) = .6695$

(c) $2 \int_0^1 (x - \frac{3}{35}x^2 - \frac{32}{35}x)\,dx = 2(\frac{3}{70}x^2 - \frac{1}{35}x^3\big|_0^1$

$= 0.02857$

7. (a)

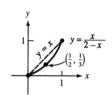

(b) $1 - I(.75) = 1 - \frac{.75}{1.25} = .4$

(c) $2 \int_0^1 (x - \frac{x}{2-x})\,dx = 2 \int_0^1 (x + 1 + \frac{2}{x-2})\,dx$

$= 2(\frac{x^2}{2} + x + 2\ln|x - 2|\big|_0^1) = 2(\frac{1}{2} + 1 - 2\ln 2)$

$= .2274$

9. (a) $C(I) = \int (.4 + .2/\sqrt{I+1})\,dI = .4I + .4(I+1)^{1/2} + K$

$C(10) = 4 + .4(11)^{1/2} + K = 8.32$

$K = 4.32 - .4(11)^{1/2}$.

$C(I) = .4I + .4(I+1)^{1/2} + 4.32 - .4(11)^{1/2}$

(b) $C(36) = .4(36) + .4(37)^{1/2} + 4.32 - .4(11)^{1/2} = \$19.82$ billion

11. (a) $C(I) = \int (.5 - .2e^{-.2I})\,dI = .5I + e^{-.2I} + K$

$C(10) = 5 + e^{-2} + K = 7.9. \quad K = 2.9 - e^{-2}$.

$C(I) = .5I - e^{-.2I} + 2.9 - e^{-2}$.

(b) $C(20) = 10 + e^{-4} + 2.9 - e^{-2} = \$12.78$ billion

13. (a) $\frac{dC}{dI} = 1 - \frac{dS}{dI} = .7 - .4e^{-.4I}$

$C(I) = \int (.7 - .4e^{-.4I})\,dI = .7I + e^{-.4I} + K$

$C(10) = 7 + e^{-4} + K = 9. \quad K = 2 - e^{-4}$

$C(I) = .7I + e^{-.4I} + 2 - e^{-4}$

(b) $C(20) = .7(20) + e^{-8} + 2 - e^{-4} = \$16.02$ billion

15. $\displaystyle\int_0^6 24000e^{-.1t}dt \;=\; -240000e^{-.1t}\Big|_0^6 \;=\; -240000(e^{-.6}-1)$

$$= \$108,285.21$$

17. $\displaystyle\int_0^5 15,600e^{-.09t}dt = \frac{-15600}{.09}(e^{-.09t})\Big|_0^5$

$= -173333.33\,e^{-.09t}\Big|_0^5 = \$62,811.12$

19. $\displaystyle\int_0^{20} 60,000e^{-.12t}dt = \frac{-60000}{.12}e^{-.12t}\Big|_0^{20} = \$454,641.02$

21. $\displaystyle\int_0^{20} 30,000e^{-.11t}dt = \frac{-30,000}{.11}e^{-.11t}\Big|_0^{20} = \$242,508.23$

23. $\displaystyle\int_{\frac{500}{20}}^{1500/20} 40x^{-.15}dx = \frac{40}{.85}x^{.85}\Big|_{25}^{75} = 1120.98 \text{ hours}$

25. $y = ax^b$, when $x = 2$, $y = 10$. So, $10 = a(2^b)$ or $a = 10(2^{-b})$.

We have a 90% learning curve so when $x = 4$, $y = .9(10) = 9$ and $9 = a(4^b)$.

Hence, $9 = 10(2^{-b})4^b = 10(2^{-b})(2^{2b})$.

$.9 = 2^b$, $\log_2(.9) = b = \dfrac{\ln.9}{\ln 2} = -.152$

$a = 10(2^{-\log_2(.9)}) = 10(\dfrac{1}{2^{\log_2.9}}) = \dfrac{10}{.9} = 11.\overline{1}$

27. $y = ax^b$. When $x = 8$, $y = 34.3$. When $x = 16$, $y = .7(34.3) = 24.01$.

$34.3 = a(8^b)$ so $a = 34.3(8^{-b}) = 34.3(2^{-3b})$.

$24.01 = a(16)^b = a(2^{4b})$.

$24.01 = 34.3(2^{-3b})(2^{4b}) = 34.3(2^b)$

$\dfrac{24.01}{34.3} = 2^b = .7$. $b = \log_2(.7) = \dfrac{\ln(.7)}{\ln 2} \doteq -.515$.

$a = \dfrac{34.3}{8^b} = 34.3(2^{-3b}) = 34.3(.7)^{-3} = 100$

29. (a) $R'(t) = C'(t)$, when $1500 - .7t^2 = 204 + .3t^2$.

$t^2 = 1296$, $t = 36$ months.

   (b) $\displaystyle -150 + \int_0^{36}\Big(R'(t)-C'(t)\Big)dt = -150 + \int_0^{36}(1500 - .7t^2 - 20t - .3t^2)\,dt$

$= -150 + \displaystyle\int_0^{36}(1296 - t^2)\,dt = 1296t - \dfrac{t^3}{3}\Big|_0^{36} - 150$

$= \$30,954,000$

31. (a) $20000 - .2t^2 = 8000 + .1t^2$, $12,000 = .3t^2$, $40000 = t^2$, $t = \sqrt{40000} = 200$ months.

(b) $-600 + \displaystyle\int_0^{200} \Big(R'(t) - C'(t)\Big)dt = -600 + \displaystyle\int_0^{200} (20000 - .2t^2 - 8000 - .1t^2)\,dt$

$= -600 + \displaystyle\int_0^{200} (12,000 - .3t^2)\,dt = 12000t - .1t^3\Big|_0^{200} - 600 = \$1,599,400,000$

33. (a) $100 - .5(\sqrt[3]{t}) = 96 + .3\sqrt[3]{t}$

$4 = .8(\sqrt[3]{t})$. $5 = \sqrt[3]{t}$, $t = 5^3 = 125$ months.

(b) $-5 + \displaystyle\int_0^{125} \Big(100 - .5\sqrt[3]{t} - 96 - .3(\sqrt[3]{t})\Big)\,dt$

$= -5 + \displaystyle\int_0^{125} (4 - .8t^{1/3})\,dt$

$= 4t - .6t^{4/3}\Big|_0^{125} - 5$

$= \$120,000,000$

35. (a) Suppose $x_1 \le x_2 \le x_3 \le ... \le x_n$ and k is an integer such that $1 \le k \le n$

$\dfrac{x_1 + x_2 + ... + x_k}{k} \le \dfrac{x_1 + x_2 + ... + x_n}{n}$ if and only if

$\dfrac{x_1 + x_2 + ... + x_k}{k} \le \dfrac{x_1 + x_2 + ... + x_k}{n} + \dfrac{x_{k+1} + ... + x_n}{n}$ if and only if

$\dfrac{(x_1 + x_2 + ... x_k)}{k} - \dfrac{(x_1 + x_2 + ... + x_k)}{n} \le \dfrac{x_{k+1} + ... + x_n}{n}$ if and only if

$\dfrac{n(x_1 + x_2 + ... + x_k) - k(x_1 + ... + x_k)}{nk} \le \dfrac{x_{k+1} + ... + x_n}{n}$ if and only if

$\dfrac{(n-k)(x_1 + ... + x_k)}{nk} \le \dfrac{x_{k+1} + ... + x_n}{n}$ if and only if

$\dfrac{x_1 + ... + x_k}{k} \le \dfrac{x_{k+1} + ... + x_n}{n-k}$.

By Exercise 34,

$\dfrac{x_1 + ... + x_k}{k} \le x_k \le x_{k+1} \le \dfrac{x_{k+1} + ... + x_n}{n-k}$

thus proving the original inequality is valid.

(b) Suppose there are n wage earners in the country with incomes of I_1, I_2, ..., I_n where $I_1 \le I_2 \le ..... \le I_n$, and k is an integer such that $1 \le k \le n$. By part(a)

$\dfrac{I_1 + ... + I_k}{k} \le \dfrac{I_1 + ... + I_n}{n}$. Hence, if we let $x = k/n$,

$I(x) = (I_1 + ... + I_k)/(I_1 + ... I_n) \le \dfrac{k}{n} = x$.

EXERCISE SET 5.5 IMPROPER INTEGRALS

1. $\displaystyle\int_1^\infty x^{-2/3}\,dx = \lim_{b\to\infty}\int_1^b x^{-2/3}\,dx = \lim_{b\to\infty}\left[3x^{1/3}\;\Big|_1^b\right]$

$\displaystyle = \lim_{b\to\infty}(3\sqrt[3]{b}-3) = \infty.$ The integral is divergent.

3. $\displaystyle\int_1^\infty 2x^{-3/4}\,dx = \lim_{b\to\infty}\int_1^b 2x^{-3/4}\,dx$

$\displaystyle = \lim_{b\to\infty}\left[24x^{1/4}\Big|_1^b\right] = \lim_{b\to\infty}(8\sqrt[4]{b}-8) = \infty$

The integral is divergent.

5. $\displaystyle\int_4^\infty \frac{2}{x}\,dx = \lim_{b\to\infty}\int_4^b \frac{2}{x}\,dx = \lim_{b\to\infty}\left[2\ln|x|\;\Big|_4^b\right]$

$\displaystyle = \lim_{b\to\infty}(2\ln b - 2\ln 4) = \infty.$ The integral diverges.

7. $\displaystyle\int_0^\infty e^{-3x}\,dx = \lim_{b\to\infty}\int_0^b e^{-3x}\,dx = \lim_{b\to\infty}\left[-\tfrac{1}{3}e^{-3x}\Big|_0^b\right]$

$\displaystyle = \lim_{b\to\infty}\left[-\tfrac{1}{3}e^{-3x}+\tfrac{1}{3}\right] = \lim_{b\to\infty}\left[-\frac{1}{3e^{3b}}+\tfrac{1}{3}\right] = \tfrac{1}{3}$

9. $\displaystyle\int_3^\infty \frac{1}{(x-2)^{3/2}}\,dx = \lim_{b\to\infty}\int_3^b (x-2)^{-3/2}\,dx$

$\displaystyle = \lim_{b\to\infty}\left[-2(x-2)^{-1/2}\;\Big|_3^b\right] = \lim_{b\to\infty}\left[-2\frac{1}{\sqrt{b-2}}+2\right] = 2$

11. $\displaystyle\int_{-\infty}^{-2}(x+1)^{-2}\,dx = \lim_{b\to-\infty}\int_b^{-2}(x+1)^{-2}\,dx = \lim_{b\to-\infty}\left[-(x+1)^{-1}\Big|_b^{-2}\right]$

$\displaystyle = \lim_{b\to-\infty}\left[1+\frac{1}{b+1}\right] = 1$

13. $\displaystyle\int_{-\infty}^{-1}3x^{-2/3}\,dx = \lim_{b\to-\infty}\int_b^{-1}3x^{-2/3}\,dx = \lim_{b\to-\infty}\left[9x^{1/3}\Big|_b^{-1}\right]$

$\displaystyle = \lim_{b\to-\infty}(-9+\sqrt[3]{b}) = -\infty.$ The integral diverges.

15. $\displaystyle\int_{-\infty}^{2} 6(3-x)^{-1/2}\,dx = \lim_{b\to-\infty}\int_{b}^{2} 6(3-x)^{-1/2}\,dx = \lim_{b\to-\infty}\left[-12(3-x)^{1/2}\Big|_{b}^{2}\right]$

$\displaystyle = \lim_{b\to-\infty}\left[-12 + 12\sqrt{3-b}\right] = \infty.$ The integral diverges.

17. $\displaystyle\int_{-\infty}^{\infty} e^{-|2x|}\,dx = \int_{-\infty}^{0} e^{-|2x|}\,dx + \int_{0}^{\infty} e^{-|2x|}\,dx$

$\displaystyle = \lim_{b\to-\infty}\int_{b}^{0} e^{2x}\,dx + \lim_{b\to\infty}\int_{0}^{b} e^{-2x}\,dx$

$\displaystyle = \lim_{b\to-\infty}\left(\tfrac{1}{2}e^{2x}\Big|_{b}^{0}\right) + \lim_{b\to\infty}\left[-\tfrac{1}{2}e^{-2x}\Big|_{0}^{b}\right]$

$\displaystyle = \lim_{b\to-\infty}\left(\tfrac{1}{2} - e^{2b}\right) + \lim_{b\to\infty}\left[-\tfrac{1}{2}e^{-2b} + \tfrac{1}{2}\right] = \tfrac{1}{2} - 0 - 0 + \tfrac{1}{2} = 1.$

19. $\displaystyle\int_{-\infty}^{\infty} e^{-|4x|}\,dx = \int_{-\infty}^{0} e^{4x}\,dx + \int_{0}^{\infty} e^{-4x}\,dx$

$\displaystyle = \lim_{b\to-\infty}\int_{b}^{0} e^{4x}\,dx + \lim_{b\to\infty}\int_{0}^{b} e^{-4x}\,dx = \lim_{b\to-\infty}\left[\tfrac{1}{4}e^{4x}\Big|_{b}^{0}\right] + \lim_{b\to\infty}\left[-\tfrac{1}{4}e^{-4x}\Big|_{0}^{b}\right]$

$\displaystyle = \lim_{b\to-\infty}\left[\tfrac{1}{4} - \tfrac{1}{4}e^{4b}\right] + \lim_{b\to\infty}\left[-\tfrac{1}{4e^{4b}} + \tfrac{1}{4}\right] = \tfrac{1}{4} + \tfrac{1}{4} = \tfrac{1}{2}$

21. $\displaystyle\int_{0}^{8} (8-x)^{-1/3}\,dx = \lim_{b\to 8^{-}}\int_{0}^{b} (8-x)^{-1/3}\,dx$

$\displaystyle = \lim_{b\to 8^{-}}\frac{-3}{2}(8-x)^{2/3}\Big|_{0}^{b} = \lim_{b\to 8}\left(\frac{-3}{2}\sqrt[3]{(8-b)^2} + 4(\tfrac{3}{2})\right) = 6$

23. $\displaystyle\int_{3}^{5} 5(x-3)^{-3}\,dx = \lim_{b\to 3^{+}}\int_{b}^{5} 5(x-3)^{-3}\,dx$

$\displaystyle = \lim_{b\to 3^{+}}\left[-\tfrac{5}{2}(x-3)^{-2}\Big|_{b}^{5}\right] = \lim_{x\to 3^{+}}\left[\frac{-5}{8} + \frac{5}{2(b-3)^2}\right] = \infty$

The integral diverges.

25. $\displaystyle\int_{-1}^{0} \frac{1}{x^3}\,dx + \int_{0}^{1} \frac{1}{x^3}\,dx = \lim_{b\to 0^{-}}\int_{-1}^{b} x^{-3}\,dx + \lim_{b\to 0^{+}}\int_{b}^{1} x^{-3}\,dx$

$\displaystyle = \lim_{b\to 0}\left[-\tfrac{1}{2}x^{-2}\Big|_{-1}^{b}\right] + \lim_{b\to 0^{+}}\left[-\tfrac{1}{2}x^{-2}\Big|_{b}^{1}\right]$

$$= \lim_{b \to 0^-} \left[\frac{-1}{2b^2} + \frac{1}{2} \right] + \lim_{b \to 0^+} \left[-\frac{1}{2} + \frac{1}{2b^2} \right] \text{ if these limits exist and are finite which they are not.}$$

Hence, $\displaystyle\int_{-1}^{1} \frac{1}{x^3} dx$ diverges.

27. $\displaystyle\int_{-1}^{1} 3x^{-4/5} dx = \int_{-1}^{0} 3x^{-4/5} dx + \int_{0}^{1} 3xe^{-4/5} dx$

$\displaystyle\lim_{b \to 0^-} \int_{-1}^{b} 3x^{-4/5} dx + \lim_{b \to 0^+} \int_{b}^{1} 3x^{-4/5} dx$

$\displaystyle= \lim_{b \to 0^-} \left[15x^{1/5} \Big|_{-1}^{b} \right] + \lim_{b \to 0^+} \left[15x^{1/5} \Big|_{b}^{1} \right]$

$\displaystyle= \lim_{b \to 0^-} \left[15(\sqrt[5]{b}) + 15 \right] + \lim_{b \to 0^+} \left[15(\sqrt[5]{b}) + 15 \right] = 30$

29. $\displaystyle\int_{1}^{2} \frac{1}{x\sqrt{\ln x}} dx = \lim_{b \to 1^+} \int_{b}^{2} \frac{1}{x\sqrt{\ln x}} dx = \lim_{b \to 1^+} \left[2\sqrt{\ln x} \Big|_{b}^{2} \right]$

$\displaystyle= \lim_{b \to 1^+} \left[2\sqrt{\ln 2} - 2\sqrt{\ln b} \right] = 2\sqrt{\ln 2}$

31. $\displaystyle\int_{1}^{4} \frac{4}{x}(\ln x)^{-2} dx = \lim_{b \to 1^+} \int_{b}^{4} \frac{4}{x}(\ln x)^{-2} dx = \lim_{b \to 1^+} \left[\frac{-4}{\ln x} \Big|_{b}^{4} \right]$

$\displaystyle= \lim_{b \to 1^+} \left[\frac{-4}{\ln 4} + \frac{4}{\ln b} \right] = -\infty. \quad \text{The integral diverges.}$

33. $\displaystyle\int_{2}^{\infty} \frac{2}{x(\ln x)^2} dx = \lim_{b \to \infty} \int_{2}^{b} \frac{2}{x}(\ln x)^{-2} dx = \lim_{b \to \infty} \left[-\frac{2}{\ln x} \Big|_{2}^{b} \right]$

$\displaystyle= \lim_{b \to \infty} \left[\frac{-2}{\ln b} + \frac{2}{\ln 2} \right] = \frac{2}{\ln 2}.$

35.

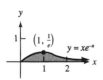

$y = xe^{-x}, \ y' = e^{-x} - xe^{-x} = (1 - x)e^{-x}.$

$y'' = -e^{-x} - e^{-x} + xe^{-x} = e^{-x}(x - 2).$

The critical value is $x = 1$. When $x = 1$, y'' is negative. Hence, there is a local maximum at $(1, \frac{1}{e})$.

Area of $R = \int_0^\infty xe^{-x}dx = \lim_{b\to\infty}\left[\int_0^b xe^{-x}dx\right] = \lim_{b\to\infty}\left[-xe^{-x} - e^{-x}\Big|_0^b\right]$

$= \lim_{b\to\infty}\left[\dfrac{-b}{e^b} - \dfrac{1}{e^b} + 0 + 1\right] = 0 - 0 + 0 + 1 = 1.$

37. $y = 3x^2 e^{-x}$

$y' = 6xe^{-x} - 3x^2 e^{-x} = 3xe^{-x}(2 - x)$

$y'' = 6e^{-x} - 6xe^{-x} - 6xe^{-x} + 3x^2 e^{-x}$

$\quad = 3e^{-x}(2 - 4x + x^2).$

$x = 2$ is the critical value.

When $x = 2$, y'' is negative.

There is a local maximum at $(2, \dfrac{12}{e^2})$.

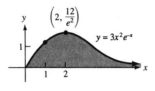

Area of $R = \int_0^\infty 3x^2 e^{-x}dx = \lim_{b\to\infty}\int_0^b 3x^2 e^{-x}dx$

$\left(\begin{matrix} u = x^2 & v = -e^{-x} \\ du = 2xdx & dv = e^{-x}dx \end{matrix}\right)$

$= 3\lim_{b\to\infty}\left[-x^2 e^{-x}\Big|_0^b + 2\int_0^b xe^{-x}dx\right] = 3\lim_{b\to\infty}\left[-x^2 e^{-x} + 2(-x-1)e^{-x}\Big|_0^b\right]$

$= 3\lim_{b\to\infty}\left[\dfrac{-b^2}{e^b} - \dfrac{-2(1+b)}{e^b} + 0 + 2\right] = 0 - 0 + 0 + 3\cdot2 = 6.$

39. $y = 2xe^{-3x}$

$y' = 2e^{-3x} - 6xe^{-3x}$

$\quad = 2e^{-3x}(1 - 3x).$

$y'' = -12e^{-3x} + 18xe^{-3x}$

$\quad = 6e^{-3x}(3x - 2).$

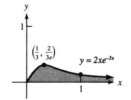

The critical value is $x = 1/3$. When $x = 1/3$, y'' is negative.

There is a local maximum at $(\dfrac{1}{3}, \dfrac{2}{3e})$.

Area of $R = \int_0^\infty 2xe^{-3x}dx = \lim_{b\to\infty}\int_0^b 2xe^{-3x}dx$ (using $\int$ table)

$= \lim_{b\to\infty}\left[\dfrac{2}{9}(-3x - 1)e^{-3x}\Big|_0^b\right] = \lim_{b\to\infty}\left[\dfrac{-6b}{9e^{3b}} - \dfrac{2}{9e^{3b}} + \dfrac{2}{9}\right] = \dfrac{2}{9}.$

41.

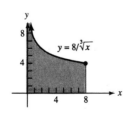

Area of $R = \int_0^8 8x^{-1/3}dx = \lim_{b\to0^+}\int_b^8 8x^{-1/3}dx$

$= \lim_{b\to0^+}\left[12x^{2/3}\Big|_b^8\right] = \lim_{b\to0^+}\left[12(4) - 12b^{2/3}\right]$

$= 48$

$-\ 220\ -$

43. $\displaystyle \int_4^\infty cx^{2/3}\,dx = \lim_{b\to\infty} \int_4^b cx^{-3/2}\,dx = \lim_{b\to\infty}\left[-2cx^{-1/2}\Big|_0^b\right]$

$\displaystyle = \lim_{b\to\infty}\left[\frac{-2c}{\sqrt{b}} + c\right] = c = 1$

45. The present value is $\displaystyle \lim_{n\to\infty} \int_0^n (12(3000)e^{-.12})\,dt$

$\displaystyle = \lim_{n\to\infty}\left(\frac{12(3000)e^{-.12t}}{-.12}\Big|_0^n\right)$

$\displaystyle = \lim_{n\to\infty}\left(-300{,}000e^{-.12t}\Big|_0^n\right)$

$\displaystyle = \lim_{n\to\infty}(300{,}000)\left(-\frac{1}{e^{.12n}}+1\right) = \$300{,}000$

EXERCISE SET 5.6 MORE ON PROBABILITY

1. (a) $\displaystyle \int_1^4 c(4-x)(x+1)\,dx = c\int_1^4 (4+3x-x^2)\,dx = c\left[4x+\frac{3}{2}x^2-\frac{x^3}{3}\Big|_1^4\right]$

$\displaystyle = c\left[16+24-\frac{64}{3}-4-\frac{3}{2}+\frac{1}{3}\right] = c\left[40-21-\frac{11}{2}\right] = c\left[\frac{27}{2}\right] = 1. \quad c = 2/27.$

(b) $\displaystyle E(X) = \frac{2}{27}\int_1^4 x(4-x)(x+1)\,dx = \frac{2}{27}\int_1^4 (4x+3x^2-x^3)\,dx$

$\displaystyle = \frac{2}{27}\left[2x^2+x^3-\frac{x^4}{4}\Big|_1^4\right] = \frac{2}{27}\left[32+64-64-2-1+\frac{1}{4}\right] = 117/54$

3. (a) $\displaystyle c\int_0^4 (-x^2+4x+5)\,dx = c\left[\frac{-x^3}{3}+2x^2+5x\Big|_0^4\right] = c\left[-\frac{64}{3}+32+20\right]$

$\displaystyle = c\left[\frac{92}{3}\right] = 1. \quad c = \frac{3}{92}.$

(b) $\displaystyle E(X) = \frac{3}{92}\int_0^4 x(-x^2+4x+5)\,dx = \frac{3}{92}\int_0^4 (-x^3+4x^2+5x)\,dx$

$\displaystyle = \frac{3}{92}\left[-\frac{x^4}{4}+\frac{4}{3}x^3+\frac{5}{2}x^2\Big|_0^4\right] = \frac{3}{92}\left[-64+\frac{256}{3}+40\right] = 2$

5. (a) $\displaystyle \int_0^5 \frac{c}{x+1}\,dx = c\ln(x+1)\Big|_0^5 = c\ln(6) = 1. \quad c = 1/\ln 6.$

(b) $\displaystyle E(X) = \frac{1}{\ln 6}\int_0^5 \frac{x}{x+1}\,dx = \frac{1}{\ln 6}\int_0^5 \left(1-\frac{1}{x+1}\right)dx = \frac{1}{\ln 6}\left[x-\ln(x+1)\Big|_0^5\right] = \frac{1}{\ln 6}[5-\ln 6].$

7. (a) $\int_0^\infty 2e^{-cx}dx = \lim_{b\to\infty}\int_0^b 2e^{-cx}dx = \lim_{b\to\infty}\left[-\frac{2}{c}e^{-cx}\Big|_0^b\right]$

$= \lim_{b\to\infty}\left[-\frac{2}{c}e^{-bc}+\frac{2}{c}\right] = \frac{2}{c} = 1.\quad c = 2.$

(b) $E(X) = \int_0^\infty 2xe^{-2x}dx = \lim_{b\to\infty}\int_0^b 2xe^{-2x}dx = 2\lim_{b\to\infty}\left(\frac{e^{-2x}}{4}(-2x-1)\Big|_0^b\right)$

$= \lim_{b\to\infty}\left(-be^{-2b}-\frac{e^{-2b}}{2}+\frac{1}{2}\right) = \frac{1}{2}$

9. (a) $\int_0^\infty ce^{-7x}dx = \lim_{b\to\infty}\left(-\frac{c}{7}e^{-7x}\Big|_0^b\right) = \lim_{b\to\infty}\left(-\frac{c}{7}e^{-7b}+\frac{c}{7}\right)$

$= \frac{c}{7} = 1.\quad c = 7.$

(b) $E(X) = \int_0^\infty 7xe^{-7x}dx = \lim_{b\to\infty}7\left(-\frac{1}{7}xe^{-7x}-\frac{1}{49}e^{-7x}\Big|_0^b\right)$

$= \lim_{b\to\infty}\left(-be^{-7b}-\frac{1}{7}e^{-7b}+\frac{1}{7}\right) = \frac{1}{7}$

11. (a) $8-\sqrt{t} \geq 0$ if $8 \geq \sqrt{t}$ or $64 \geq t \geq 0.$ Thus, $f(t) \geq 0$ for all $t.$

$\int_0^{36}\frac{1}{144}(8-\sqrt{t})dt = \frac{1}{144}(8t-\frac{2}{3}t^{3/2}\Big|_0^{36}) = \frac{1}{144}(288-144) = 1.$

Thus, f is a pdf.

(b) $E(T) = \int_0^{36}\frac{1}{144}(8t-t^{3/2})dt = \frac{1}{144}(4t^2-\frac{2}{5}t^{5/2}\Big|_0^{36})$

$= \frac{1}{144}(4(36)^2-.4(6)^5) = 14.4$

13. $f(t) = \frac{1}{90},\ 0 \leq t \leq 90$ seconds

$\quad = 0,$ elsewhere.

If T is the number of seconds the motorist must wait,

$P(T \geq 30) = \int_{30}^{90}\frac{1}{90}dt = \frac{2}{3}.$

15. $f(t) = \frac{1}{135},\ 0 \leq t \leq 135$ minutes

$\quad = 0,$ elsewhere

where T is the number of minutes before the next show begins.

$P(T \leq 15) = \int_0^{15}\frac{1}{135}dt = \frac{15}{135} = 1/9$

17. (a) $f(t) \geq 0$ for all t since $21 - t^{1/3} \geq 0$ if $21 \geq \sqrt[3]{t}$ or $t \leq 9261$.

$$\int_0^{8000} \frac{1}{48000}(21 - t^{1/3})\,dt = \frac{1}{48000}(21t - \frac{3}{4}t^{4/3})\Big|_0^{8000}$$

$$= \frac{1}{48000}(168,000 - 120,000) = 1. \quad \text{Thus, } f \text{ is a pdf.}$$

(b) $E(T) = \int_0^{8000} \frac{1}{48000}(21t - t^{4/3})\,dt = \frac{1}{48000}(\frac{21t^2}{2} - \frac{3}{7}t^{7/3})\Big|_0^{8000}$

$$\frac{1}{48000}(672,000,000 - \frac{3}{7}(1,280,000,000) = 2571.43 \text{ minutes}$$

19. (a) $31 - \sqrt[3]{t} \geq 0$ if $\sqrt[3]{t} \leq 31$ or $t \leq 29{,}791$.
Thus, t is never negative.

$$\int_0^{21,952} \frac{1}{219,520}(31 - \sqrt[3]{t})\,dt = \frac{1}{219,520}(31t - \frac{3}{4}t^{4/3}\Big|_0^{21,952})$$

$$= \frac{1}{219,520}(680512 - 460992) = 1. \quad f \text{ is a pdf.}$$

(b) $E(T) = \int_0^{21952} \frac{1}{219520}(31t - t^{4/3})\,dt$

$$= \frac{1}{219520}(\frac{31t^2}{2} - \frac{3}{7}t^{7/3}\Big|_0^{21952})$$

$$= \frac{31(21952)}{20} - \frac{3(21952)(614656)}{7(21951)(10)} = 7683.2 \text{ hours}$$

21. (a) $220 - \sqrt{t} \geq 0$ if $220 \geq \sqrt{t}$ or $0 \leq t \leq 48{,}400$.
Thus, $f(t)$ is never negative.

$$\int_0^{900} \frac{1}{180,000}(220 - \sqrt{t}\,)\,dt = 220t - \frac{2}{3}t^{3/2}\Big|_0^{900}(\frac{1}{180,000})$$

$$= (198,000 - 18,000)(\frac{1}{180,000}) = 1. \quad \text{Thus, } f \text{ is a pdf.}$$

(b) $E(T) = \frac{1}{180,000}\int_0^{900}(220t - t^{3/2})\,dt$

$$= \frac{1}{180,000}(110t^2 - .4t^{5/2}\Big|_0^{900}) = \frac{1}{180,000}(89,100,000 - 9,720,000)$$

$$= 441 \text{ hours}$$

23. (a) $360 - \sqrt{t} \geq 0$ if $360 \geq \sqrt{t}$ or $129600 \geq t \geq 0$.
Thus, $f(t)$ is never negative.

$$\frac{1}{2,430,000}\int_0^{8100}(360 - \sqrt{t}\,)\,dt = \frac{1}{2,430,000}(360t - \frac{2}{3}t^{3/2}\Big|_0^{8100})$$

$$= \frac{1}{2,430,000}(2916000 - 486000) = 1. \quad f \text{ is a pdf.}$$

(b) $E(T) = \frac{1}{2,430,000} \int_0^{8100} (360t - t^{3/2}) dt =$

$\frac{1}{2,430,000} \left[180t^2 - .4t^{5/2} \Big|_0^{8100} \right] = 3,888$ hours.

25. (a) $9 - \sqrt[3]{t} \geq 0$ if $9 \geq \sqrt[3]{t}$ or $729 \geq t$. Thus, $f(t)$ is never negative.

$\int_0^{343} \frac{4}{5145}(9 - \sqrt[3]{t}) dt = \frac{4}{5145}(9t - \frac{3}{4}t^{4/3}) \Big|_0^{343}$

$= \frac{4}{5145}(3087 - 1800.75) = 1$. f is a pdf.

(b) $E(T) = \int_0^{343} \frac{4}{5145}(9t - t^{4/3}) dt = \frac{4}{5145}(\frac{9}{2}t^2 - \frac{3}{7}t^{7/3} \Big|_0^{343})$

$= \frac{4}{5145}(529420.5 - 352947) = 137.2$ hours.

27. (a) f is never negative and $\int_0^{\infty} \frac{1}{120} e^{-t/120} dt$

$= \lim_{b \to \infty} \int_0^b \frac{1}{120} e^{-t/120} dt = \lim_{b \to \infty} (-e^{-t/120} \Big|_0^b$

$= \lim_{b \to \infty} (-e^{-b/120} + 1) = 1$. Thus, f is a pdf.

(b) $E(T) = \int_0^{\infty} \frac{t}{120} e^{-t/120} dt = \lim_{b \to \infty} \int_0^b \frac{t}{120} e^{-t/120} dt$

$= \lim_{b \to \infty} \frac{1}{120}(120)^2 (-\frac{1}{120}t - 1)e^{-t/120} \Big|_0^b$

$= \lim_{b \to \infty} (120(-\frac{b}{120} - 1)e^{-b/120} + 120) = 120$ months

(c) $P(24 \leq T \leq 36) = \int_{24}^{30} \frac{1}{120} e^{-t/120} dt = -e^{-t/120} \Big|_{24}^{30}$

$= -e^{-.25} + e^{-.2} = .0399$

29. (a) $f(t)$ is never negative and $\int_0^{\infty} \frac{1}{6000} e^{-t/6000} dt$

$= \lim_{b \to \infty} (-e^{-t/6000} \Big|_0^b) = \lim_{b \to \infty} (-e^{-b/6000} + 0) = 1$.

Thus, f is a pdf.

(b) $E(T) = \int_0^{\infty} \frac{t}{6000} e^{-t/6000} dt = \lim_{b \to \infty} \left(6000(\frac{-t}{6000} - 1)e^{-t/6000} \Big|_0^b \right)$

$= \lim_{b \to \infty} (-be^{-b/6000} - 6000e^{-b/6000} + 6000) = 6000$ hours.

(c) $P(1460 \leq T \leq 2190) = -e^{-t/6000} \Big|_{1460}^{2190}$

$= -e^{-2190/6000} + e^{-1460/6000} = -.69419665 + .784010133 = .09$

31. $P(T < 4) = \frac{1}{4500} \int_0^4 (30 - \sqrt{t})\, dt = \frac{1}{4500}(30t - \frac{2}{3}t^{3/2})\big|_0^4 = .02548$

$P(4 \leq T \leq 16) = \frac{1}{4500}(30t - \frac{2}{3}t^{3/2})\big|_4^{16} = .09718 - .02548 = .071$

The profit is $-\$14.00$ for a dryer lasting less than 60 days, $-\$2.50$ for a dryer lasting 60 to 240 days and $\$9.00$ otherwise.

$E(\text{profit}) = -14(.02548) - 2.5(.0717) + 9(1 - .02548 - .0717) = \$7.59$

33. A circuit is used 270 hours in 180 days and 540 hours in 360 days.

$P(T < 270) = \frac{1}{2430000} \int_0^{270} (360 - \sqrt{t})\, dt$

$= \frac{1}{2,430,000}(360t - \frac{2}{3}t^{3/2})\big|_0^{270} = .03878$

$P(270 \leq T \leq 540) = .03778$
$P(T > 540) = 1 - .03878 - .03778 = .92344$
$E(\text{profit}) = (-4.50)(.03878) + (-.25)(.03778) + 4(.92344) = \$3.51$

EXERCISE SET 5.7 CHAPTER REVIEW

1. If $y = f_1(x) = e^{-4x}$, $y' = -4e^{-4x}$, $y'' = 16e^{-4x}$ and

$y'' - y' - 20y = 16e^{-4x} + 4e^{-4x} - 20e^{-4x} = 0$

If $y = f_2(x) = e^{5x}$, $y' = 5e^{5x}$, $y'' = 25e^{5x}$ and

$y'' - y' - 20y = 25e^{5x} - 5e^{5x} - 20e^{5x} = 0$

3. If $y = x^2 + 3x + 1$, $y' = 2x + 3$, $y'' = 2$ and
$x^2 y'' + 2xy' - 6y + 12x + 6 =$
$2x^2 + 4x^2 + 6x - 6x^2 - 18x - 6 + 12x + 6 = 0$

5. (a) $\int \frac{1}{x}dx = \int 4t^3\, dt.$

$\ln|x| = t^4 + C$. If $t = -1$, $x = e^4$ and $4 = 1 + C$ $C = 3$
$\ln|x| = t^4 + 3$.

$x = e^{t^4 + 3}$ is the relevant solution

(b) $\int e^{2y}dy = \int 3xe^x dx$

$\frac{1}{2}e^{2y} = 3(xe^x - e^x) + C$. If $x = 0$, $y = 2$ and $\frac{1}{2}e^4 = -3 + C$. $C = 3 + e^4/2$.

$e^{2y} = 6xe^x - 6e^x + 6 + e^4$

(c) $(x^2 + 1)\ln(y + e)\frac{dy}{dx} = x(y + e)$. $\int \frac{\ln(y + e)dy}{y + e} = \int \frac{x}{x^2 + 1}dx.$

$\frac{1}{2}(\ln(y + e))^2 = \frac{1}{2}\ln(x^2 + 1) + \ln C.$

If $x = 0$, and $y = 0$ and $\ln C = 1/2$

$\left(\ln(y + e)\right)^2 = 1 + \ln(x^2 + 1)$ is the solution.

7. $P = P_0 e^{.08t}$

    (a) $3P_0 = P_0 e^{.08t}$

           $3 = e^{.08t}$

           $\dfrac{\ln 3}{.08} = t = 13.7327$ years

    (b) $9000 = 6000 e^{.08t}$

           $1.5 = e^{.08t}$

           $\dfrac{\ln 1.5}{.08} = t = 5.068$ years.

9. $\eta = \dfrac{p\,dq}{q\,dp} = \dfrac{-3p^2}{2500 - p^2}$, $0 < p < 50$.

$\displaystyle\int \dfrac{dq}{q} = \int \dfrac{-3p}{2500 - p^2}dp$

$\ln q = \frac{3}{2}\ln(2500 - p^2) + \ln C$.

If $p = 30$, $q = 192{,}000$ and $\ln 192{,}000 = \ln C(64{,}000)$ $C = 3$.

$\ln q = \ln(2500 - p^2)^{3/2} \cdot 3$

$q = 3(2500 - p^2)^{3/2}$, $0 < p < 50$.

11. (a)

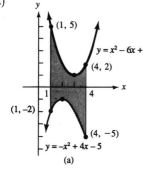

(a)

(b) $A = \displaystyle\int_1^4 (x^2 - 6x + 10) - (-x^2 + 4x - 5)\,dx$

$\qquad = \displaystyle\int_1^4 (2x^2 - 10x + 15)\,dx$

$\qquad = \frac{2}{3}x^3 - 5x^2 + 15x\Big|_1^4$

$\qquad = \dfrac{128}{3} - 20 - \left(\frac{2}{3} + 10\right)$

$\qquad = 12$

13. (a)

(b) $A = \displaystyle\int_{-8}^1 (3 - \sqrt{1 - x})\,dx + 2(3) + \int_3^{12}(3 - \sqrt{x - 3}$

$\qquad = 3x + \frac{2}{3}(1 - x)^{3/2}\Big|_{-8}^1 + 6 + (3x - \frac{2}{3}(x - 3)^{3/2}\big|$

$\qquad = 3 - (-24 + 18) + 6 + 18 - 9 = 24$

15. (a)

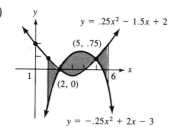

$y = .25x^2 - 1.5x + 2$

$(5, .75)$

1 6

$(2, 0)$

$y = -.25x^2 + 2x - 3$

(b) $A_1 = \displaystyle\int_1^2 .25x^2 - 1.5x + 2 - (-.25x^2 + 2x - 3)\,dx = \int_1^2 (.5x^2 - 3.5x + 5)\,dx = 11/12$

$A_2 = \displaystyle\int_2^5 (-.5x^2 + 3.5x - 5)\,dx = 2.25$

$A_3 = \displaystyle\int_5^6 (.5x^2 - 3.5x + 5)\,dx = 11/12$

$A = A_1 + A_2 + A_3 = 49/12$

17. (a) $S(q) = D(q)$ if $q/3 = -q/4 + 14;\ 0 \le q \le 56$
$4q = -3q + 168,\ 7q = 168,\ q = 24$ and $p = 8.$

The equilibrium quantity is 24,000 units.
The equilibrium price is \$8/unit.

(b) $C.S. = \displaystyle\int_0^{24} \left(-\frac{q}{4} + 14 - 8\right)dq = -q^2/8 + 6q\Big|_0^{24} = \$72{,}000$

(c) $P.S. = \displaystyle\int_0^{24} (8 - q/3)\,dq = 8q - q^2/6\Big|_0^{24} = \$96{,}000$

19. (a) $D(q) = mq + b.\quad m = \dfrac{45 - 35}{40 - 80} = \dfrac{10}{-40} = -1/4.\quad D(q) = -\frac{1}{4}q + 55.$

$D(q) = S(q)$ if $q = 100$ and $p = 30.$

The equilibrium quantity is 100,000 units.
The equilibrium price is \$30/unit.

(b) $C.S. = \displaystyle\int_0^{100} (55 - q/4 - 30)\,dq = 25q - q^2/8\Big|_0^{100} = \$1{,}250{,}000.$

(c) $P.S. = \displaystyle\int_0^{100} (30 - 3\sqrt{q})\,dq = 30q - 2q^{3/2}\Big|_0^{100} = \$1{,}000{,}000$

21. (a)

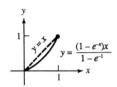

$y = \frac{3x^2 + 8x}{11}$

(b) $\quad I(.25) \doteq .1989$

(c) $\quad 2\displaystyle\int_0^1 (x - \frac{3}{11}x^2 - \frac{8}{11}x)\,dx$

$\qquad = 2(\frac{3}{22}x^2 - \frac{1}{11}x^3\big|_0^1) = \overline{.09}$

23. (a)

y

$y = x$

$y = \frac{(1 - e^{-x})x}{1 - e^{-1}}$

(b) $\quad I(.25) = \dfrac{.055299804}{.632120558} \doteq .087$

(c) $\quad 2\displaystyle\int_0^1 (x - (\frac{1}{1 - e^{-1}})x + \frac{1}{1 - e^{-1}} \cdot e^{-x}x\,dx$

$\qquad = 2\left[\dfrac{x^2}{2} - \dfrac{x^2}{2}(\dfrac{1}{1 - \frac{1}{e}}) + \dfrac{1}{1 - \frac{1}{e}}(-xe^{-x} - e^{-x})\right]\bigg|_0^1$

$\qquad = 2\left[\dfrac{1}{2} - \dfrac{1}{2}(\dfrac{1}{1 - \frac{1}{e}}) + \dfrac{1}{1 - \frac{1}{e}}(-2e^{-1}) + \dfrac{1}{1 - \frac{1}{e}}\right]$

$\qquad = 2\left[\dfrac{1}{2} + \dfrac{1}{2}(\dfrac{1}{1 - \frac{1}{e}}) - \dfrac{2}{e - 1}\right] \doteq .254$

25. $\displaystyle\int_0^{15} 54{,}000e^{-.08t}\,dt = \dfrac{-54{,}000}{.08}e^{-.08t}\big|_0^{15} = \$471{,}694$

27. $y = ax^b$. When $x = 8$, $y = 30$ and when $x = 16$, $y = 30(.9) = 27$.

$30 = a(8^b) = a(2^{3b}).\quad a = \dfrac{30}{2^{3b}} = \dfrac{30}{(2^b)^3}$

$27 = a(16^b) = a(2^{4b}) = \dfrac{30}{2^{3b}}(2^{4b}) = 30(2^b)$

$2^b = \dfrac{27}{30} = .9.\quad b = \log_2(.9) = \dfrac{\ln(.9)}{\ln 2} \doteq -.152$

$a = \dfrac{30}{(.9)^3} \doteq 41.15.$

29. (a) $\displaystyle\int_1^\infty \dfrac{4}{x^2}\,dx = \lim_{b \to \infty}\int_1^b 4x^{-2}\,dx = \lim_{b \to \infty}\left[-\dfrac{4}{x}\bigg|_1^b\right]$

$\qquad = \lim_{b \to \infty}\left[-\dfrac{4}{b} + 4\right] = 4$

(b) $\displaystyle\int_{-\infty}^{2} e^{3x}\,dx = \lim_{b\to-\infty}\int_{b}^{2} e^{3x}\,dx = \lim_{b\to-\infty}\left[\tfrac{1}{3}e^{3x}\Big|_{b}^{2}\right]$

$= \lim_{b\to-\infty}\left[\tfrac{e^6}{3} - \tfrac{e^{3b}}{3}\right] = \tfrac{e^6}{3}$

(c) $\displaystyle\int_{2}^{\infty}\frac{3x}{x^2+1}\,dx = \lim_{b\to\infty}\left[\int_{2}^{b}\frac{3x}{x^2+1}\,dx\right]$

$= \lim_{b\to\infty}\left[\tfrac{3}{2}\ln(x^2+1)\Big|_{2}^{b}\right] = \lim_{b\to\infty}\tfrac{3}{2}\ln(b^2+1) - \tfrac{3}{2}\ln(5)$

$= \infty$. The integral diverges.

(d) $\displaystyle\int_{-\infty}^{0}\frac{e^x}{e^x+5}\,dx = \lim_{b\to-\infty}\int_{b}^{0}\frac{e^x}{e^x+5}\,dx = \lim_{b\to-\infty}\left[\ln(e^x+5)\Big|_{b}^{0}\right]$

$= \lim_{b\to-\infty}\left[\ln 6 - \ln(e^b+5)\right] = \ln(1.2)$

(e) $\displaystyle\int_{-\infty}^{\infty}\frac{3x^2}{(x^3+1)^2}\,dx = \int_{-\infty}^{-1}\frac{3x^2\,dx}{(x^3+1)^2} + \int_{-1}^{\infty}\frac{3x^2}{(x^2+1)^2}\,dx$

$= \lim_{a\to-\infty}\int_{a}^{-1}\frac{3x^2}{(x^3+1)^2}\,dx + \lim_{a\to\infty}\int_{-1}^{a}\frac{3x^2}{(x^3+1)^2}\,dx$

$= \lim_{b\to-1^-}\left[\lim_{a\to-\infty}\int_{a}^{b}\frac{3x^2}{(x^3+1)^2}\,dx\right] + \lim_{b\to-1^+}\left[\lim_{a\to\infty}\int_{b}^{a}\frac{3x^2}{(x^3+1)^2}\,dx\right]$

$= \lim_{b\to-1^-}\left[\lim_{a\to-\infty}\frac{-1}{x^3+1}\Big|_{a}^{b}\right] + \lim_{b\to-1^+}\left[\lim_{a\to\infty}\frac{-1}{x^3+1}\Big|_{b}^{a}\right]$

$= \lim_{b\to-1^-}\left[\frac{-1}{b^3+1}\right] + \lim_{b\to-1^+}\left[\frac{1}{b^3+1}\right] = \infty$ The integral diverges.

(f) $\displaystyle\int_{0}^{3} 3x^{-1/2}\,dx = \lim_{b\to0^+}\int_{b}^{3} 3x^{-1/2}\,dx$

$= \lim_{b\to0^+}\left[6x^{1/2}\Big|_{b}^{3}\right] = \lim_{b\to0^+}\left[6\sqrt{3} - 6\sqrt{b}\right] = 6\sqrt{3}$

31. (a) $13 - \sqrt{t} \geq 0$ if $\sqrt{13} \geq t$ or $0 \leq t \leq 169$.

$\displaystyle\int_{0}^{121}\frac{3}{2057}(13-\sqrt{t})\,dt = \frac{3}{2057}\left(13t - \tfrac{2}{3}t^{3/2}\Big|_{0}^{121}\right)$

$= \frac{3}{2057}\left(1573 - \tfrac{2662}{3}\right) = 1$.

(b) $E(T) = \displaystyle\int_{0}^{121}\frac{3}{2057}(13t - t^{3/2})\,dt$

$= \frac{3}{2057}\left(\tfrac{13t^2}{2} - \tfrac{2}{5}t^{5/2}\Big|_{0}^{121}\right) = 44.84$ weeks.

33. (a) $f(t)$ is never negative.

$$\int_0^\infty \frac{1}{23} e^{-t/23} \, dt = \lim_{b \to \infty} \left(-e^{-t/23} \Big|_0^b \right)$$

$$= \lim_{b \to \infty} (e^{-b/23} + 1) = 1. \quad \text{Thus, } f \text{ is a pdf.}$$

(b) $E(T) = \displaystyle\int_0^\infty \frac{t}{23} e^{-t/23} \, dt = \lim_{b \to \infty} \left(t e^{-t/23} - 23 e^{-t/23} \Big|_0^b \right)$

$$\lim_{b \to \infty} (b e^{-b/23} - 23 e^{-b/23} - 0 + 23) = 23 \text{ years.}$$

(c) $P(T \geq 15) = \displaystyle\int_{15}^\infty \frac{1}{23} e^{-t/23} \, dt$

$$= \lim_{b \to \infty} \left(-e^{-t/23} \Big|_{15}^b \right) = e^{-15/23} = .52$$

CHAPTER 6

FUNCTIONS OF SEVERAL VARIABLES

EXERCISE SET 6.1 FUNCTIONS OF SEVERAL VARIABLES

1. The domain of f is the set of all ordered pairs of real numbers

    $f(1,2) = 1 + 3(2) + 4 = 11$
    $f(-2,3) = 4 + 3(-6) + 9 = -5$
    $f(3,-4) = 9 + 3(-12) + 16 = -11$

3. The domain of f is the set of all ordered pairs of real numbers except for those of the form $(5,c)$ or $(d,7)$.

    $f(0,-1) = 0/40 = 0$
    $f(-2,-5) = -6/84 = -1/14$
    $f(3,2) = 9/10$

5. The domain of f is the set of all ordered pairs of real numbers except those of the form $(-2,c)$ or (d,d).

    $f(3,4) = 11/-5 = -2.2$
    $f(-1,6) = -11/7$
    $f(3,-2) = -1/25$

7. The domain of f is the set of all ordered pairs of real numbers except those of the form (x,y) where $4 - x^2 < 0$. The latter occurs if $x > 2$ or $x < -2$

    $f(2,4) = 16 + \sqrt{4-4} = 16$
    $f(1,5) = 10 + \sqrt{3}$
    $f(0,3) = 0 + \sqrt{4-0} = 2$

9. The domain of f is the set of all ordered pairs of real numbers except those of the form (x,y) where $25 - x^2 - y^2 < 0$ or $x^2 + y^2 > 25$. The domain consists of points inside or on the circle with center $(0,0)$ and radius 5.

    $f(1,2) = \sqrt{25-1-4} = 2\sqrt{5}$
    $f(3,4) = \sqrt{25-9-16} = 0$
    $f(-1,2) = \sqrt{25-1-4} = 2\sqrt{5}$

11. The domain of f is the set of all ordered triples of real numbers.

    $f(1,1,2) = 1 + 1 + 8 = 10$
    $f(-1,2,-3) = 1 - 2 + 18 = 17$
    $f(2,4,0) = 4 + 8 + 0 = 12$

13. The domain of f is the set of all ordered triples or real numbers except for those of the form (x,y,z) where $x^2 + y^2 > 9$.

$$f(1,1,3) = 9 + \sqrt{9-1-1} = 9 + \sqrt{7}$$
$$f(0,2,5) = 15\sqrt{9-0-4} = 15 + \sqrt{5}$$
$$f(-1,2,-3) = -9 + \sqrt{9-1-4} = -9 + 2 = -7$$

15. The domain of f is the set of all ordered triples of real numbers except for those of the form (x,y,z) where $z > 4$ or $z < -4$.

$$f(1,3,2) = 1 + 27 - \sqrt{12} = 28 - 2\sqrt{3}$$
$$f(0,3,-4) = 0 + 27 - \sqrt{0} = 27$$
$$f(-2,4,-3) = 4 + 48 - \sqrt{7} = 52 - \sqrt{7}$$

17, 19, 21.

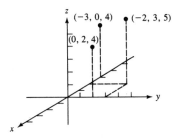

23, 25.

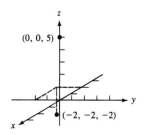

27. $f(x,y) = \dfrac{4}{x-y}$

$f(1,2) = -4$
$f(2,1) = 4$
$f(2,0) = 2$

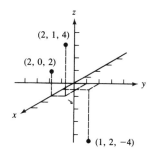

29. $f(x,y) = \dfrac{2y}{(x+3)(y+2)}$

$f(1,0) = 0$
$f(-2,2) = 1$
$f(0,3) = .4$

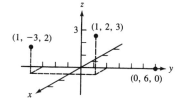

31. $f(x,y) = 5x - \sqrt{6-y}$
$f(1,2) = 3$
$f(1,-3) = 2$
$f(0,6) = 0$

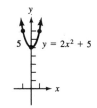

33. $5 = y - 2x^2$
$y = 2x^2 + 5$

x	y
0	5
1	7
-1	7

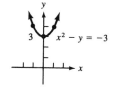

35. $-3 = x^2 - y, \; y = x^2 + 3$

x	y
0	3
1	4
-1	4

37. $6 = x^2 + y^2 + z^2$
$4 = x^2 + y^2$

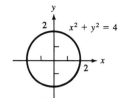

39. $16 = x^2 + y^2$

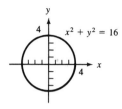

41. The revenue for brand A is $p_A(300 - 10 \ p_A + 20p_B)$
The revenue for brand B is $p_B(450 - 15p_B + 25p_A)$
The total revenue is

$$R(p_A, p_B) = p_A(300 \ - \ 10 \ p_A + 20p_B) + p_B(450 - 15p_B + 25p_A)$$
$$= 300p_A - 10p_A{}^2 + 450p_B - 15p_B{}^2 + 45p_Ap_B$$

43. $C(x,y) =$ cost of top $+$ cost of base $+$ cost of sides

$xyh = 3000$

$h = \dfrac{3000}{xy}$

$$C(x,y) \ = 3xy + 20xy + 2x\left(\dfrac{3000}{xy}\right)12 + 2y\left(\dfrac{3000}{xy}\right)12$$
$$= 23xy + 72{,}000/y + 72{,}000/x.$$

EXERCISE SET 6.2 PARTIAL DIFFERENTIATION

1. $f_x(1,3)$

$= \lim\limits_{h \to 0} \dfrac{f(1 + h, 3) - f(1,3)}{h}$

$= \lim\limits_{h \to 0} \dfrac{9(1 + h)^2 - 9}{h}$

$= \lim\limits_{h \to 0} \dfrac{9 + 18h + 9h^2 - 9}{h}$

$= \lim\limits_{h \to 0} \dfrac{9h^2 + 18h}{h}$

$= (9h + 18) = 18$

$f_y(1,3)$

$= \lim\limits_{h \to 0} \dfrac{f(1, 3 + h) - f(1,3)}{h}$

$= \lim\limits_{h \to 0} \dfrac{(3 + h)^2 - 9}{h}$

$$= \lim_{h \to 0} \frac{6h + h^2}{h}$$
$$= \lim_{h \to 0} \frac{h(6 + h)}{h} = 6$$

3. $f_x(-1, -3)$ $= \lim_{h \to 0} \frac{f(-1+h, -3) - f(-1, -3)}{h}$
$$= \lim_{h \to 0} \frac{-3(-1+h)^2(81) + 243}{h}$$
$$= \lim_{h \to 0} \frac{-243(1 - 2h + h^2) + 243}{h}$$
$$= \lim_{h \to 0} \frac{(486 - 243h)h}{h} = 486$$

$f_y(-1, -3)$ $= \lim_{h \to 0} \frac{f(-1, -3+h) + 243}{h}$
$$= \lim_{h \to 0} \frac{-3(-3+h)^4 + 243}{h}$$
$$= \lim_{h \to 0} \frac{-3(81 - 108h + 54h^2 - 12h^3 + h^4) + 243}{h}$$
$$= \lim_{h \to 0} \frac{-3h(-108 + 54h - 12h^2 + h^3)}{h}$$
$$= -3(-108) = 324$$

5. $f_x(5,2)$ $= \lim_{h \to 0} \frac{f(5+h,2) - f(5,2)}{h}$
$$= \lim_{h \to 0} \frac{2(5+h)(8) - 80}{h}$$
$$= \lim_{h \to 0} \frac{16h}{h} = 16$$

$f_y(5,2)$ $= \lim_{h \to 0} \frac{f(5,2+h) - 80}{h}$
$$= \lim_{h \to 0} \frac{10(2+h)^3 - 80}{h}$$
$$= \lim_{h \to 0} \frac{80 + 120h + 60h^2 + 10h^3 - 80}{h}$$
$$= 120$$

7. $f_s(s,t) = -3t^7 D_s(s^3) = -3t^7(3s^2) = -9t^7s^2$
$f_t(s,t) = -3s^3 D_t(t^7) = -3s^3(7t^6) = -21s^3t^6$

9. $f_x(x,y) = -3y^5 D_x(x^2) + 4yD_x(x^3) = -6xy^5 + 12x^2y$
$f_y(x,y) = -3x^2 D_y(y^5) + 4x^3 D_y(y) = -15x^2y^4 + 4x^3$

11. $h(x,y) = \dfrac{3}{2x^3 y^2} + \dfrac{5y}{2x} = \dfrac{3}{2}x^{-3}y^{-2} + \dfrac{5}{2}yx^{-1}$
$h_x(x,y) = \dfrac{3}{2}y^{-2}D_x(x^{-3}) + \dfrac{5}{2}yD_x(x^{-1}) = \dfrac{-9}{2}y^{-2}x^{-4} - \dfrac{5}{2}yx^{-2}$

$$= \frac{-4.5}{x^4 y^2} - \frac{5y}{2x^2}$$

$$h_y(x,y) \; = 1.5x^{-3} D_y(y^{-2}) + \frac{5}{2}x^{-1} \cdot 1 = \frac{-3}{x^3 y^3} + \frac{5}{2x}$$

13. $f(x,y) = 3e^{x^2 y^5}$

$$f_1(x,y) = 3e^{x^2 y^5} D_y(x^2 y^5) = 6xy^5 e^{x^2 y^5}$$

$$f_2(x,y) = 3e^{x^2 y^5} D_y(x^2 y^5) = 15x^2 y^4 e^{x^2 y^5}$$

15. $h(u,v) = 5e^{-u^2 v^6}$

$$h_1(u,v) = 5e^{-u^2 v^6} D_u(-u^2 v^6) = -10uv^6 e^{-u^2 v^6}$$

$$h_2(u,v) = 5e^{-u^2 v^6} D_v(-u^2 v^6) = -30u^2 v^5 e^{-u^2 v^6}$$

17. $g(x,y) = -4x^3 y^2 e^{x^3 y}$

$$g_1(x,y) = -4y^2 e^{x^3 y} D_x(x^3) - 4x^3 y^2 D_x(e^{x^3 y})$$

$$= -12x^2 y^2 e^{x^3 y} - 12x^5 y^3 e^{x^3 y}$$

$$g_2(x,y) = -4x^3 e^{x^3 y} D_y(y^2) - 4x^3 y^2 D_y(e^{x^3 y})$$

$$= -8x^3 y e^{x^3 y} - 4x^6 y^2 e^{x^3 y}$$

19. $f(x,y) = \dfrac{5e^{3xy}}{2x^2 y^3}$

$$f_1(x,y) \; = \frac{2x^2 y^3 D_x(5e^{3xy}) - 5e^{3xy} D_x(2x^2 y^3)}{(2x^2 y^3)^2}$$

$$= \frac{30x^2 y^4 e^{3xy} - 20xy^3 e^{3xy}}{4x^4 y^6}$$

$$= \frac{(15xy - 10)e^{3xy}}{2x^3 y^3}$$

$$f_2(x,y) \; = \frac{2x^2 y^3 D_y(5e^{3xy}) - 5e^{3xy} D_y(2x^2 y^3)}{4x^4 y^6}$$

$$= \frac{2x^2 y^3 (15xe^{3xy}) - 5e^{3xy}(6x^2 y^2)}{4x^4 y^6}$$

$$= \frac{e^{3xy}(30x^3 y^3 - 30x^2 y^2)}{4x^4 y^6}$$

$$= \frac{15e^{3xy}(xy - 1)}{2x^2 y^4}$$

21. $h(x,y) = 7xy(5^{x^4y^3})$

$h_1(x,y) = 7xyD_x(5^{x^4y^3}) + 5^{x^4y^3}D_x(7xy)$

$= 7xy5^{x^4y^3}\ln5 D_x(x^4y^3) + 5^{x^4y^3}(7y)$

$= 5^{x^4y^3}(7xy(4x^3y^3)\ln5 + 7y)$

$= (28x^4y^4\ln5 + 7y)5^{x^4y^3}$

$h_2(x,y) = 7xyD_y(5^{x^4y^3}) + 5^{x^4y^3}D_y(7xy)$

$= 7xy(5^{x^4y^3})\ln5 D_y(x^4y^3) + 5^{x^4y^3}(7x)$

$= 5^{x^4y^3}(21x^5y^3\ln5 + 7x)$

23. $g_1(x,y) = \dfrac{D_x(3x^2 + 5y^6)}{3x^2 + 5y^6} = \dfrac{6x}{3x^2 + 5y^6}$

$g_2(x,y) = \dfrac{D_y(3x^2 + 5y^6)}{3x^2 + 5y^6} = \dfrac{30y^5}{3x^2 + 5y^6}$

25. $f(x,y) = e^{-x^2y^4}\ln(3x^{10} + 5x^2y^4)$

$f_1(x,y) = -2xy^4e^{-x^2y^4}\ln(3x^2 + 5x^2y^4) + \dfrac{e^{-x^2y^4}(30x^9 + 10xy^4)}{3x^{10} + 5x^2y^4}$

$f_2(x,y) = -4x^2y^3e^{-x^2y^4}\ln(3x^{10} + 5x^2y^4) + \dfrac{e^{-x^2y^4}(20x^2y^3)}{3x^{10} + 5x^2y^4}$

27. $h_1(x,y) = \dfrac{2xy^6}{(\ln3)(x^2y^6 + 6)}$

$h_2(x,y) = \dfrac{6x^2y^5}{(\ln3)(x^2y^6 + 6)}$

29. $g(x,y) = (-3\log_7(4x^2 + y^2))(5x^3y^2 + 1)^{-1}$

$g_1(x,y) = \dfrac{-24x}{(\ln7)(4x^2 + y^2)(5x^3y^2 + 1)} + 3(5x^3y^2 + 1)^{-2}(\log_7(4x^2 + y^2))(15x^2y^2)$

$g_2(x,y) = \dfrac{-6y}{(\ln7)(4x^2 + y^2)(5x^3y^2 + 1)} + 3(5x^3y^2 + 1)^{-2}(\log_7(4x^2 + y^2))(10x^3y)$

31. $f_x(x,y) = 10xy^5 \qquad\qquad f_y(x,y) = 25x^2y^4$

$f_{xx}(x,y) = 10y^5 \qquad\qquad f_{yy}(x,y) = 100x^2y^3$

33. $f_x(x,y) \quad = 5x^4y^33^{x^5y^3}\ln 3 \qquad f_y(x,y) = 3x^5y^23^{x^5y^3}\ln 3$

$\quad f_{xx}(x,y) \quad = 20x^3y^3(3^{x^5y^3})\ln 3 + 5x^4y^3\ln 3(5x^4y^3 \cdot 3^{x^5y^3}\ln 3)$

$\qquad\qquad = (20x^3y^3 + 25x^8y^6\ln 3)(\ln 3)3^{x^5y^3}$

$\quad f_{yy}(x,y) \quad = (6x^5y)3^{x^5y^3}\ln 3 + 3x^5y^2\ln 3(3y^2x^5)3^{x^5y^3}\ln 3$

$\qquad\qquad = (\ln 3)3^{x^5y^3}(6x^5y + 9x^{10}y^4\ln 3)$

35. $f_x(x,y) \quad = \dfrac{6x}{(3x^2 + 5y^4)\ln 5} \qquad f_y(x,y) \quad = \dfrac{20y^3}{(3x^2 + 5y^4)\ln 5}$

$\quad f_{xx}(x,y) \quad = \dfrac{6}{\ln 5}(3x^2 + 5y^4 - 6x^2) \cdot \dfrac{1}{(3x^2 + 5y^4)^2} = \dfrac{6(5y^4 - 3x^2)}{(\ln 5)(3x^2 + 5y^4)^2}$

$\quad f_{yy}(x,y) \quad = \dfrac{20\left(3y^2(3x^2 + 5y^4) - y^3(20y^3)\right)}{(\ln 5)(3x^2 + 5y^4)^2}$

$\qquad\qquad = \dfrac{20(9x^2y^2 - 5y^6)}{(\ln 5)(3x^2 + 5y^4)^2}$

37. $f(x,y) \quad = e^{-3xy}x^{-2}y^{-1}$

$\quad f_x(x,y) \quad = -3ye^{-3xy}(x^{-2}y^{-1}) - 2x^{-3}y^{-1}e^{-3xy}$

$\qquad\qquad = -3x^{-2}e^{-3xy} - 2x^{-3}y^{-1}e^{-3xy}$

$\quad f_{xx}(x,y) \quad = 6x^{-3}e^{-3xy} - 3x^{-2}(-3y)e^{-3xy} + 6x^{-4}y^{-1}e^{-3xy} + 6x^{-3}yy^{-1}e^{-3xy}$

$\qquad\qquad = e^{-3xy}(6x^{-3} + 9x^{-2}y + 6x^{-4}y^{-1} + 6x^{-3})$

$\quad f_y(x,y) \quad = -3xe^{-3xy}x^{-2}y^{-1} - y^{-2}x^{-2}e^{-3xy} = -3x^{-1}y^{-1}e^{-3xy} - x^{-2}y^{-2}e^{-3xy}$

$\quad f_{yy}(x,y) \quad = 3x^{-1}y^{-2}e^{-3xy} + 9y^{-1}e^{-3xy} + 2x^{-2}y^{-3}e^{-3xy} + 3x^{-1}y^{-2}e^{-3xy}$

39. $f_x(x,y) \quad = \dfrac{e^5(2xe^3)}{(e^3x^2 + y^e)\ln 3} = \dfrac{2e^8}{\ln 3}(\dfrac{x}{e^3x^2 + y^e})$

$\quad f_{xx}(x,y) \quad = \dfrac{2e^8}{\ln 3}\dfrac{\left(e^3x^2 + y^e - xe^3(2x)\right)}{(e^3y^2 + y^e)^2}$

$\qquad\qquad = \dfrac{2e^8(y^e - e^3x^2)}{(e^3x^2 + y^e)^2\ln 3}$

$\quad f_y(x,y) \quad = e^5\dfrac{ey^{e-1}}{(e^3x^2 + y^e)\ln 3} = \dfrac{e^6}{\ln 3}\dfrac{y^{e-1}}{(e^3x^2 + y^e)}$

$\quad f_{yy}(x,y) = \dfrac{e^6}{\ln 3}\left((e-1)y^{e-2}(e^3x^2 + y^e) - y^{e-1}ey^{e-1}\right) \cdot \dfrac{1}{(e^3x^2 + y^e)^2}$

41. $f_x(x,y) \quad = 25x^4y^7$

$\quad f_{xy}(x,y) \quad = 175x^4y^6$

$\quad f_y(x,y) \quad = 35x^5y^6$

$\quad f_{yx}(x,y) \quad = 175x^4y^6 = f_{xy}(x,y)$

43. $f_x(x,y) \quad = 6xe^{(3x^2+5y^3)}$

$\quad f_{xy}(x,y) \quad = 6x(15y^2)e^{(3x^2+5y^3)} = 90xy^2e^{(3x^2+5y^3)}$

$\quad f_y(x,y) \quad = 15y^2e^{(3x^2+5y^3)}$

$\quad f_{yx}(x,y) \quad = 15y^2(6x)e^{(3x^2+5y^3)} = 90xy^2e^{(3x^2+5y^3)} = f_{xy}(x,y)$

45. $f_x(x,y) \quad = 2xe^{5y}$

$\quad f_{xy}(x,y) \quad = 10xe^{5y}$

$\quad f_y(x,y) \quad = 5x^2e^{5y}$

$\quad f_{yx}(x,y) \quad = 10xe^{5y} = f_{xy}(x,y)$

47. $f_x(x,y) \quad = 2xe^{xy} + (x^2 + y^3)ye^{xy}$

$\qquad\qquad\quad = (2x + x^2y + y^4)e^{xy}$

$\quad f_{xy}(x,y) \quad = (x^2 + 4y^3)e^{xy} + (2x + x^2y + y^4)xe^{xy}$

$\qquad\qquad\quad = (3x^2 + 4y^3 + x^3y + xy^4)e^{xy}$

$\quad f_y(x,y) \quad = 3y^2e^{xy} + (x^2 + y^3)xe^{xy}$

$\qquad\qquad\quad = (3y^2 + x^3 + xy^3)e^{xy}$

$\quad f_{yx}(x,y) \quad = (3x^2 + y^3)e^{xy} + (3y^2 + x^3 + xy^3)ye^{xy}$

$\qquad\qquad\quad = (3x^2 + 4y^3 + yx^3 + xy^4)e^{xy} = f_{xy}(x,y)$

49. $f(x,y) \quad = e^{-5xy}\ln(x^2 + 6y^8)$

$\quad f_x(x,y) \quad = -5ye^{-5xy}\ln(x^2 + 6y^8) + \dfrac{2xe^{-5xy}}{x^2 + 6y^8}$

$\quad f_y(x,y) \quad = -5xe^{-5xy}\ln(x^2 + 6y^8) + \dfrac{48y^7e^{-5xy}}{x^2 + 6y^8}$

$\quad f_{xy}(x,y) \quad = e^{-5xy}\Big(-5\ln(x^2 + 6y^8) + 25xy\ln(x^2 + 6y^8)$

$\qquad\qquad - \dfrac{240y^8 + 10x^2}{x^2 + 6y^8} - \dfrac{96xy^7}{(x^2 + 6y^8)^2}\Big) = f_{yx}(x,y)$

EXERCISE SET 6.3 CHAIN RULE

1. (a) We substitute $76.2 + .3t$ for p_x and $22.2 + 1.2\sqrt{t}$ for p_y to obtain
$$q_x = 30{,}000 - 430\sqrt{76.2 + .3t} - 350 \ \sqrt[3]{22.2 + 1.2\sqrt{t}}.$$

(b) When $t = 16$,
$$q_x = 30{,}000 - 430\sqrt{76.2 + .3(16)} - 350 \ \sqrt[3]{22.2 + 1.2\sqrt{16}}$$
$$= 30{,}000 - 430(9) - 350(3) = 25{,}080 \text{ units.}$$

3. (a) Replacing p_A by $20 + .2t$ and p_B by $14.75 + .002t^2$, we have
$$q_A = 500 - 12\sqrt{20 + .2t} - 123 \ \sqrt[4]{14.75 + .002t^2}$$

(b) When $t = 25$,
$$q_A = 500 - 12\sqrt{20 + 5} - 123 \ \sqrt[4]{14.75 + 1.25}$$
$$= 500 - 60 - 123(2) = 194 \text{ bottles}$$

5. (a) $z = 4x^3y^5 = 4(e^{-2t})^3(e^{3t})^5 = 4e^{-6t}e^{15t} = 4e^{9t}$
$$\frac{dz}{dt} = 4D_t(9t)e^{9t} = 36e^{9t}$$

(b) $\dfrac{\partial z}{\partial x} = 12x^2y^5$ $\qquad\qquad$ $\dfrac{\partial z}{\partial y} = 20x^3y^4$

$\dfrac{dx}{dt} = -2e^{-2t}$ $\qquad\qquad$ $\dfrac{dy}{dt} = 3e^{3t}$

$\dfrac{dz}{dt} = 12x^2y^5(-2e^{-2t} + 60x^3y^4e^{3t}$

$\qquad = -24(e^{-2t})^2(e^{3t})^5e^{-2t} + 60(e^{-2t})^3(e^{3t})^4e^{3t}$

$\qquad = -24e^{-4t+15t-2t} + 60^{-6t+12t+3t} = 36e^{9t}$

7. (a) $z = \ln(x^2 + y^4) = \ln\left((2^t)^2 + (t^3)^4\right) = \ln(2^{2t} + t^{12})$
$$\frac{dz}{dt} = \frac{2(\ln 2)2^{2t} + 12t^{11}}{2^{2t} + t^{12}}$$

(b) $\dfrac{\partial z}{\partial x} = \dfrac{2x}{x^2 + y^4}$ $\qquad\qquad$ $\dfrac{\partial z}{\partial y} = \dfrac{4y^3}{x^2 + y^4}$

$\dfrac{dx}{dt} = (\ln 2)2^t$ $\qquad\qquad$ $\dfrac{dy}{dt} = 3t^2$

$\dfrac{dz}{dt} = \dfrac{2x}{x^2 + y^2}(\ln 2)(2^t) + \dfrac{4y^3}{x^2 + y^4}(3t^2)$

$\qquad = \dfrac{1}{(2^t)^2 + (t^3)^4}\left((2 \cdot 2^t(\ln 2)(2^t) + 4(t^3)^3(3t^2)\right)$

$\qquad = (2(2^{2t})\ln 2 + 12t^{11})/(2^{2t} + t^{12})$

9. $\dfrac{\partial z}{\partial x}=\dfrac{\partial z}{\partial u}\cdot\dfrac{\partial u}{\partial x}+\dfrac{\partial z}{\partial v}\cdot\dfrac{\partial v}{\partial x}$

$$=\frac{2u}{u^2+5v^2}(2x)+\frac{10v}{u^2+5v^2}(2xye^{x^2y})$$

$$=\frac{4ux+20xyve^{x^2y}}{u^2+5v^2}=\frac{4x(x^2+3y^6)+20xye^{2x^2y}}{(x^2+3y^6)^2+5e^{2x^2y}}$$

$\dfrac{\partial z}{\partial y}=\dfrac{\partial z}{\partial u}\cdot\dfrac{\partial u}{\partial y}+\dfrac{\partial z}{\partial v}\cdot\dfrac{\partial v}{\partial y}$

$$=\frac{2u}{u^2+5v^2}(18y^5)+\frac{10v}{u^2+5v^2}x^2e^{x^2y}$$

$$=36\left((x^2+3y^6)y^5+10x^2e^{2x^2y}\right)\bigg/\left((x^2+3y^6)^2+5e^{2x^2y}\right)$$

11. $\dfrac{\partial z}{\partial u}=10ue^{uv}+5u^2ve^{uv}$ $\qquad\qquad\dfrac{\partial u}{\partial x}=\dfrac{2x}{x^2+y^2}$

$\dfrac{\partial z}{\partial v}=5u^3e^{uv}$ $\qquad\qquad\qquad\qquad\dfrac{\partial v}{\partial x}=12xy^5$

$\dfrac{\partial u}{\partial y}=\dfrac{2y}{x^2+y^2}$ $\qquad\qquad\qquad\dfrac{\partial v}{\partial y}=30x^2y^4$

$\dfrac{\partial z}{\partial x}=(10ue^{uv}+5u^2ve^{uv})(2x)/(x^2+y^2)+5u^3e^{uv}(12xy^5)$

$\qquad=10e^{uv}u(2+uv)x/(x^2+y^2)+60u^3xy^5e^{uv}$

$\dfrac{\partial z}{\partial y}=5ue^{uv}(2+uv)(2y)/(x^2+y^2)+5u^3e^{uv}(30x^2y^4)$

Finally, replace u by $\ln(x^2+y^2)$ and v by $6x^2y^5$

13. $\dfrac{\partial z}{\partial u}=15u^2v^2w^4e^{5u^3v^2w^4}$

$\dfrac{\partial z}{\partial v}=10vu^3w^4e^{5u^3v^2w^4}$

$\dfrac{\partial z}{\partial w}=20w^3u^3v^2e^{5u^3v^2w^4}$

$\dfrac{\partial u}{\partial x}=3x^2\qquad\dfrac{\partial u}{\partial y}=6\qquad\qquad\dfrac{\partial v}{\partial x}=10x$

$\dfrac{\partial v}{\partial y}=12y\qquad\quad\dfrac{\partial w}{\partial x}=30x^4\qquad\dfrac{\partial w}{\partial y}=12y^3$

$\dfrac{\partial z}{\partial x}=15u^2v^2w^4e^{5u^3v^2w^4}(3x^2)+10vu^3w^4e^{5u^3v^2w^4}(10x)+$

$\qquad\qquad 20w^3u^3v^2e^{5u^3v^2w^4}(30x^5)$

$\qquad=5u^2vw^3e^{5u^3v^2w^4}(9x^2vw+20uwx+120uvx^5)$

$\dfrac{\partial z}{\partial y}=15u^2v^2w^4e^{5u^3v^2w^4}(6)+10vu^3w^4e^{5u^3v^2w^4}(12y)+$

$\qquad\qquad 20w^3u^3v^2e^{5u^3v^2w^4}(12y^3)$

$\qquad=u^2vw^3e^{5u^3v^2w^4}(90vw+120uwy+240uvy^3)$

Now replace u by x^3+6y, v by $5x^2+6y^2$ and w by $5x^6+3y^4$

15. $\quad\dfrac{\partial z}{\partial r}=\dfrac{r}{\sqrt{r^2+s^2+t^4}},\qquad \dfrac{\partial z}{\partial s}=\dfrac{s}{\sqrt{r^2+s^2+t^4}},\qquad \dfrac{\partial z}{\partial t}=\dfrac{2t^3}{\sqrt{r^2+s^2+t^4}}$

$$\dfrac{\partial r}{\partial x}=ye^{xy}\qquad\qquad \dfrac{\partial s}{\partial x}=2xy^3 \qquad\qquad \dfrac{\partial t}{\partial x}=1$$

$$\dfrac{\partial r}{\partial y}=xe^{xy}\qquad\qquad \dfrac{\partial s}{\partial y}=3x^2y^2 \qquad\qquad \dfrac{\partial t}{\partial y}=3$$

$$\dfrac{\partial z}{\partial x}=\dfrac{rye^{xy}}{\sqrt{r^2+s^2+t^4}}+\dfrac{2xy^3s}{\sqrt{r^2+s^2+t^4}}+\dfrac{2t^3}{\sqrt{r^2+s^2+t^4}}$$

$$=(rye^{xy}+3xy^3s+2t^3)/\sqrt{r^2+s^2+t^4}$$

$$\dfrac{\partial z}{\partial y}=(rxe^{xy}+3x^2y^2s+6t^3)/\sqrt{r^2+s^2+t^4}$$

Now replace r by e^{xy}, s by x^2y^3 and t by $x+3y$.

17. $\quad\dfrac{\partial C}{\partial p_X}=\dfrac{\partial C}{\partial q_X}\cdot\dfrac{\partial q_X}{\partial p_X}+\dfrac{\partial C}{\partial q_Y}\cdot\dfrac{\partial q_Y}{\partial p_X}$

$$=(16+.00003q_Y^2)(-15)+.00006q_Xq_Y+17)(-.06p_X)$$

When $p_X=15$ and $p_Y=36$, $q_X=5000-225-26(36)=3839$

$q_Y=4500-(.03)(225)-540(6)=1253.25$, and the total cost is changing at the rate of $(16+.00003(1253.25)^2)(-15)+(.00006(3839)(1253.25)+17)(-.06)(15)$
$=-\$1221.89$ per dollar

19. $\quad\dfrac{\partial C}{\partial p_X}=\dfrac{\partial C}{\partial q_X}\cdot\dfrac{\partial q_X}{\partial p_X}+\dfrac{\partial C}{\partial q_Y}\cdot\dfrac{\partial q_Y}{\partial p_X}=$
$$\dfrac{175}{\sqrt{q_X}}(-10)+.04q_Y\Big(\dfrac{-50}{\sqrt{p_X}}\Big).$$

When $p_X=25$ and $p_Y=32$,
$q_X=2970-250-35(32)=1600$
and $q_Y=1950-100(5)-.25(32)^2=1194$.
The total cost is changing at the rate of

$$\dfrac{-1750}{40}-\dfrac{.04(1194)(50)}{40}=-\$103.45 \text{ per dollar}$$

21.

$$\dfrac{dx}{dt}=-88\text{ft/sec}$$

$$\dfrac{dy}{dt}=73\text{ ft/sec}$$

where x and y are the distances of A and B from the intersection, respectively. l, the distance between A and B, is given by

$$l = \sqrt{x^2 + y^2}$$

$$\frac{dl}{dt} = \frac{\partial l}{\partial x} \cdot \frac{dx}{dt} + \frac{\partial l}{\partial y} \cdot \frac{dy}{dt} = \frac{x}{\sqrt{x^2 + y^2}}(-88) + \frac{y}{\sqrt{x^2 + y^2}}(73).$$

When $x = 4000$ and $y = 5000$

$$\frac{dl}{dt} = (4000(-88) + 5000(73))/\sqrt{41,000,000}$$

$$= 13,000/6403 = 2.03$$

The distance between the cars is increasing at the rate of approximately 2.03 ft/sec.

23. Let $F(x,y) = 7x^2 y^3 + e^x + 3 - 4x^3 - e^{xy} = 0$

$$\frac{dy}{dx} = \frac{-\dfrac{\partial F(x,y)}{\partial x}}{\dfrac{\partial F(x,y)}{\partial y}} = \frac{-(14xy^3 + e^x - 12x^2 - ye^{xy})}{21x^2 y^2 - xe^{xy}}$$

$$= \frac{12x^2 + ye^{xy} - 14xy^3 - e^x}{21x^2 y^2 - xe^{xy}}$$

25. Let $F(x,y,z) = x^2 + y^2 + z^2 - 36 = 0$

$$\frac{\partial z}{\partial x} = \frac{-\partial x}{\partial z} = \frac{-x}{z}$$

$$\frac{\partial z}{\partial y} = \frac{-\partial y}{\partial z} = \frac{-y}{z}$$

27. Let $F(x,y,z) = e^{xy} + e^{xz} + e^{yz} - 50$

$$\frac{\partial z}{\partial x} = \frac{-(ye^{xy} + ze^{yz})}{xe^{xz} + ye^{yz}}$$

$$\frac{\partial z}{\partial y} = \frac{-xe^{xy} - ze^{yz}}{xe^{xz} + ye^{yz}}$$

29. $u = F(x,y,z) = 0.$

$$\frac{\partial u}{\partial x} = \frac{\partial F(x,y,z)}{\partial x} \cdot \frac{\partial x}{\partial x} + \frac{\partial F(x,y,z)}{\partial y} \cdot \frac{\partial y}{\partial x} + \frac{\partial F(x,y,z)}{\partial z} \cdot \frac{\partial z}{\partial x}$$

$$\frac{\partial F(x,y,z)}{\partial x} \cdot 1 + \frac{\partial F(x,y,z)}{\partial y} \cdot 0 + \frac{\partial F(x,y,z)}{\partial z} \cdot \frac{\partial z}{\partial x} = 0$$

$$\frac{\partial F(x,y,z)}{\partial z} \cdot \frac{\partial z}{\partial x} = -\frac{\partial F(x,y,z)}{\partial x} \quad \text{and} \quad \frac{\partial z}{\partial x} = \frac{-\dfrac{\partial F(x,y,z)}{\partial x}}{\dfrac{\partial F(x,y,z)}{\partial z}}$$

Similarly,

$$\frac{\partial u}{\partial y} = \frac{\partial F(x,y,z)}{\partial x} \cdot \frac{\partial x}{\partial y} + \frac{\partial F(x,y,z)}{\partial y} \cdot \frac{\partial y}{\partial y} + \frac{\partial F(x,y,z)}{\partial z} \cdot \frac{\partial z}{\partial y} = 0$$

$$\frac{\partial F x,y,z)}{\partial x} \cdot 0 + \frac{\partial F(x,y,z)}{\partial y} \cdot 1 + \frac{\partial F(x,y,z)}{\partial z} \cdot \frac{\partial z}{\partial y} = 0$$

and $\dfrac{\partial z}{\partial y} = \dfrac{-\dfrac{\partial F(x,y,z)}{\partial y}}{\dfrac{\partial F(x,y,z)}{\partial z}}$

EXERCISE SET 6.4 APPLICATIONS OF PARTIAL DIFFERENTIATION

1. (a) $f(4,3) = \sqrt{16+9} = \sqrt{25} = 5$

   (b) $f_x(x,y) = \dfrac{x}{\sqrt{x^2+y^2}}$, $f_x(4,3) = 4/5$

   $f_y(x,y) = \dfrac{y}{\sqrt{x^2+y^2}}$, $f_y(4,3) = 3/5$

   $\triangle x = .1$, $\triangle y = -.2$

   $f(4.1,2.8) \doteq .8(.1) + (.6)(-.2) + 5 = 4.96$

3. (a) $h(27,4) = 9/2$

   (b) $h_x(x,y) = \dfrac{2}{3\sqrt[3]{x}\sqrt{y}}$, $h_x(27,4) = \dfrac{2}{18} = 1/9$

   $h_y(x,y) = \dfrac{-1}{2}x^{2/3}y^{-3/2}$, $h_y(27,4) = -\dfrac{9}{2}\cdot\dfrac{1}{8} = -\dfrac{9}{16}$

   $\triangle x = -1$, $\triangle y = .5$

   $h(26,4.5) \doteq -\dfrac{1}{9} - \dfrac{9}{32} + \dfrac{9}{2}$ or 4.1

5. (a) $g(13,5) = \sqrt{169-25} = 12$

   (b) $g_x(x,y) = \dfrac{x}{\sqrt{x^2-y^2}}$, $g_x(13,5) = \dfrac{13}{12}$

   $g_y(x,y) = \dfrac{-y}{\sqrt{x^2-y^2}}$, $g_y(13,5) = -5/12$

   $\triangle x = -.1$, $\triangle y = -.8$

   $g(12.9,4.2) \doteq \dfrac{-1.3}{12} + \dfrac{1}{3} + 12 = 12.225$

7. (a) $L(4,2.25) = 6000 - 400(2) + 800(1.5) = 6,400$ pounds

   (b) $\triangle p_L = -.05$, $\triangle p_P = .05$

   $L_1(p_L,p_P) = -\dfrac{200}{\sqrt{p_L}}$ $L_2(p_L,p_P) = \dfrac{400}{\sqrt{p_P}}$

   $L(3.95,2.30) \doteq \dfrac{-200}{2}(-.05) + \dfrac{400}{1.5}(.05) + 6400 = 6418.33$ pounds

9. (a) $Q = 250\sqrt{900}\sqrt{8100} = 675,000$ units

(b) $\dfrac{2Q}{2K} = 125K^{-1/2}L^{1/2}$ $\qquad$ $\dfrac{2Q}{2L} = 125K^{1/2}L^{-1/2}$

$\Delta Q = -2, \; \Delta L = 40$

When $Q = 898$ and $L = 8140$,

$Q \doteq 125(3)(-2) + \dfrac{125}{90}\sqrt{900}(40) + 675,000 = 675,917$

11. (a) $P_1(p_A,p_B) = -8p_A + 8p_B + 57$

$P_1(10,8) = -80 + 64 + 57 = 41$

When the price of brand B is held fixed at \$8 per pound, the daily profit increases at the rate of \$41 per dollar increase in the price of brand A, when the price of brand A is \$10.

(b) $P_2(p_A,p_B) = 8p_A - 12p_B + 88$
$P_2(10,8) = 80 - 96 + 88 = 72$

When the price of brand A is held fixed at \$10 per pound, the daily profit increases at the rate of \$72 per dollar increase in the price of brand B, when the price of brand B is \$8.

13. $f(5.2,6.9) \doteq f_{p_A}(5,7)(.2) + f_{p_B}(5,7)(-.1) + 8000$

$= 120(.2) + 180(-.1) + 8000 = 8006$ units

15. $\dfrac{\partial D_A}{\partial p_A} = -10p_A$ $\qquad\qquad$ $\dfrac{\partial D_A}{\partial p_B} = 8/\sqrt{p_B}$

$\dfrac{\partial D_B}{\partial p_A} = 12$ $\qquad\qquad$ $\dfrac{\partial D_B}{\partial p_B} = -\dfrac{15}{2\sqrt{p_B}}$

When $p_A = 13$ and $p_B = 16$

$\dfrac{\partial D_A}{\partial p_A} = -130, \;\; \dfrac{\partial D_A}{\partial p_B} = 2, \;\; \dfrac{\partial D_B}{\partial p_A} = 12, \;\; \dfrac{\partial D_B}{\partial p_B} = -\dfrac{15}{8}$

When the price of brand B is held fixed at \$16 per gallon, the demand for brand A decreases by 130 gallons, and the demand for brand B increases by 12 gallons, approximately, as the price of a gallon of brand A increases from \$13 to \$14.

When the price of brand A is held fixed at \$13 per gallon, the demand for brand A increases by 2 gallons and the demand for brand B decreases by 15/8 gallons, approximately, as the price of a gallon of brand B increases from \$16 to \$17.

17. $\dfrac{\partial D_x}{\partial p_x} = -5, \;\; \dfrac{\partial D_x}{\partial p_y} = 4, \;\; \dfrac{\partial D_y}{\partial p_y} = -5, \;\; \dfrac{\partial D_y}{\partial p_x} = 6$

The commodities are substitutes since $\dfrac{\partial D_x}{\partial p_y} > 0$ and $\dfrac{\partial D_y}{\partial p_x} > 0.$

19. $\dfrac{\partial D_x}{\partial p_x} = -6(p_x)^{1/2}$ $\qquad\qquad$ $\dfrac{\partial D_x}{\partial p_y} = -10p_y$

$\dfrac{\partial D_y}{\partial p_x} = -8p_x$ $\qquad\qquad$ $\dfrac{\partial D_y}{\partial p_y} = -\dfrac{35}{2}p_y^{3/2}$

Since p_x and p_y are positive, $-10p_y$ and $-8p_x$ are negative and the commodities are complements.

21. $\dfrac{\partial D_x}{\partial p_x} = -14p_x^{4/3}$ $\qquad\qquad$ $\dfrac{\partial D_y}{\partial p_y} = 20p_y^{3/2}$

$\dfrac{\partial D_x}{\partial p_x} = 5p_x^{2/3}$ $\qquad\qquad$ $\dfrac{\partial D_y}{\partial p_y} = -6_y^{2}$

The commodities are substitutes.

23. $\dfrac{\partial D_x}{\partial p_x} = -1/p_x$ $\qquad\qquad$ $\dfrac{\partial D_x}{\partial p_y} = 6p_y$

$\dfrac{\partial D_y}{\partial p_x} = \dfrac{1}{p_x+1}$ $\qquad\qquad$ $\dfrac{\partial D_y}{\partial p_y} = -6p^2y$

The commodities are substitutes.

25. $\dfrac{\partial D_x}{\partial p_x} = \dfrac{-150y}{2p_x^{3/2}}$ $\qquad\qquad$ $\dfrac{\partial D_x}{\partial p_y} = \dfrac{150}{p_x^{1/2}}$

$\dfrac{\partial D_y}{\partial p_x} = \dfrac{175}{p_y^{1/4}}$ $\qquad\qquad$ $\dfrac{\partial D_y}{\partial p_y} = \dfrac{-175p_x}{4p_y^{5/4}}$

The commodities are substitutes.

27. (a) $\dfrac{\partial D_X}{\partial p_X} = -3.$ $\quad D_X(6,4) = 450 - 18 + 20 = 452$

$\eta_{XX} = -3(\frac{6}{452}) = .039823.$

When the price of commodity Y is held fixed at \$4 and the price of X is \$6, the demand for commodity X decreases by approximately .040% as the price of commodity X increases by 1%.

(b) $\dfrac{\partial D_Y}{\partial p_Y} = -4.$ $\quad D_Y(6,4) = 600 + 30 - 16 = 614.$ $\quad \eta_{YY} = -4(\frac{4}{614}) = -.026$

When the price of X is \$6 and is held fixed, and the price of Y is \$4, the demand for Y decreases by approximately .026% as the price of Y increases by 1%.

(c) $\eta_{XY} = \dfrac{\partial D_X}{\partial p_Y} \cdot \dfrac{p_Y}{D_X} = \dfrac{5(4)}{452} = .044$

When the price of X is fixed at \$6 and the price of Y is \$4, an increases of 1% in the price of Y causes an increase of about .044% in the demand for X.

(d) $\eta_{YX} = \dfrac{\partial D_Y}{\partial p_X} \cdot \dfrac{p_X}{D_Y} = \dfrac{5(6)}{614} = .049$

When the price of Y is fixed at \$4 and the price of X is \$6, an increase of 1% in the price of X will cause an approximate .049% increase in the demand for Y.

29. (a) $\eta_{XX} = \dfrac{-6p_X^{3/2}}{3176} = \dfrac{-6(27)}{3176} = -.051$

When the price of Y s fixed at \$6, and the price of X is \$9, an increase of 1% in the price of X causes a decrease of approximately .051% in the demand for X.

(b) $\eta_{YY} = \dfrac{-12p_Y^2}{2812} = \dfrac{-12(36)}{2812} = -.154$

When the price of X is fixed at \$9 and the price of Y is \$6, an increase of 1% in the price of Y causes an approximate .154% decrease in the demand for Y.

(c) $\eta_{XY} = \dfrac{-12p_Y^2}{3176} = \dfrac{-12(36)}{3176} = -.136$

When the price of X is fixed at \$9 and the price of Y is \$6, an increase of 1% in the price of Y causes an approximate .136% decrease in the demand for X.

(d) $\eta_{YX} = \dfrac{-10(27)(9)}{2812} = -.864$

When the price of Y is fixed at \$6 and the price of X is \$9, an increase of 1% in the price of X causes an approximate .864% decrease in the demand for Y.

31. (a) $\eta_{XX} = \dfrac{-14(8)^{7/3}}{15924} = -.113$

When the price of Y is fixed at \$16 and the price of X is \$8, an increase of 1% in the price of X causes an approximate .113% decrease in the demand for X.

(b) $\eta_{YY} = \dfrac{-6(16)^{3/2}}{8840} = -.043$

When the price of X is fixed at \$8 and the price of Y is \$16, an increase of 1% in the price of Y causes an approximate .043% decrease in the demand for Y.

(c) $\eta_{XY} = \dfrac{20(16)^{5/2}}{15924} = 1.286$

When the price of X is fixed at \$8 and the price of Y is \$16, an increase of 1% in the price of Y causes an approximate 1.286% increase in the demand for X.

(d) $\eta_{YX} = \dfrac{5(8)^{5/3}}{8840} = .018$

When the price of Y is fixed at \$16 and the price of X is \$8, an increase of 1% in the price of X causes an approximate .018% increase in the demand for Y.

33. $\eta_{XX} = \dfrac{-50p_Y}{(p_X)^{1/2}(900)} = \dfrac{-50(27)}{3(900)} = -.5$

When the price of Y is fixed at \$27 and the price of X is \$9, and increase of 1% in the price of X causes a decrease of approximately .5% in the demand for X.

$$\eta_{YY} = \frac{-40p_X}{(p_Y)^{1/3}(360)} = \frac{-40(9)}{3(360)} = -.333$$

When the price of X is fixed at \$9 and the price of Y is \$27, an increase of 1% in the price of Y will cause an approximate .333% decrease in the demand for Y.

$$\eta_{XY} = \frac{100p_Y}{(p_X)^{1/2}(900)} = \frac{100(27)}{3(900)} = 1$$

When the price of X is fixed at \$9 and the price of Y is \$27, an increase of 1% in the price of Y will cause an approximate 1% decrease in the demand for X.

$$\eta_{YX} = \frac{120p_X}{(p_Y)^{1/3})(360)} = \frac{120(9)}{3(360)} = 1$$

When the price of Y is fixed at \$27 and the price of X is \$9, an increase of 1% in the price of X will cause an approximate 1% increase in the demand for Y.

EXERCISE SET 6.5 OPTIMIZATION

1. $f(x,y) \quad = x^2 + y^2 - 25$

   $f_x(x,y) \quad = 2x = 0$ if $x = 0$

   $f_y(x,y) \quad = 2y = 0$ if $y = 0$

   $(0,0)$ is the critical point

   $f_{xx}(x,y) = 2$

   $f_{yy}(x,y) = 2, \quad f_{xy}(x,y) = 0$

   $A = 2, \quad B = 2, \quad C = 0, \quad AB - C^2 = 4$, There is a local minimum at $(0,0)$

3. $f_x(x,y) = 2x - 2 = 0$ if $x = 1$

   $f_y(x,y) = 2y - 6 = 0$ if $y = 3$

   $(1,3)$ is the critical point

   $f_{xx}(x,y) = 2$

   $f_{yy}(x,y) = 2$

   $f_{xy}(x,y) = 0$

   $A = 2, \quad B = 2, \quad C = 0, \quad D = 4$. there is a local minimum at $(1,3)$

5. $f_x(x,y) = 2x - 8 = 0$ if $x = 4$

   $f_y(x,y) = -2y + 6 = 0$ if $y = 3$

   $(4,3)$ is the critical point

   $f_{xx}(x,y) = 2$

   $f_{yy}(x,y) = -2$

   $D = -4$. There is a saddle point when $x = 4$ and $y = 3$.

7. $f_x(x,y) = 2x - 5y + 3$

$f_y(x,y) = -5x + 12y - 6$

solving $\begin{matrix} 2x - 5y + 3 = 0 \\ -5x + 12y - 6 = 0 \end{matrix}\Big\} \rightarrow \begin{cases} 10x - 25y + 15 = 0 \\ -10x + 24y - 12 = 0 \end{cases}\quad \begin{matrix} y = 3 \\ x = 6 \end{matrix}$

The critical point is (6,3).

$f_{xx}(x,y) = 2,\ f_{yy}(x,y) = 12,\ f_{xy}(x,y) = -5$

$D = AB - C^2 = 24 - 25 < 0.$ There is a saddle point when $x = 6$ and $y = 3$.

9. $f_x(x,y) = 4x - 9y + 7$

$f_y(x,y) = -9x + 20y - 5$

Solving the system

$\begin{cases} 4x - 9y = -7 \\ -9x + 20y = 5 \end{cases} \rightarrow \begin{cases} 36x - 81y = -63 \\ 36x + 80y = 20 \end{cases}\quad \begin{matrix} y = 43 \\ x = 95 \end{matrix}$

The critical point is (95,43).

$f_{xx}(x,y) = 4,\ f_{yy}(x,y) = 20,\ f_{xy}(x,y) = 20,\ f_{xy}(x,y) = -9$

$AB - C^2 = 80 - 81 < 0.$ There is a saddle point when $x = 95$ and $y = 43$.

11. $f_x(x,y) = 2x + 7y - 7$

$f_y(x,y) = 7x + 24y + 5$

Solving the system

$\begin{cases} 2x + 7y = 7 \\ 7x + 24y = -5 \end{cases} \rightarrow \begin{cases} 14x + 49y = 49 \\ -14x - 48y = 10 \end{cases}\quad \begin{matrix} y = 59 \\ x = -203 \end{matrix}$

The critical point is $(-203, 59)$.

$f_{xx}(x,y) = 2,\ f_{yy}(x,y) = 24,\ f_{xy} = 7.$

$AB - C^2 = 48 - 49 < 0.$ There is a saddle point when $x = -203$ and $y = 59$.

13. $f_x(x,y) = 2x - 4 = 0$ if $x = 2$

$f_y(x,y) = 3y^2 + 9 \neq 0.$

No critical points.

15. $f_x(x,y) = 6x^2 + 6x - 12 = 6(x^2 + x - 2) = 6(x + 2)(x + 1) = 0$ if $x = 1$ or -2

$f_y(x,y) = 2y - 6 = 0$ if $y = 3$

The critcal points are (1,3) and $(-2,3)$.

$f_{xy}(x,y) = 12x + 6$

$f_{yy}(x,y) = 2,\ f_{xy}(x,y) = 0$

Point	A	B	C	D	conclusion
(1,3)	18	2	0	+	local minimum
$(-2,3)$	-18	2	0	$-$	saddle point

17. $f_x(x,y) = 2x + 6 = 0$ if $x = -3$

$f_y(x,y) = 3y^2 + 12y - 15 = 3(y^2 + 4y - 5) = 3(y+5)(y-1) = 0$ if $y = 1$ or -5

The critical points are $(-3,1)$ and $(-3,-5)$

$f_{xx}(x,y) = 2$, $f_{yy}(x,y) = 6y + 12$, $f_{xy}(x,y) = 0$

Point	A	B	C	D	conclusion
$(-3,1)$	2	18	0	+	local minimum
$(-3,-5)$	2	-18	0	-	saddle point

19. $f_x(x,y) = 3x^2 + 12x - 15 = 3(x^2 + 4x - 5) = 3(x+5)(x-1) = 0$ if $x = 1$ or -5

$f_y(x,y) = 3y^2 - 6y - 9 = 3(y^2 - 2y - 3) = 3(y-3)(y+1) = 0$ if $y = 3$ or -1

Critical points on $(1,3)$, $(1,-1)$, $(-5,3)$, $(-5,-1)$.

$f_{xx}(x,y) = 6x + 12$, $f_{yy}(x,y) = 6y - 6$, $f_{xy}(x,y) = 0$

Point	A	B	C	D	conclusion
$(1,3)$	18	12	0	+	local minimum
$(1,-1)$	18	-12	0	-	saddle point
$(-5,3)$	-18	12	0	-	saddle point
$(-5,-1)$	-18	-12	0	+	local maximum

21. $f_x(x,y) = 2x - 4y$

$f_y(x,y) = 3y^2 - 4x + 4$

Solving the system

$$\begin{cases} 2x - 4y = 0 \\ 3y^2 - 4x + 4 = 0 \end{cases} \rightarrow \begin{cases} x = 2y \\ 3y^2 - 4x + 4 = 0 \end{cases} \rightarrow \begin{cases} 3y^2 - 8y + 4 = 0 \\ (3y-2)(y-2) = 0 \end{cases}$$

$y = 2/3$ or $y = 2$. If $y = 2$, $x = 4$. If $y = 2/3$, $x = 4/3$.

The critical points are $(4,2)$ and $(4/3,2/3)$.

$f_{xx}(x,y) = 2$, $f_{yy}(x,y) = 6y$, $f_{xy}(x,y) = -4$.

Point	A	B	C	D	conclusion
$(4,2)$	2	12	-4	+	local minimum
$(4/3,2/3)$	2	4	-4	-	saddle point

23. $f_x(x,y) = 2x - 6y$

$f_y(x,y) = 3y^2 - 6x + 24$

Solving the system

$$\begin{cases} 2x - 6y = 0 \\ 3y^2 - 6x + 24 = 0 \end{cases} \rightarrow \begin{cases} x = 3y \\ y^2 - 2x + 8 = 0 \end{cases} \quad \begin{array}{l} y^2 - 6y + 8 = 0 \\ (y-4)(y-2) = 0, \end{array}$$

$y = 2$ or $y = 4$.

If $y = 2$, $x = 6$. If $y = 4$, $x = 12$.

The critical points are $(6,2)$ and $(12,4)$.

$f_{xx}(x,y) = 2$, $f_{yy}(x,y) = 6y$, $f_{xy}(x,y) = -6$.

Point	A	B	C	D	conclusion
$(6,2)$	2	12	-6	$-$	saddle point
$(12,4)$	2	24	-6	$+$	local minimum

25. $f_x(x,y) = 18xy - 6y^2 + 12x - 144$.

$f_y(x,y) = 3y^2 + 9x^2 - 12xy = 3(y^2 - 4xy + 3x^2) = 3(y-x)(y-3x)$

If $x = y$, $18x^2 - 6x^2 + 12x - 144 = 12x^2 + 12x - 144 = 12(x^2 + x - 12) = 12(x+4)(x-3) = 0$

if $x = 3$ or $x = -4$.

If $y = 3x$, $54x^2 - 54x^2 + 12x - 144 = 0$ if $x = 12$.

The critical points are $(3,3)$, $(-4,-4)$ and $(12,36)$.

$f_{xx}(x,y) = 18y + 12$, $f_{yy}(x,y) = 6y - 12x$, $f_{xy}(x,y) = 18x - 12y$

Point	A	B	C	D	conclusion
$(3,3)$	66	-18		$-$	saddle point
$(-4,-4)$	-60	24		$-$	saddle point
$(12,36)$	660	72	-216	$+$	local minimum

27. $f_x(x,y) = y - \dfrac{8}{x^2} = 0$ if $y = \dfrac{8}{x^2}$

$f_y(x,y) = x - \dfrac{1}{y^2} = 0$ if $x = 1/y^2$

$x = \dfrac{x^4}{64}$, $64x = x^4$, $x^4 - 64x = x(x^3 - 64) = 0$, $x = 0$ or $x = 4$.

The critical point is $(4,1/2)$.

$f_{xx}(x,y) = 16/x^3$, $f_{yy}(x,y) = 2/y^3$, $f_{xy}(x,y) = 1$

Point	A	B	C	D	conclusion
$(4,1/2)$	1/4	16	1	$+$	local minimum

29. $f_x(x,y) = 2ye^{2xy} = 0$ if $y = 0$

$f_y(x,y) = 2xe^{2xy} = 0$ if $x = 0$.

The critical point is $(0,0)$

$f_{xx}(x,y) = 4y^2 e^{2xy}$

$f_{yy}(x,y) = 4x^2 e^{2xy}$

$f_{xy}(x,y) = 2e^{2xy} + 4xye^{2xy}$

Point	A	B	C	D	conclusion
$(0,0)$	0	0	2	$-$	saddle point

31. $f_x(x,y) = \ln(xy) + \dfrac{(x-2)}{x} = \ln x + \ln y + 1 - 2/x$

$f_y(x,y) = \dfrac{(x-2)}{y} = 0$ if $x = 2$

$\ln(2y) = 0$ if $y = 1/2$.

The critical point is $(2, \frac{1}{2})$

$f_{xx}(x,y) = \frac{1}{x} + 2/x^2, \quad f_{yy}(x,y) = \dfrac{2-x}{y^2}, \quad f_{xy}(x,y) = \frac{1}{y}$

Point	A	B	C	D	conclusion
(2,1/2)	1	0	2	$-$	saddle point

33. $f_x(x,y) = y - 4x = 0$ if $y = 4x$

$f_y(x,y) = x - \dfrac{1}{y}$.

$x - \dfrac{1}{y} = x - \dfrac{1}{4x} = \dfrac{4x^2-1}{4x} = 0$ if $x = \pm 1/2$.

The critical point is $(\frac{1}{2}, 2)$ as $(-\frac{1}{2}, -2)$ is not in the domain

$f_{xx}(x,y) = -4$

$f_{yy}(x,y) = 1/y^2$

$f_{xy}(x,y) = 1$

Point	A	B	C	D	conclusion
(1/2,2)	-4	1/4	1	$-$	saddle point

35. $P(q_A, q_B) = -7q_A{}^2 + 5q_A q_B - q_B{}^2 + 120q_A + 180q_B - 16400$

$\dfrac{\partial P}{\partial q_A} = -14q_A + 5q_B + 120$
$\quad \begin{cases} -28q_A + 10q_B = -240 \\ 25q_A - 10q_B = -900 \end{cases}$
$\quad \begin{aligned} -3q_A &= -1140 \\ q_A &= 380 \\ q_B &= 1040 \end{aligned}$

$\dfrac{\partial P}{\partial q_B} = 5q_A - 2q_B + 180$

The critical point is $(380, 1040)$

$\dfrac{\partial^2 P}{\partial q_A} = -14, \quad \dfrac{\partial^2 P}{\partial q_B} = -2, \quad \dfrac{\partial^2 P}{\partial q_A \partial q_B} = 5$

A	B	C	D
-14	-2	5	3

The maximum occurs when $q_A = 380$ and $q_B = 1040$.

The maximum profit is \$100,000

37. $P(q_S, q_D) = 30q_S + 45q_D - 3q_S^2 + 3q_S q_D - q_D^2 - 1125$

$\dfrac{\partial P}{\partial q_S} = 30 - 6q_S + 3q_D$
$\quad \begin{cases} -4q_S + 2q_D = -20 \\ 3q_S - 2q_D = -45 \end{cases}$
$\quad \begin{aligned} -q_S &= -65 \\ q_S &= 65 \\ q_D &= 120 \end{aligned}$

$\dfrac{\partial P}{\partial q_D} = 45 + 3q_S - 2q_D$

The critical point is $(65,120)$

$$\frac{\partial^2 P}{\partial q_S} = -6, \quad \frac{\partial^2 P}{\partial q_D} = -2, \quad \frac{\partial^2 P}{\partial q_S \partial q_D} = 3$$

A	B	C	D
-6	-2	3	3

The maximum profit occurs when $q_S = 65$ and $q_D = 120$

The maximum profit is \$2,550,000

39. $P(p_A, p_B) = (p_A - 2)(820 - 1800p_A + 1000p_B) + (p_B - 3)(1100 + 1000p_A - 1000p_B)$

$$= 1420p_A + 2100p_B - 1800p_A^2 - 1000p_B^2 + 2000p_A p_B - 4940$$

$$\frac{\partial P}{\partial p_A} = 1420 - 3600p_B + 2000p_B$$

$$\frac{\partial P}{\partial p_B} = 2100 - 2000p_B + 2000p_A$$

The critical point occurs when $p_A = 2.2$ and $p_B = 3.25$

A	B	C	D
-3600	-2000	2000	$+$

The maximum weekly profit occurs when brand A sells for \$2.20 and brand B sells for \$3.25.

The maximum profit is \$22 + \$12.50 = \$34.50

41.

$lwh = 1152$

$$h = \frac{1152}{lw}$$

$$C(l,w) = 8lw + 2(6)hw + 2(6)hl$$

$$= 8lw + \frac{12(1152)}{l} + \frac{12(1152)}{w}$$

$$\frac{\partial C}{\partial l} = 8w - \frac{13824}{l^2} \qquad \frac{\partial C}{\partial w} = 8l - \frac{13824}{w^2}.$$

$$8w = \frac{13824}{l^2} \text{ when } w = \frac{1728}{l^2} \text{ and } 8l - \frac{13824}{w^2} = 8l - \frac{13824 l^4}{(1728)^2}$$

$$= 8l\left(1 - \frac{l^3}{1728}\right) = 0 \text{ if } l = 12.$$

The critical point occurs when $l = 12$ and $w = 12$

$$\frac{\partial^2 C}{\partial l^2} = \frac{13824(2)}{l^3} \qquad \frac{\partial^2 C}{\partial w^2} = \frac{13824(2)}{w^3}$$

A	B	C	D
16	16	8	$+$

The minimum cost occurs when $l = 12$ feet, $w = 12$ feet and $h = 8$ feet.

43. $P(p_A, p_B) = (p_A - 12)(456 - 30p_A + 5p_B) + (p_B - 15)(666 + 10p_A - 40p_B)$

$$= 666p_A - 30p_A^2 + 15p_A p_B - 40p_B^2 + 1206p_B - 15{,}462$$

$\dfrac{\partial P}{\partial p_A} = 666 - 60p_A + 15p_B$

$\dfrac{\partial P}{\partial p_B} = 1206 + 15p_A - 80p_B$

The critical point occurs when $p_A = 15.60$ and $p_B = 18$

$\dfrac{\partial^2 P}{\partial p_A^2} = -60, \quad \dfrac{\partial^2 p}{\partial p_B^2} = -80, \quad \dfrac{\partial^2 P}{\partial p_A^2 p_B} = 15$

$AB - C^2 > 0$. The maximum profit occurs when $p_A = \$15.60$ and $p_B = \$18$

The maximum profit is \$586.80

45. $P(q_A, q_B) = \Big(196 - .05(q_A + q_B)\Big)(q_A + q_B) - .02q_A^2 - 80q_A - .03q^2 - 60q_B - 24{,}000$

$$= -.07q_A^2 - .08q_B^2 - .1q_A q_B + 116q_A + 136q_B - 24{,}000$$

$\dfrac{\partial P}{\partial q_A} = -.14q_A - .10q_B + 116$

$\dfrac{\partial P}{\partial q_B} = -.10q_A - .16q_B + 136$

The critical point occurs when $q_A = 400$ and $q_b = 600$

$AB - C^2 = (.14)(.16) - (.10)(.10) > 0.$

The maximum profit occurs when 400 units are produced at location A and 600 at location

EXERCISE SET 6.6 LAGRANGE MULTIPLIERS

1. $3x + 4y = 25$ and $d(x,y) = \sqrt{x^2 + y^2}$

$F(x,y,\lambda) = \sqrt{x^2 + y^2} - \lambda(3x + 4y - 25)$

$F_x(x,y,\lambda) = \dfrac{x}{\sqrt{x^2 + y^2}} - 3\lambda = 0$

$F_y(x,y,\lambda) = \dfrac{y}{\sqrt{x^2 + y^2}} - 4\lambda = 0$

$F_\lambda(x,y,\lambda) = -(3x + 4y - 25) = 0$

$\dfrac{x}{3\sqrt{x^2 + y^2}} = \dfrac{y}{4\sqrt{x^2 + y^2}} \rightarrow 4x = 3y \text{ or } x = \dfrac{3}{4}y$

$3x + 4y - 25 = \dfrac{9}{4}y + 4y - 25 = 0 \text{ if } \dfrac{25}{4}y = 25 \text{ or } y = 4, \ x = 3 \text{ and } \lambda = \dfrac{1}{5}$

The critical point for d is $(3,4)$ and since a minimum distance must occur, it will be at the point $(3,4)$.

The minimum distance is 5 units.

3. $d = \sqrt{(x+1)^2 + (y-3)^2}$ subject to $x + y = 8$.

$d^2 = (x+1)^2 + (y-3)^2$.

Let $F(x,y,\lambda) = (x+1)^2 + (y-3)^2 - \lambda(x + y - 8)$

$F_x(x,y,\lambda) = 2(x+1) - \lambda = 0$

$F_y(x,y,\lambda) = 2(y-3) - \lambda = 0$

$F_\lambda(x,y,\lambda) = -x - y + 8 = 0$

$2(x+1) = 2(y-3)$

$x = y - 4$

$-x - y + 8 = -y + 4 - y + 8 = 0$

$-2y = -12$

$y = 6$

$x = 2$

The minimum distance is $\sqrt{9 + 9} = \sqrt{18} = 3\sqrt{2}$.

5. $d = \sqrt{x^2 + y^2 + z^2}$ subject to $2x - y + 3z = 14$.

$d^2 = x^2 + y^2 + z^2$

Let $F(x,y,z,\lambda) = x^2 + y^2 + z^2 - \lambda(2x - y - 3z - 14)$

$F_x(x,y,z,\lambda) = 2x - 2\lambda$

$F_y(x,y,z,\lambda) = 2y + \lambda$

$F_z(x,y,z,\lambda) = 2x + 3\lambda$

$F_\lambda(x,y,z,\lambda) = -2x + y + 3z + 14$

The system of equations is

$$
\begin{array}{rcrcl}
x & & & - & \lambda = 0 \\
& 2y & & + & \lambda = 0 \\
& & 2z & + & 3\lambda = 0 \\
-2x + y + 3z & & & = & -14
\end{array}
$$

$$
\left\{
\begin{array}{rcrcl}
x & & & - & \lambda = 0 \\
& 2y & & + & \lambda = 0 \\
& & 2z & + & 3\lambda = 0 \\
-2x + y + 3z & & & = & -14
\end{array}
\right.
\qquad 2E_1 + E_4 \to E_4
$$

$$
\left\{
\begin{array}{rcrcl}
x & & & - & \lambda = 0 \\
& 2y & & + & \lambda = 0 \\
& & 2z & + & 3\lambda = 0 \\
& y + 3z & - & 2\lambda & = -14
\end{array}
\right.
\qquad
\begin{array}{l}
(-1/2)E_2 + E_4 \to E_4 \\
(1/2)E_2 \to E_2
\end{array}
$$

$$
\left\{
\begin{array}{rcrcl}
x & & & - & \lambda = 0 \\
& y & & + & \lambda/2 = 0 \\
& & 2z & + & 3\lambda = 0 \\
& 3z & - & (5/2)\lambda & = -14
\end{array}
\right.
\qquad
\begin{array}{l}
3E_3 - 2E_4 \to E_4 \\
E_3/2 \to E_3
\end{array}
$$

$$
\left\{
\begin{array}{rcrcl}
x & & & - & \lambda = 0 \\
& y & & + & \lambda/2 = 0 \\
& & z & + & 3\lambda/2 = 0 \\
& & 14\lambda & = & 28
\end{array}
\right.
\qquad
\begin{array}{l}
\lambda = 2 \\
x = 2 \\
y = -1 \\
z = -3
\end{array}
$$

The minimum distance is $\sqrt{4 + 1 + 9} = \sqrt{14}$

7. $d = \sqrt{(x+1)^2 + (y-2)^2 + (z-3)^2}$ subject to $x + 2y + 2z = 27$

$d^2 = (x+1)^2 + (y-2)^2 + (z-3)^2$

$F(x,y,z,\lambda) = (x+1)^2 + (y-2)^2 + (z-3)^2 - \lambda(x + 2y + 2z - 27)$

$F_x(x,y,z,\lambda) = 2(x+1) - \lambda = 2x - \lambda + 2 = 0 \to x = \frac{\lambda}{2} - 1$

$F_y(x,y,z,\lambda) = 2(y-2) - 2\lambda = 2y - 2\lambda - 4 = 0 \to y = \lambda + 2$

$F_z(x,y,z,\lambda) = 2(z-3) - 2\lambda = 2z - 2\lambda - 6 = 0 \to z = \lambda + 3$

$F_\lambda(x,y,z,\lambda) = -x - 2y - 2z + 27 = 0$

$-x - 2y - 2z + 27 = \frac{-\lambda}{2} + 1 - 2\lambda - 4 - 2\lambda - 6 + 27 = 0$

$\frac{-9\lambda}{2} = -18$

$\lambda = 4$

$x = 1, \quad y = 6, \quad z = 7$

The minimum distance is $\sqrt{4 + 16 + 16} = 6$

9. Let $f(x,y.z) = xyz$ subject to $x + y + z = K$, $K > 0$, and $x > 0$, $y > 0$, $z > 0$.

$F(x,y,z,\lambda) = xyz - \lambda(x + y + z - K)$

$F_x(x,y,z,\lambda) = yz - \lambda = 0$ (1)

$F_y(x,y,z,\lambda) = xz - \lambda = 0$ (2)

$F_z(x,y,z,\lambda) = xy - \lambda = 0$ (3)

$F_\lambda(x,y,z,\lambda) = -x - y - z + K = 0$ (4)

From (1), $\lambda = yz$. Hence,

$xz - yz = 0 \to x = y$
$xy - yz = 0 \to x = z$

The relevant critical point occurs when $x = y = z = K/3$

The maximum value of the product is $(\frac{K}{3})^3 = \frac{K^3}{27}$. If the arithmetic mean is $\dfrac{x + y + z}{3} = K$,

$x + y + z = 3K$, and $xyz \leq \dfrac{(3K)^3}{27} = K^3$. Hence, $\sqrt[3]{xyz} \leq K$ and the geometric mean of x, y and z is no larger than the arithmetic mean of x, y and z.

11.

$lwh = 125$

$f(l,w,h) = 2(lw + wh + lh)$ which is a minimum when $lw + wh + lh$ is a minimum

$F(l,w,h,\lambda) = lw + wh + lh - \lambda(lwh - 125)$

$F_l(l,w,h,\lambda) = w + h - \lambda wh = 0$

$F_w(l,w,h,\lambda) = h + l - \lambda lh = 0$

$F_h(l,w,h,\lambda) = w + l - \lambda lw = 0$

$F_\lambda(l,w,h,\lambda) = -lwh + 125 = 0 \to h = \dfrac{125}{lw}$

$$w + \frac{125}{lw} - \frac{125\lambda}{l} = 0 \Rightarrow lw^2 + 125 - 125\lambda w = 0$$

$$\frac{125}{lw} + l - \frac{125\lambda}{w} = 0 \Rightarrow 125 + l^2 w - 125\lambda l = 0$$

$$w + l - \lambda lw = 0 \Rightarrow \lambda = \frac{w+l}{lw}$$

$$lw^2 + 125 - \frac{125(w+l)}{l} = 0 \Rightarrow l^2 w^2 + 125l - 125w - 125l = 0$$

$$125 + l^2 w - 125\frac{(w+l)}{w} = 0 \Rightarrow l^2 w^2 + 125w - 125w - 125l = 0$$

$$\begin{matrix} l^2 w = 125 \\ w^2 l = 125 \end{matrix} \Rightarrow \frac{l^2 w}{w^2 l} = \frac{l}{w} = 1 \text{ and } l = w.$$

$$l^3 = 125, \ l = 5, \ w = 5 \text{ and } h = \frac{125}{25} = 5.$$

The critical point occurs when $l = w = h = 5$ in. and the minimum of plywood is $2(75) = 150$ in^2.

13.

$$lw + 2hl + 2hw = 108$$
$$v(l,w,h) = lwh$$

$$V(l,w,h,\lambda) = lwh - \lambda(lw + 2hl + 2hw - 108)$$

(1) $\quad \frac{\partial V}{\partial l} = wh - \lambda(w + 2h) = 0 \Rightarrow \lambda = \frac{wh}{w+2h}$

(2) $\quad \frac{\partial V}{\partial w} = lh - \lambda(l + 2h) = 0 \Rightarrow \lambda = \frac{lh}{l+2h}$

(3) $\quad \frac{\partial V}{\partial h} = lw - \lambda(2l + 2w) = 0$

(4) $\quad \frac{\partial V}{\partial \lambda} = -lw - 2hl - 2hw + 108 = 0$

$$\frac{w}{w+2h} = \frac{l}{l+2h} \Rightarrow wl + 2wh = lw + 2lh \text{ and } w = l$$

From (3) $\lambda = \frac{l^2}{4l} = \frac{l}{4}$

From (1) $lh - \frac{l}{4}(l + 2h) = 0$ or $4h = l + 2h, \ h = \frac{l}{2}$

From (4) $-l^2 - l^2 - l^2 + 108 = 0$

$$3l^2 = 108, \ l^2 = 36, \ l = 6, \ w = 6 \text{ and } h = 3$$

The critical point occurs when $l = w = 6$ inches and $h = 3$ inches, and these are the dimensions for maximum volume.

15.

$$1.5\, lw + 2(hl + wh) = 288$$
$$v(l,w,h) = lwh$$

$$V(l,w,h,\lambda) = lwh - \lambda(1.5lw + 2hl + 2wh - 288)$$

(1) $\quad \frac{\partial V}{\partial l} = wh - \lambda(1.5w + 2h) = 0 \Rightarrow \lambda = \frac{wh}{1.5w + 2h}$

(2) $\quad \dfrac{\partial V}{\partial w} = lh - \lambda(1.5l + 2h) = 0 \Rightarrow \lambda = \dfrac{lh}{1.5l + 2h}$

(3) $\quad \dfrac{\partial V}{\partial h} = lw - \lambda(2l + 2w) = 0$

(4) $\quad \dfrac{\partial V}{\partial \lambda} = -1.5lw - 2hl - 2wh + 288 = 0$

From (1) and (2) $\qquad \dfrac{w}{1.5w + 2h} = \dfrac{l}{1.5l + 2h}$

$\qquad\qquad\qquad\qquad 1.5wl + 2wh = 1.5wl + 2hl, \ \ w = l$

From (4) $\qquad\qquad\qquad 4wh = 288 - 1.5w^2$

$\qquad\qquad\qquad\qquad h = \dfrac{72}{w} - \dfrac{3}{8}w$

From (3) $\qquad\qquad\qquad \lambda = \dfrac{w^2}{4w} = \dfrac{w}{4}$

From (1) $\qquad\qquad\qquad w(\dfrac{72}{w} - \dfrac{3}{8}w) - \dfrac{w}{4}(\dfrac{3w}{2} + \dfrac{144}{w} - \dfrac{3}{4}w) = 0$

$\qquad\qquad\qquad\qquad 72 - \dfrac{3}{8}w^2 - \dfrac{3}{8}w^2 - 36 + \dfrac{3}{16}w^2 = 0$

$\qquad\qquad\qquad\qquad 36 = \dfrac{9}{16}w^2, \ \ w = \dfrac{6(4)}{3} = 8$

$l = 8 \quad h = 9 - 3 = 6$

The maximum volume is $8(8)(6) = 384 \text{ ft}^3$

17. $\quad V(l, w, h, \lambda) = lwh - \lambda(3lw + 4lh + 4wh - 144)$

$\quad \dfrac{\partial V}{\partial l} = wh - \lambda(3w + 4h) = 0 \Rightarrow \lambda = \dfrac{wh}{3w + 4h} \qquad\qquad (1)$

$\quad \dfrac{\partial V}{\partial w} = lh - \lambda(3l + 4h) = 0 \Rightarrow \lambda = \dfrac{lh}{3l + 4h} \qquad\qquad (2)$

$\quad \dfrac{\partial V}{\partial h} = lw - \lambda(4l + 4w) = 0 \Rightarrow \lambda = \dfrac{lw}{4l + 4w} \qquad\qquad (3)$

$\quad \dfrac{\partial V}{\partial \lambda} = -3lw - 4lh - 4lw + 144 = 0 \qquad\qquad\qquad\quad (4)$

From (1) and (2) $3lw + 4wh = 3lw + 4lh$

$\qquad\qquad\qquad\qquad w = l$

From (3) $\quad \lambda = \dfrac{w^2}{8w} = \dfrac{w}{8}$

From (4) $\quad 3w^2 + 8wh = 144$

$\qquad\qquad\quad 8wh = 144 - 3w^2$

$\qquad\qquad\qquad h = \dfrac{18}{w} - \dfrac{3w}{8}$

From (1) $\quad 18 - \dfrac{3}{8}w^2 - \dfrac{w}{8} \ (3w + \dfrac{72}{w} - \dfrac{3}{2}w) = 0$

$\qquad\qquad\quad 18 - \dfrac{3}{4}w^2 - 9 + \dfrac{3w^2}{16} = 0$

$\qquad\qquad\qquad 9 = \dfrac{9}{16}w^2$

$$w = 4, \; l = 4, \; h = \frac{9}{2} - \frac{3}{2} = 3.$$

The maximum volume is $4(4)(3) = 48 \text{ ft}^3$.

19. $lwh = 144$

$$a(l,w,h) = 2lh + 6hw + lw$$

$$A(l,w,h,\lambda) = 2lh + 6hw + lw - \lambda(lwh - 144)$$

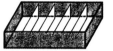

(1) $\dfrac{\partial A}{\partial l} = 2h + w - wh\lambda = 0 \Rightarrow \lambda = \dfrac{2h+w}{wh} = \dfrac{2}{w} + \dfrac{1}{h}$

(2) $\dfrac{2A}{\partial w} = 6h + l - lh\lambda = 0 \Rightarrow \lambda = \dfrac{6h+l}{lh} = \dfrac{6}{l} + \dfrac{1}{h}$

(3) $\dfrac{\partial A}{\partial h} = 2l + 6w - lw\lambda = 0$

(4) $\dfrac{\partial A}{\partial \lambda} = -lwh + 144 = 0$

From (1) and (2) $\dfrac{2}{w} = \dfrac{6}{l}, \;\; 2l = 6w, \;\; l = 3w$

From (3) $6w + 6w - 3w^2\left(\dfrac{2}{w} + \dfrac{1}{h}\right) = 6w - \dfrac{3w^2}{h} = 0.$

$h = \dfrac{w}{2}$

From (4) $3w(w)(w/2) = 144$

$$3w^3 = 288$$
$$w^3 = 288/3 = 96$$
$$w = \sqrt[3]{96} = 2\sqrt[3]{12}$$
$$l = 6\sqrt[3]{12}$$
$$h = \sqrt[3]{12}$$

The area is minimized when

$$w = 2 \cdot \sqrt[3]{12} \text{ inches}, \; l = 6\sqrt[3]{12} \text{ inches and } h = \sqrt[3]{12} \text{ inches}$$

21.

River

x [   ] x

y

$2x + y = 1600$

$A(x,y,\lambda) = xy - \lambda(2x + y - 1600)$

$\dfrac{\partial A}{\partial x} = y - 2x = 0 \Rightarrow \lambda = y/2$

$\dfrac{\partial A}{\partial y} = x - \lambda = 0 \Rightarrow \lambda = x$

$\left. \right\} \Rightarrow y = 2x$

$\dfrac{\partial A}{\partial \lambda} = -2x - y + 1600 = 0 \Rightarrow -2x - 2x + 1600 = 0$

$$4x = 1600$$
$$x = 400, \;\; y = 800$$

The area is maximized when the ends of the pasture are 400 feet and the side adjacent the river is 800 feet.

23.

$\pi r^2 h = 25$. Let K be the cost per sqaure inch of cardboard.

$$C(r, h\lambda) = 2\pi r^2 K + \frac{2\pi r h K}{3} - x(\pi r^2 h - 25).$$

$$\frac{\partial C}{\partial r} = 4\pi r K + \frac{2\pi}{3} h K - 2\pi \lambda r h = 0 \qquad (1)$$

$$\frac{\partial C}{\partial h} = \frac{2\pi r K}{3} - \lambda \pi r^2 = 0 \Rightarrow \lambda = \frac{2K}{3r}$$

$$\frac{\partial C}{\partial \lambda} = -\pi r^2 h + 25 \Rightarrow h = \frac{25}{\pi r^2}$$

from (1)

$$4\pi r K + \frac{2\pi}{3}\left(\frac{25}{\pi r^2}\right)K - 2\pi\left(\frac{2K}{3r}\right)r\left(\frac{25}{\pi r^2}\right) =$$

$$4\pi r K + \frac{50K}{3r^2} - \frac{100K}{3r^2} = 0$$

$$12\pi r^3 K = 50K$$

$$r^3 = \frac{25}{6\pi}$$

$$r = \sqrt[3]{\frac{25}{6\pi}}$$

$$h = \frac{25(6\pi)^{2/3}}{\pi(25)^{2/3}} = \sqrt[3]{\frac{900}{\pi}}$$

The minimum cost occurs when the radius is $\sqrt[3]{\frac{25}{6\pi}}$ in. and the height is $\sqrt[3]{\frac{900}{\pi}}$ in.

25. $30L + 70K = 210,000$

$$F(L, K, \lambda) = 320 L^{.4} K^{.6} - \lambda(30L + 70K - 210,000)$$

$$\frac{\partial F}{\partial L} = \frac{128}{L^{.6}} K^{.6} - 30\lambda = 0 \Rightarrow \lambda = \frac{64}{15}\left(\frac{K}{L}\right)^{.6} \qquad (1)$$

$$\frac{\partial F}{\partial K} = \frac{192 L^{.4}}{K^{.4}} - 70\lambda = 0 \Rightarrow \lambda = \frac{192}{70}\left(\frac{L}{K}\right)^{.4} \qquad (2)$$

$$\frac{\partial F}{\partial \lambda} = -30L - 70K + 210,000 = 0 \qquad (3)$$

From (1) and (2)

$$\frac{64}{15}\left(\frac{K}{L}\right)^{.6} = \frac{192}{70}\left(\frac{L}{K}\right)^{.4}$$

$$\frac{64}{15}K = \frac{192}{70}L$$

$$K = \frac{2880L}{4480} = \frac{9}{14}L$$

From (3)

$$30L + 70\left(\frac{9}{14}\right)L = 210,000$$

$$75L = 210,000$$

$$L = 2800$$

$$K = \frac{9}{14}(2800) = 1800$$

The maximum production level is $320(2800)^{.4}(1800)^{.6} = 687348$ units.

27. $50L + 120K = 600,000$

$$F(L,K,\lambda) = 750L^{.6}K^{.4} - \lambda(50L + 120K - 600,000)$$

$$\frac{\partial F}{\partial L} = \frac{450K^{.4}}{L^{.4}} - 50\lambda = 0 \Rightarrow \lambda = \frac{9K^{.4}}{L^{.4}}$$

$$\frac{\partial F}{\partial K} = \frac{300L^{.6}}{K^{.6}} - 120\lambda = 0 \Rightarrow \lambda = \frac{5L^{.6}}{2K^{.6}}$$

$$\frac{\partial F}{\partial \lambda} = -50\ L - 120K + 600,000 = 0$$

$$18K = 5L$$
$$L = \frac{18K}{5}$$
$$50(\tfrac{18K}{5}) + 120K = 600,000$$
$$300K = 600,000$$
$$K = 2000$$
$$L = 7200$$

The maximum production is $750(7200)^{.6}(2000)^{.4} = 3,234,989$ units

29. $400L^{.25}K^{.75} = 100,000$

$$C(L,K,\lambda) = 20L + 960K - \lambda(400L^{.25}K^{.75} - 100,000)$$

(1) $\frac{\partial C}{\partial L} = 20 - \frac{100\lambda K^{.75}}{L^{.75}} = 0 \Rightarrow \lambda = \frac{1}{5}(\frac{L}{K})^{.75}$

(2) $\frac{\partial C}{\partial K} = 960 - \frac{300}{K^{.25}}\lambda L^{.25} = 0 \Rightarrow \lambda = \frac{16}{5}(\frac{K}{L})^{.25}$

(3) $\frac{\partial C}{\partial \lambda} = -400L^{.25}K^{.75} + 100,000 = 0$

From (1) and (2)
$$(\tfrac{L}{K})^{.75} = 16(\tfrac{K}{L})^{.25}$$
$$L = 16K$$
From (3) $400(16K)^{.25}K^{.75} = 100,000$
$$800K = 100,000$$
$$K = 125$$
$$L = 2000$$

The minimum total cost is $20(2000) + 960(125) = \$160,000$

31. $420L^{2/3}K^{1/3} = 336,000$

$$C(L,K,\lambda) = 30L + 405K - \lambda(420L^{2/3}K^{1/3} - 336,000)$$
$$\frac{\partial C}{\partial L} = 30 - 280\frac{K^{1/3}\lambda}{L^{1/3}} = 0 \Rightarrow \lambda = \frac{3}{28}\frac{L^{1/3}}{K^{1/3}}$$

$$\frac{\partial C}{\partial K} = 405 - 140\frac{L^{2/3}}{K^{2/3}}\lambda = 0 \Rightarrow \lambda = \frac{81}{28}\frac{K^{2/3}}{L^{1/3}}$$

$$\frac{\partial C}{\partial \lambda} = -420L^{2/3}K^{1/3} + 336,000 = 0$$

$$\frac{3L}{28} = \frac{81}{28}K, \quad L = 27K$$

$$420(27K)^{2/3}K^{1/3} = 336{,}000$$
$$K = 88.\overline{8}$$
$$L = 2400$$

The minimum total cost is $30(2400) + 405(88.\overline{8}\ ) = \$108{,}000$

33. From Exercise 25 $\lambda_0 = \frac{64}{15}\left(\frac{1800}{2800}\right)^{.6} = 3.27$

The approximate change in maximum production for each dollar spent on labor and capital is 3.27 units.

35. From Exercise 27 $\lambda_0 = 9\left(\frac{2000}{7200}\right)^{.4} = 5.39$

For each additional dollar spent on labor and capital, the maximum production increases by about 5.39 units.

37. $x + y = 120$

$$F(x,y,\lambda) = 3x^{5/2}y^{3/2} - \lambda(x + y - 120)$$

$$\frac{\partial F}{\partial x} = \frac{15}{2}(xy)^{3/2} - \lambda = 0 \Rightarrow \lambda = \frac{15}{2}x^{3/2}y^{3/2}$$

$$\frac{\partial F}{\partial y} = \frac{9}{2}x^{5/2}y^{1/2} - \lambda = 0 \Rightarrow \lambda = \frac{9}{2}x^{5/2}y^{1/2}$$

$$\frac{\partial F}{\partial \lambda} = -x - y + 120 = 0$$

$$15x^{3/2}y^{3/2} = 9x^{5/2}y^{1/2}$$
$$15y = 9x$$

$y = \frac{3}{5}x,\ x + \frac{3}{5}x = \frac{8x}{5} = 120,\ x = 75,\ y = 45,$ and $\lambda = \frac{2}{15}(75 \cdot 45)^{1.5} = 1{,}470{,}523$

If \$75,000 is spent on development and \$45,000 is spent on promotion, the number of books sold is maximized. If an additional \$1000 is spent on development and promotion, the number of books sold increases by approximately 1,470,000.

39. $q_A + q_B = 600$

$$F(q_A, q_B, \lambda) = .6q_A^2 + .5q_Aq_B + .4q_B^2 + 150q_A + 100q_B + 20{,}000 - \lambda(q_A + q_B - 600)$$

$$\frac{\partial F}{\partial q_A} = 1.2q_A + .5q_B + 150 - \lambda = 0 \Rightarrow \lambda = 1.2q_A + .5q_B + 150$$

$$\frac{\partial F}{\partial q_B} = .5q_A + .8q_B + 100 - \lambda = 0 \Rightarrow \lambda = .5q_A + .8q_B + 100$$

$$\frac{\partial F}{\partial \lambda} = -q_A - q_B + 600 = 0$$

$$1.2q_A + .5q_B + 150 = .5q_A + .8q_B + 100$$

$$.7q_A = .3q_B - 50$$

$$q_A = \frac{3q_B - 500}{7}$$

$$\frac{3q_B - 500}{7} + q_B = 600$$

$$10q_B - 500 = 4200$$
$$10q_B = 4700$$
$$q_B = 470$$
$$q_A = 130$$

The critical point occurs when $q_A = 130$ and $q_B = 470$. At the critical point $C = 215{,}550$. When $q_A = 0$ and $q_B = 600$, $C = 224{,}000$. When $q_A = 600$ and $q_B = 0$, $C = 326{,}000$. Thus, the minimum costs occurs when 130 units are produced at site A and 470 are produced at site B.

41. (a)

Wall

$$30x + 20x + 20y = C$$
$$a(x,y) = xy$$

$$A(x,y,\lambda) = xy - \lambda(50x + 20y - C)$$

$$\frac{\partial A}{\partial x} = y - 50\lambda = 0 \Rightarrow \lambda = y/50$$

$$\frac{\partial A}{\partial y} = x - 20\lambda = 0 \Rightarrow \lambda = \frac{x}{20}$$

$$\Rightarrow y = \frac{5}{2}x$$

$$\frac{\partial A}{\partial x} = -50x - 20y + C = 0$$

$$50x + 50x = C$$
$$x = C/100, \quad y = \frac{1}{40}C$$

The maximum area is
$$A_M = C^2/4000$$

(b) $\lambda = \dfrac{y}{50} = \dfrac{C}{2000}$

(c) $\dfrac{dA_M}{dC} = \dfrac{C}{2000}$ which is the answer to part (b).

(d) If $C = 800$, the maximum area is $\dfrac{800(800)}{4000} = 160$ ft^2

(e) $\lambda = \dfrac{800}{2000} = .4$. The maximum area is increased by approximately .4 ft^2.

(f) When $C = 801$, the maximum area is $\dfrac{(801)^2}{4000} = 160.40025$ and the increase in area is

.40025 ft^2 which is nearly the same as the answer to part (e).

EXERCISE SET 6.7 THE METHOD OF LEAST SQUARES

1.

i	x_i	y_i	$mx_i + b$	$mx_i + b - y_i$
1	1	5	$m + b$	$m + b - 5$
2	2	7	$2m + b$	$2m + b - 7$
3	3	8	$3m + b$	$3m + b - 8$
4	4	9	$4m + b$	$4m + b - 9$

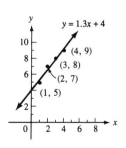

$$f(m,b) = (m+b-5)^2 + (2m+b-7)^2 + (3m+b-8)^2 + (4m+b-9)^2$$
$$f_m(m,b) = 2(m+b-5) + 4(2m+b-7) + 6(3m+b-8) + 8(4m+b-9)$$
$$= 60m + 20b - 158$$
$$f_b(m,b) = 2(m+b-5) + 2(2m+b-7) + 2(3m+b-8) + 2(4m+b-9)$$
$$= 20m + 8b - 58$$

Solving the system
$$60m + 20b = 158$$
$$20m + 8b = 58$$

we find $m = 1.3$ and $b = 4$ at the critical point.

$f_{mm} = 60$, $f_{bb} = 8$, $f_{bm} = 20$, $60(8) - 400 > 0$. The minimum occurs when $m = 1.3$ and $b = 4$

The line is $y = 1.3x + 4$

3.

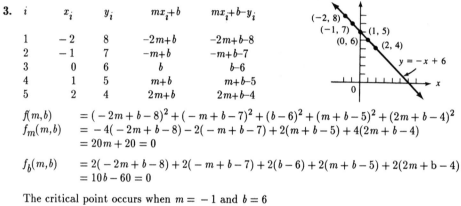

i	x_i	y_i	mx_i+b	mx_i+b-y_i
1	-2	8	$-2m+b$	$-2m+b-8$
2	-1	7	$-m+b$	$-m+b-7$
3	0	6	b	$b-6$
4	1	5	$m+b$	$m+b-5$
5	2	4	$2m+b$	$2m+b-4$

$$f(m,b) = (-2m+b-8)^2 + (-m+b-7)^2 + (b-6)^2 + (m+b-5)^2 + (2m+b-4)^2$$
$$f_m(m,b) = -4(-2m+b-8) - 2(-m+b-7) + 2(m+b-5) + 4(2m+b-4)$$
$$= 20m + 20 = 0$$

$$f_b(m,b) = 2(-2m+b-8) + 2(-m+b-7) + 2(b-6) + 2(m+b-5) + 2(2m+b-4)$$
$$= 10b - 60 = 0$$

The critical point occurs when $m = -1$ and $b = 6$

$f_{mm} = 20$, $f_{bb} = 10$, $f_{mb} = 0$

The minimum occurs at the critical point and the line has equation

$$y = -x + 6$$

5.

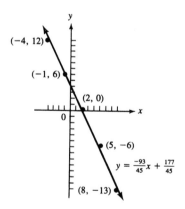

i	x_i	y_i	x_iy_i	x_i^2
1	-4	12	-48	16
2	-1	6	-6	1
3	2	0	0	4
4	5	-6	-30	25
5	8	-13	-104	64
	X=10	Y=-1	P=-188	S=110

$$m = \frac{5(-188) - 10(-1)}{5(110) - (10)(10)} = \frac{-930}{450} = \frac{-31}{15}$$

$$b = \frac{110(-1) + 188(10)}{5(110) - 100} = \frac{1770}{450} = \frac{59}{15} = 3.93$$

The equation of the line is $y = \dfrac{-31}{15}x + \dfrac{59}{15}$

7. $\dfrac{\sum x_i}{n} = \dfrac{1}{2}$ $x' = x - \dfrac{1}{2}$

i	x_i	x_i'	y_i	$x_i'y_i$	$(x_i')^2$
1	-2	-2.5	7	-17.5	6.25
2	-1	-1.5	5	-7.5	2.25
3	0	$-.5$	3	-1.5	.25
4	1	.5	1	.5	.25
5	2	1.5	-1	-1.5	2.25
6	3	2.5	-3	-7.5	6.25
	$X=0$		$Y=12$	$P=-35$	$S=17.5$

$$m = \frac{P}{S} = \frac{-35}{17.5} = -2 \qquad b = \frac{Y}{n} = \frac{12}{6} = 2$$

The equation of the line is $y = -2x' + 2$

or $y = -2\left(x - \dfrac{1}{2}\right) + 2$

$y = -2x + 3$

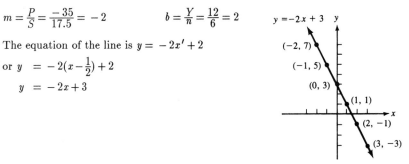

9.

i	x_i	y_i	x_iy_i	x_i^2	$x_i-.6$	$y_i+2.2$	$(x_i-.6)(y_i+2.2)$	$(x_i-.6)^2$
1	-4	-17	68	16	-4.6	-14.8	68.08	21.16
2	-3	-14	42	9	-3.6	-11.8	42.48	12.46
3	-1	-7	7	1	-1.6	-4.8	7.68	2.56
4	3	6	18	9	2.4	8.2	19.68	5.67
5	8	21	168	64	7.4	23.2	171.68	54.76
Sums:	$n=5$ $X=3$	$Y=-11$	$P=303$	$S=99$			309.6	97.2

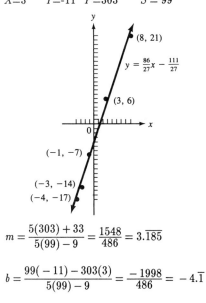

$$m = \frac{5(303) + 33}{5(99) - 9} = \frac{1548}{486} = 3.\overline{185}$$

$$b = \frac{99(-11) - 303(3)}{5(99) - 9} = \frac{-1998}{486} = -4.\overline{1}$$

The equation of the regression line is

$$y = 3.\overline{185}x - 4.\overline{1}$$

Using the alternative method

$$m = \frac{309.6}{97.2} = 3.\overline{185}$$

$$b = -2.2 - 3.\overline{185}(.6) = -4.\overline{1}$$

11. (a)

year	x_i	y_i	$x_i y_i$	x_i^2
1983	-4	42	-168	16
1984	-3	40	-120	9
1985	-2	43	-86	4
1986	-1	45	-45	1
1987	0	44	0	0
1988	1	48	48	1
1989	2	48	96	4
1990	3	51	153	9
1991	4	57	228	16
$n=9$	X=0	Y=418	P=106	S=60

$$m = \frac{106}{60}$$

$$b = \frac{418}{9}$$

The equation of regression line is

$$y = \frac{53}{30}x + \frac{418}{9} \text{ where } x \text{ is the number of years after 1987}$$

(b) Letting $x = 5$, $\frac{53}{30}(5) + \frac{418}{9} = 55.3$ orders

13. (a)

x_i	y_i	$x_i y_i$	x_i^2
15	9	135	225
16	9.5	152	256
16.5	10	165	272.25
17	10.7	181.9	289
19	12	228	361
X=83.5	Y=51.2	P=861.9	S=1403.25

$$m = \frac{5(861.0) - 83.5(51.2)}{5(1403.25) - 83.5(83.5)} = \frac{34.3}{44} \sim .78$$

$$b = \frac{51.2(1403.25) - 861.9(83.5)}{5(1403.25) - 83.5(83.5)} = \frac{-122.25}{44} \sim -2.78$$

The equation of the regression line is $y = \frac{343}{440}x - \frac{12225}{4400}$

(b) $\frac{343}{440}(23) - \frac{12225}{4400} \doteq 15.15$ or 1515 students

15. (a)

x_i	y_i	$x_i y_i$
85	3.1	263.5
75	2.7	202.5
60	2.2	132
72	3.0	216
90	3.5	315
70	2.8	196
75	3.0	225
88	3.0	264

$X=615$ $Y=23.3$ $P=1814$ $S=\sum x_i^2 = 48003$

$$m = \frac{8(1814) - 615(23.3)}{8(48003) - 615(615)} = \frac{182.5}{5799} \doteq .0315$$

$$b = \frac{48003(23.3) - 1814(615)}{8(48003) - 615(615)} = \frac{28599}{5799} \doteq .493$$

The equation of the regression line is
$$y = \frac{1825}{57990}x + \frac{28599}{57990}$$

(b) $\frac{1825}{57990}(79) + \frac{28599}{57990} \doteq 2.98$

17. (a) $X = \sum x_i = 510$

$Y = \sum y_i = 522$

$P = \sum x_i y_i = 33662$

$S = \sum x_i^2 = 31290$

$$m = \frac{10(33662) - 510(522)}{10(31290) - 510(510)} = \frac{70400}{52800} = \frac{4}{3}$$

$$b = \frac{31290(522) - 33662(510)}{10(31290) - 510(510)} = \frac{-834240}{52800} = \frac{-869}{55} = -15.8$$

The equation of the regression line is $y = \frac{4}{3}x - 15.8$

(b) If $x = 35$, $y = \frac{4}{3}(35) - 15.8 \doteq 31$ deaths.

19. (a) $X = \sum x_i = 36$

$Y = \sum y_i = 3140$

$P = \sum x_i y_i = 18240$

$S = \sum x_i^2 = 204$

$$m = \frac{8(18240) - 36(3140)}{8(204) - 36(36)} = \frac{32880}{336} = \frac{685}{7} \sim 97.9$$

$$b = \frac{204(3140) - 18240(36)}{8(204) - 36(36)} = \frac{-16080}{336} = \frac{-335}{7} \sim -47.9$$

The equation of the regression line is

$$y = \frac{685}{7}x - \frac{335}{7}.$$

(b) When $x = 10$, $y = \frac{6850 - 335}{7} \doteq \$930.71$

21. $\dfrac{nP - XY}{nS - X^2}(\overline{x}) + \dfrac{SY - PX}{nS - X^2}$

$= \dfrac{nP - XY}{nS - X^2}\left(\dfrac{X}{n}\right) + \dfrac{Sy - PX}{nS - X^2}$

$= \dfrac{PX - X^2\overline{y} + SY - PX}{nS - X^2} \quad = \dfrac{-X^2\overline{y} + SY}{nS - X^2}$

$\qquad\qquad = \dfrac{-X^2\overline{y} + nS\overline{y}}{nS - X^2} = \dfrac{\overline{y}(nS - X^2)}{nS - X^2} = \overline{y}$

Thus, $(\overline{x}, \overline{y})$ lies on the regression line

23. $\sum (a_i - \overline{a}) = \sum a_i - \sum \overline{a}$

$\qquad = \sum a_i - n\overline{a} = \sum a_i - \dfrac{n \sum a_i}{n} = \sum a_i - \sum a_i = 0.$

25. $m = \dfrac{nP - XY}{nS - X^2} = \dfrac{\dfrac{nP - XY}{n}}{\dfrac{nS - X^2}{n}} = \dfrac{\sum (x_i - \overline{x})(y_i - \overline{y})}{\sum (x_i - \overline{x})^2}$

EXERCISE SET 6.8 DOUBLE INTEGRALS

1. $\displaystyle\int (24x^2 y^3 + 20xy^4)\,dy$

$= 24x^2 \displaystyle\int y^3\,dy + 20x \displaystyle\int y^4\,dy$

$= 24x^2 y^4/4 + 20xy^5/5 + C(x)$

$= 6x^2 y^4 + 4xy^5 + C(x)$

3. $\displaystyle\int 6x^2(x^3 + 1)^4 y\,dx = y \displaystyle\int 6x^2(x^3 + 1)^4\,dx$

(let $u = x^3 + 1$), $du = 3x^2\,dx$)

$= y \displaystyle\int 2u^4\,du = \tfrac{2}{5}yu^5 + C(y) = \tfrac{2}{5}y(x^3 + 1)^5 + C(y)$

5. $\displaystyle\int 6x^2e^{3y}\,dx = 6e^{3y}\int x^2\,dx = 6e^{3y}\cdot\frac{x^3}{3} + C(y)$

$\qquad = 2x^3e^{3y} + C(y)$

7. $\displaystyle\int_1^2 (24x^2y^3 + 20xy^4)\,dy = 6x^2y^4 + 4xy^5\big|_{y=1}^2$

$\qquad = (6x^2(16) + 4x(32)) - (6x^2\cdot 1 + 4x\cdot 1)$

$\qquad = 96x^2 + 128x - 6x^2 - 4x = 90x^2 + 124x$

9. $\displaystyle\int_0^1 6x^2(x^3+1)^4 y\,dx = \tfrac{2}{5}(x^3+1)^5 y\big|_{x=0}^1 = \tfrac{2}{5}(2)^5 y - \tfrac{2}{5}(1)^5 y$

$\qquad = \tfrac{64}{5}y - \tfrac{2}{5}y = \tfrac{62}{5}y$

11. $\displaystyle\int_{3y}^{y^2} (24x^2y^3 + 20xy^4)\,dx = 8x^3y^3 + 10x^2y^5\;\big|_{x=3y}^{y^2}$

$\qquad = (8(y^2)^3y^3 + 10(y^2)^2y^5) - (8(3y)^3y^3 + 10(3y)^2y^5)$

$\qquad = 8y^9 + 10y^9 - 216y^6 - 90y^7$

$\qquad = 18y^9 - 90y^7 - 216y^6$

13. $\displaystyle\int_{\sqrt{y}}^{3y} \frac{8xy}{x^2+1}\,dx = 4y\ln(x^2+1)\big|_{x=\sqrt{y}}^{3y}$

$\qquad = 4y\ln(9y^2+1) - 4y\ln(y+1)$

15. $\displaystyle\int_1^2\int_{x^3}^{8x^6}(5x^2\;^3\!\sqrt{y}\;dy)\,dx = \int_1^2 5x^2(\tfrac{3}{4}y^{4/3})\big|_{y=x^3}^{8x^6}\,dx$

$\qquad = \int_1^2 \frac{15x^2}{4}(16x^8 - x^4)\,dx = \int_1^2 (60x^{10} - \tfrac{15}{4}x^6)\,dx$

$\qquad = \frac{60}{11}x^{11} - \frac{15}{28}x^7\big|_1^2 = \frac{60}{11}(2)^{11} - \frac{15}{28}(2)^7 - \frac{60}{11} + \frac{15}{28} = 11097.4$

17. $\displaystyle\int_1^2\int_0^{\sqrt{y}} \frac{8xy}{x^2+1}\,dx\,dy = \int_1^2 \frac{8y\ln(x^2+1)}{2}\big|_0^{\sqrt{y}}\,dx =$

$\qquad \displaystyle\int_1^2 4(y\ln(y+1)\,dy.$ To evaluate this integral, let $u = y+1$ and use the integral tables to obtain

$\qquad \displaystyle\int_{u=2}^3 4(u-1)\ln u\,du = \int_2^3 4u\ln u - 4\ln u\,du$

$\qquad = 2u^2\ln u - u^2 - 4(u\ln u - u)\big|_2^3$

$$= 2u^2 \ln u - u^2 - 4u \ln u + 4u|_2^3$$
$$= (18\ln 3 - 9 - 12\ln 3 + 12) - (8\ln 2 - 4 - 8\ln 2 + 8)$$
$$= 6\ln 3 - 1$$

19. (a) $\displaystyle\int_1^3 \left[\int_2^5 (24x^2 y^3 + 20xy^4)\, dx \right] dy$

$$= \int_1^3 \left[8x^3 y^3 + 10x^2 y^4 \big|_{x=2}^5 \right] dy$$

$$= \int_1^3 (1000y^3 + 250y^4) - (64y^3 + 40y^4)\, dy$$

$$= \int_1^3 (936y^3 + 210y^4)\, dy = 234y^4 + 42y^5 \big|_1^3$$

$$= 234(81) + 42(243) - (234 + 42) = 28{,}884$$

(b) $\displaystyle\int_2^5 \left[\int_1^3 (24x^2 y^3 + 20xy^4)\, dy \right] dx$

$$= \int_2^5 (6x^2 y^4 + 4xy^5)/\big|_{y=1}^3\, dx = \int_2^5 (486x^2 + 972x) - (6x^2 + 4x)\, dx$$

$$= \int_2^5 (480x^2 + 968x)\, dx = 160x^3 + 484x^2 \big|_2^5$$

$$= (160(125) + 484(25)) - (160(8) + 484(4))$$

$$= 28{,}884$$

The answers to parts (a) and (b) are the same.

21. (a) $\displaystyle\int_2^4 \left[\int_0^2 \frac{8xy}{x^2 + 1}\, dy \right] dx$

$$= \int_2^4 \left(\frac{4xy^2}{x^2 + 1} \big|_{y=0}^2 \right) dx = \int_2^4 \frac{16x}{x^2 + 1}\, dx$$

$$= 8\ln(x^2 + 1)\big|_2^4 = 8\ln(17) - 8\ln(5)$$

$$= 8\ln(17/5) = 8\ln(3.4)$$

(b) $\displaystyle\int_0^2 \left[\int_2^4 \frac{8xy}{x^2 + 1}\, dx \right] dy = \int_0^2 y(4\ln(x^2 + 1)\big|_{x=2}^4\, dy$

$$= \int_0^2 4y(\ln 17 - \ln 5)\, dy = 2y^2 \ln(17/5)\big|_{y=0}^2$$

$$= 8\ln(3.4)$$

23.

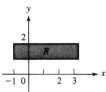

$$\iint\limits_R f(x,y)\,dA = \int_{-1}^{3}\int_{1}^{2}(3x^2+4xy)\,dy\,dx$$

$$= \int_{-1}^{3}\left[3x^2 y + 2xy^2\Big|_{y=1}^{2}\right]dx = \int_{-1}^{3}(6x^2+8x)-(3x^2+2x)\,dx$$

$$= \int_{-1}^{3}(3x^2+6x)\,dx = x^3+3x^2\Big|_{-1}^{3} = (27+27)-(-1+3)$$

$$= 54-2 = 52$$

25.

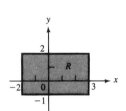

$$\int_{-2}^{3}\left[\int_{-1}^{2}\frac{x^2 y^3}{y^4+4}\,dy\right]dx$$

$$= \int_{-2}^{3}\frac{x^2}{4}\ln(y^4+4)\Big|_{y=-1}^{2}\,dx$$

$$= \int_{-2}^{3}\frac{x^2}{4}(\ln 20 - \ln 5)\,dx = \int_{-2}^{3}\frac{x^2}{4}\ln 4\,dx$$

$$= \frac{\ln 4}{12}x^3\Big|_{-2}^{3} = \frac{\ln 4}{12}(27+8) = \frac{35\ln 4}{12}$$

27. R is a Type II region

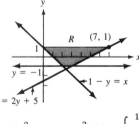

$$\iint\limits_R (3x^2-5xy+3y^3)\,dA = \int_{-1}^{1}\int_{1-y}^{2y+5}(3x^2-5xy+3y^3)\,dx\,dy$$

$$= \int_{-1}^{1} x^3 - \frac{5}{2}x^2 y + 3xy^3\Big|_{1-y}^{2y+5}\,dy =$$

$$\int_{-1}^{1}(2y+5)^3 - \frac{5}{2}(2y+5)^2 y + 3(2y+5)y^3 - (1-y)^3 + \frac{5}{2}(1-y)^2 y + 3(1-y)y^3\,dy$$

- 271 -

$$= \frac{1}{8}(2y+5)^4 - \frac{5}{2}\left(y^4 + \frac{20}{3}y^3 + \frac{25}{2}y^2\right) + \frac{6}{5}y^5 + \frac{15}{4}y^4 + \frac{1}{4}(1-y)^4$$
$$+ \frac{5}{2}\left(\frac{y^2}{2} - \frac{2}{3}y^3 + \frac{y^4}{4}\right) + \frac{3}{4}y^4 - \frac{3}{5}y^5\Big|_{-1}^{1} = 250.53$$

29. R is a Type I region

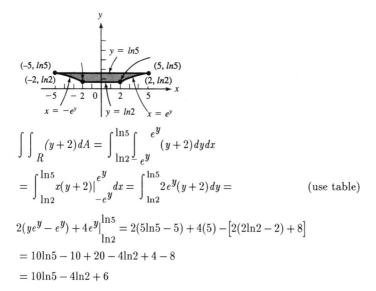

$$\iint_R (y+2)\,dA = \int_{\ln2}^{\ln5}\int_{-e^y}^{e^y} (y+2)\,dy\,dx$$
$$= \int_{\ln2}^{\ln5} x(y+2)\Big|_{-e^y}^{e^y}\,dx = \int_{\ln2}^{\ln5} 2e^y(y+2)\,dy = \qquad \text{(use table)}$$

$$2(ye^y - e^y) + 4e^y\Big|_{\ln2}^{\ln5} = 2(5\ln5 - 5) + 4(5) - \big[2(2\ln2 - 2) + 8\big]$$
$$= 10\ln5 - 10 + 20 - 4\ln2 + 4 - 8$$
$$= 10\ln5 - 4\ln2 + 6$$

31. (a)

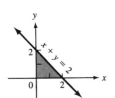

$$\{(x,y)|0 \le x \le 2,\, 0 \le y \le 2 - x\}$$

(b) $\{(x,y)|0 \le y \le 2,\, 0 \le x \le 2 - y\}$

(c) $\displaystyle\iint_R xe^{(2-y)^3}\,dA = \int_{y=0}^{2}\Big[\int_{x=0}^{2-y} xe^{(2-y)^3}\,dx\Big]dy$

$$= \int_0^2 \Big[\frac{x^2}{2}e^{(2-y)^3}\Big|_0^{2-y}\Big]dy = \int_0^2 \frac{1}{2}(2-y)^2 e^{(2-y)^3}\,dy$$

$$(\text{let } u = 2 - y) = -\frac{1}{6}e^{(2-y)^3}\Big|_{y=0}^{2}$$

$$= -\frac{1}{6}e^0 - \left(-\frac{1}{6}e^8\right) = \frac{1}{6}e^8 - \frac{1}{6}$$

33.

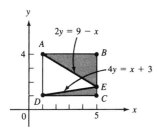

$$\iint\limits_{R} f(x,y)\,dA = \int_{1}^{2}\int_{4y-3}^{5}(5x^2y + 2x - 3y)\,dx\,dy + \int_{2}^{4}\int_{9-2y}^{5}(5x^2y + 2x - 3y)\,dx\,dy = 1120$$

35. The average value is given by

$$\frac{1}{\left(3-(-1)\right)(2-1)}\int_{-1}^{3}\int_{1}^{2}(3x^2 + 4xy)\,dy\,dx$$

$$= \frac{1}{4}\int_{-1}^{3} 3x^2y + 2xy^2\Big|_{1}^{2}\,dx$$

$$= \frac{1}{4}\int_{-1}^{3}(6x^2 + 8x - 3x^2 - 2x)\,dx$$

$$= \frac{1}{4}\int_{-1}^{3}(3x^2 + 6x)\,dx = \frac{1}{4}(x^3 + 3x^2\Big|_{-1}^{3}$$

$$= \frac{1}{4}(54 - 2) = \frac{52}{4} = 13$$

37. The average value is

$$\frac{1}{5(3)}\int_{y=-1}^{2}[\int_{x=-2}^{3}\frac{x^2y^3}{y^4+4}\,dx]\,dy$$

$$= \frac{1}{15}\int_{-1}^{2}\frac{x^3}{3}(\frac{y^3}{y^4+4})\Big|_{x=-2}^{3}\,dy$$

$$= \frac{1}{45}\int_{-1}^{2} 35(\frac{y^3}{y^4+4})\,dy$$

$$= \frac{35}{45(4)}\ln(y^4+4)\Big|_{-1}^{2}$$

$$= \frac{7}{36}(\ln 20 - \ln 5) = \frac{7}{36}\ln 4$$

39. $z = 2 - x - y$

Volume $= \iint\limits_{R}(2 - x - y)\,dA$

$$= \int_{0}^{1}[\int_{0}^{1-x^2}(2 - x - y)\,dy]\,dx$$

$$= \int_0^1 2y - xy - \frac{y^2}{2}\Big|_0^{1-x^2} dx$$

$$= \int_0^1 2(1-x^2) - x(1-x^2) - \frac{(1-x^2)^2}{2} dx$$

$$= \int_0^1 \left(2 - 2x^2 - x + x^3 - \frac{1}{2} + x^2 - \frac{x^4}{2}\right) dx$$

$$= \int_0^1 \left(\frac{3}{2} - x^2 + x^3 - x - \frac{x^4}{2}\right) dx$$

$$= \frac{3}{2}x - \frac{x^3}{3} + \frac{x^4}{4} - \frac{x^2}{2} - \frac{x^5}{10}\Big|_0^1$$

$$= \frac{3}{2} - \frac{1}{3} + \frac{1}{4} - \frac{1}{2} - \frac{1}{10}$$

$$= \frac{15}{4} - \frac{1}{3} - \frac{1}{10} = \frac{75 - 20 - 6}{60} = \frac{49}{60}$$

41. The average is

$$\frac{1}{(.41)(.74)} \int_8^{8.74} \left[\int_4^{4.41} 2000 - 350\sqrt{p_C} + 75\sqrt[3]{p_T} \ dp_C \right] dp_T$$

$$= \frac{1}{.3034} \int_8^{8.74} 2000 p_C - 350\left(\frac{2}{3}\right)p_C^{3/2} + 75 \sqrt[3]{p_T}p_C\Big|_4^{4.41} dp_T$$

$$= \frac{1}{.3034} \int_8^{8.74} 820 - 350\left(\frac{2}{3}\right)(1.261) + 75(.41)\sqrt[3]{p_T} \ dp_T$$

$$= \frac{1}{.3034} \int_8^{8.74} (525.7\overline{6} + 30.75\sqrt[3]{p_T} \ dp_T$$

$$= \frac{1}{.3034}\left(525.7\overline{6}p_T + 30.75(.75)p_T^{4/3}\Big|_8^{8.74}\right)$$

$$= \frac{1}{.3034}(389.06 + 46.197) \doteq 1435 \text{ lb.}$$

43. $P(x + y \le 250) = \iint_R f(x,y) \, dA$

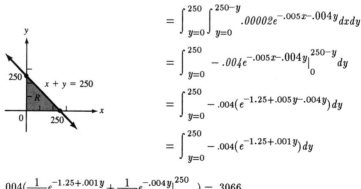

$$= \int_{y=0}^{250} \int_{y=0}^{250-y} .00002e^{-.005x-.004y} dx dy$$

$$= \int_{y=0}^{250} -.004e^{-.005x-.004y}\Big|_0^{250-y} dy$$

$$= \int_{y=0}^{250} -.004\left(e^{-1.25+.005y-.004y}\right) dy$$

$$= \int_{y=0}^{250} -.004\left(e^{-1.25+.001y}\right) dy$$

$$= -.004\left(\frac{1}{.001}e^{-1.25+.001y} + \frac{1}{.004}e^{-.004y}\Big|_0^{250}\right) = .3066$$

1. (a) The domain is the set of all ordered pairs of real numbers

$f(1,3) = 1 + 18 - 9 = 10$

$f(-2, -1) = 4 + 12 - 1 = 15$

(b) The domain is the set of all ordered pairs of real numbers except those of the form (x,y) where $y = 2x$

$f(1,3) = 4/(-1) = -4$

$f(0,1) = 1/(-1) = -1$

(c) The domain is the set of all ordered pairs of real numbers except for those of the form (x,y) where $x = 2$ or $y = -3$.

$f(-1,4) = -3/\left((-3)(7)\right) = 1/7$

$f(3, -2) = 9/1 = 9$

(d) The domain consits of all ordered pairs of real numbers except for those of the form (x,y) where $x^2 + y^2 > 100$. $x^2 + y^2 > 100$ occurs at points outside the circle with center $(0,0)$ and radius 10.

$f(2,5) = \sqrt{100 - 4 - 25} = \sqrt{71}$

$f(8,5) = \sqrt{100 - 64 - 25} = \sqrt{11}$

(e) The domain consists of all ordered pairs of real numbers except for those of the form (x,y) where $2x - y = 0$ or $y = 2x$.

$f(2, -1) = 0/(1 - e^5) = 0$

$f(1,1) = 3/(1 - e) = \dfrac{3}{1 - e}$

(f) The domain consists of all ordered triples of real numbers

$f(-1,1,3) = 3 - 18 - 27 = -42$

$f(0,2, -1) = 0 + 0 + 1 = 1$

(g) The domain is the set of all ordered triples of real numbers except for those of the form (x,y,z) where $x = 3$, $y = -2$ or $z = 5$.

$f(2, -1,4) = (2 - 2 - 12)/\left((-1)(1)(-1)\right) = -12$

$f(5,2,3) = (5 + 4 - 9)/\left((2)(4)(-2)\right) = 0$

(h) The domain is the set of all ordered triples of real numbers except for those of the form (x,y,z) where $x > -3$ and $y < 5$, or $x < -3$ and $y > 5$.

$f(-2,6,0) = \sqrt{1(1)(9)} = 3$

$f(2,7, -1) = \sqrt{5(2)(10)} = 10$

3. (a) $f(0, -1) = 1$
$f(2,0) = 8$
$f(1,2) = -3$.   The points are $(0, -1,1)$, $(2,0,8)$, $(1,2, -3)$.

(b) $f(0,1) = -1$
$f(2,0) = 1/3$
$f(1, -1) = 0$.   Three points are $(0,1, -1)$, $(2,0,1/3)$ and $(1, -1,0)$.

(c) $f(0,3) = 0$
$f(3,0) = -1.5$
$f(0,0) = 0.$ Three points are $(0,3,0)$, $(3,0,-1.5)$ and $(0,0,0)$

(d) $f(0,0) = 6$
$f(0,6) = 0$
$f(6,0) = 0.$ Three points are $(0,0,6)$, $(0,6,0)$ and $(6,0,0)$.

(e) $f(1,0) = 1/(1-e^3)$
$f(0,1) = -3/(1-e^{-2})$
$f(2,1) = -1/(1-e^4).$ Three points are $(1,0,1/(1-e^3))$, $(0,1,-3/(1-e^{-2}))$, $(2,1,-1/(1-e^4))$

5. Revenue $= R(p_A, p_B) =$ revenue for brand A + revenue for brand B.

$= p_A(500 - 8p_A + 12p_B) + p_B(400 + 15p_A - 9p_B)$

$= 500p_A - 8p_A^2 + 400p_B - 9p_B^2 + 27p_Ap_B$

7. (a) $f_x(x,y) = y^6 D_x(3x^4) = y^6(12x^3) = 12x^3y^6$

$f_y(x,y) = 3x^4 Dy(y^6) = 3x^4(6y^5) = 18x^4y^5$

(b) $f_x(x,y) = \dfrac{(x^2+y^4)D_x(x^3y^2) - x^3y^2 D_x(x^2+y^4)}{(x^2+y^4)^2}$

$= \dfrac{(x^2+y^4)(3x^2y^2) - x^3y^2(2x)}{(x^2+y^4)^2} = \dfrac{x^4y^2 + 3x^2y^6}{(x^2+y^4)^2}$

$= \dfrac{x^4y^2 + 3x^2y^6}{(x^2+y^4)^2}$

$f_y(x,y) = \dfrac{(x^2+y^4)D_y(x^3y^2) - (x^3y^2)Dy(x^2+y^4)}{(x^2+y^4)^2}$

$= \dfrac{(x^2+y^4)(2yx^3) - x^3y^2(4y^3)}{(x^2+y^4)^2}$

$= \dfrac{2x^5y - 2y^5x^3}{(x^2+y^4)^2}$

(c) $f_x(x,y) = 2xe^{xy} + (x^2+y^5)ye^{xy}$

$f_y(x,y) = 5y^4 e^{xy} + (x^2+y^5)xe^{xy}$

(d) $f_x(x,y) = (6xy)\ln(x^2+y^4) + (3x^2+5y)(\dfrac{2x}{x^2+y^4})$

$f_y(x,y) = (3x^2+5)\ln(x^2+y^4) + (3x^2y+5y)(\dfrac{4y^3}{x^2+y^4})$

(e) $f_x(x,y) = 5(4xy + 3y^2)e^{2x^2y + 3xy^2}$

   $f_y(x,y) = 5(2x^2 + 6xy)e^{2x^2y + 3xy^2}$

(f) $f_x(x,y) \quad = \dfrac{\frac{1}{x+y}(x+y) - (\ln(x+y) + 5)(1)}{(x+y)^2}$

   $= \dfrac{-4 - \ln(x+y)}{(x+y)^2}$

   $f_y(x,y) \quad = \dfrac{\frac{1}{x+y}(x+y) - (\ln(x+y) + 5)(1)}{(x+y)^2} = \dfrac{-4 - \ln(x+y)}{(x+y)^2}$

(g) $f(x,y) = (5x^2 + 6y^4)^{1/2}$

   $f_x(x,y) \quad = \frac{1}{2}(5x^2 + 6y^4)^{-1/2}D_x(5x^2 + 6y^4)$

   $= 5x(5x^2 + 6y^4)^{-1/2}$

   $f_y(x,y) \quad = \frac{1}{2}(5x^2 + 6y^4)^{-1/2}Dy(5x^2 + 6y^4)$

   $= 12y^3(5x^2 + 6y^4)^{-1/2}$

9. (a) $\dfrac{\partial z}{\partial x} = 35x^6y^3$ $\qquad\qquad$ $\dfrac{\partial z}{\partial y} = 15x^7y^2$

   $\dfrac{\partial^2 z}{\partial x^2} = 210x^5y^3$ $\qquad\qquad$ $\dfrac{\partial^2 z}{\partial y^2} = 30x^7y$

   $\dfrac{\partial^2 z}{\partial y \partial x} = 105x^6y^2$ $\qquad\qquad$ $\dfrac{\partial^2 z}{\partial x \partial y} = 105x^6y^2$

(b) $\dfrac{\partial z}{\partial x} = 10xy^3e^{5x^2y^3}$

   $\dfrac{\partial z}{\partial y} = 15x^2y^2e^{5x^2y^3}$

   $\dfrac{\partial^2 z}{\partial x^2} = 10y^3e^{5x^2y^3} + 100x^2y^6e^{5x^2y^3}$

   $\dfrac{\partial^2 z}{\partial y^2} = 30x^2ye^{5x^2y^3} + 225x^4y^4e^{5x^2y^3}$

   $\dfrac{\partial^2 z}{\partial y \partial x} = 30xy^2e^{5x^2y^3} + 150x^3y^5e^{5x^2y^3}$

   $\dfrac{\partial^2 z}{\partial x \partial y} = 30xy^2e^{5x^2y^3} + 150x^3y^5e^{5x^2y^3}$

(c) $z = \ln x + \ln y + \ln(5x + 6y)$

   $\dfrac{\partial z}{\partial x} = \frac{1}{x} + \dfrac{5}{5x + 6y}$

$$\frac{\partial^2 z}{\partial x^2} = \frac{-1}{x^2} - \frac{25}{(5x+6y)^2}$$

$$\frac{\partial z}{\partial y \partial x} = \frac{-30}{(5x+6y)^2}$$

$$\frac{\partial z}{\partial y} = \frac{1}{y} + \frac{6}{5x+6y}$$

$$\frac{\partial^2 z}{\partial y^2} = \frac{-1}{y^2} - \frac{36}{(5x+6y)^2}$$

$$\frac{\partial^2 z}{\partial x \partial y} = \frac{-30}{(5x+6y)^2}$$

11. (a) $f(2.98,3.01) = 3(2.98)^2(3.01)^2 = 241.3719361$

(b) $f(3,3) = 3(9)(9) = 243$

$$\frac{\partial f}{\partial x} = 6xy^2, \ \frac{\partial f}{\partial y} = 6x^2 y, \ \Delta x = -.02, \ \Delta y = .01$$

$$f(2.98,3.01) \doteq 162(-.02) + 162(.01) + 243 = 241.38.$$

(c) The percentage error is

$$\frac{241.3719361 - 241.38}{241.3719361} = -.0033408\%$$

13. (a) $\dfrac{\partial D_x}{\partial p_x} = -10p_x, \ \dfrac{\partial D_y}{\partial p_y} = 9p_y^2$

$$\frac{\partial D_y}{\partial p_x} = 18p_x^2, \ \frac{\partial D_y}{\partial p_y} = -10p_y$$

When $p_x = 8$ and $p_y = 7$

$$\frac{\partial D_x}{\partial p_x} = -80, \ \frac{\partial D_x}{\partial p_y} = 441, \ \frac{\partial D_y}{\partial p_x} = 1152, \ \frac{\partial D_y}{\partial p_y} = -70.$$

When the price of y is held fixed at \$7, the demand for x decreases by approximately 80 units and the demand for y increases by 1152 units as the price of x increases from \$8 to \$9.

When the price of x is held fixed at \$8, the demand for x increases by approximately 441 units and the demand for y decreases by 70 units, as the price of y increases from \$7 to \$8.

(b) The commodities are substitutes since $9p_y^2 > 0$ and $12p_x > 0$.

15. (a) $f_x(x,y) = 4x + 6y + 4$
$f_y(x,y) = 6x + 10y + 8$

$$\begin{cases} 4x + 6y = -4 \\ 6x + 10 = -8 \end{cases} \rightarrow \begin{cases} -6x - 9y = 6 \\ 6x + 10y = -8 \end{cases} \ y = -2, \ x = 2$$

The critical point is $(2, -2)$.
$f_{xx}(x,y) = 4, \ f_{yy}(x,y) = 10, \ f_{xy}(x,y) = 6$
$AC - B^2 = 40 - 36 > 0.$ there is a local minimum at $(2, -2)$

(b) $f_x(x,y) = 3x^2 - 6 = 0$ if $x = \pm\sqrt{2}$
$f_y(x,y) = 3y^2 + 4y - 7 = (3y + 7)(y - 1) = 0$ if $y = 1$ or $-7/3$.

The critical points are $(\sqrt{2},1)$, $(\sqrt{2}, -7/3)$, $(-\sqrt{2},1)$, $(-\sqrt{2}, -7/3)$.

$f_{xx}(x,y) = 6x$, $f_{yy}(x,y) = 6y + 4$, $f_{xy}(x,y) = 0$.

point	A	B	C	D	conclusion
$(\sqrt{2},1)$	$6\sqrt{2}$	10	0	+	local minimum
$(\sqrt{2}, -7/3)$	$6\sqrt{2}$	-10	0	$-$	saddle point
$(-\sqrt{2},1)$	$-6\sqrt{2}$	10	0	$-$	saddle point
$(-\sqrt{2}, -7/3)$	$-6\sqrt{2}$	-10	0	+	local maximum

(c) $f_x(x,y) = 3y - 18x = 0$ if $y = 6x$
$f_y(x,y) = 3x - 8/y$

$$3x - \frac{8}{y} = 3x - \frac{8}{6x} = 3x - \frac{4}{3x} = \frac{9x^2 - 4}{3x} = 0 \text{ if } x = \pm 2/3.$$

The critical points are $(2/3,4)$, $(-2/3, -4)$
$f_{xx}(x,y) = -18$, $f_{yy}(x,y) = 8/y^2$, $f_{xy}(x,y) = 3$

point	A	B	C	D	conclusion
$(2/3,4)$	-18	+	3	$-$	saddle point
$(-2/3, -4)$	-18	+	3	$-$	saddle point

(d) $f_x(x,y) = (2x + y - 6)e^{(x^2+y+y^2 - 6x)}$

$f_y(x,y) = (x + 2y)e^{(x^2+y+y^2 - 6x)}$

$x + 2y = 0$ if $x = -2y$
$2x + y - 6 = -4y + y - 6 = 0$ if $y = -2$

The critical point is $(4, -2)$

$f_{xx}(x,y) = 2e^{(x^2+xy+y^2 - 6x)} + (2x + y - 6)e^{(x^2+xy+y^2 - 6x)}$

$f_{yy}(x,y) = 2e^{(x^2+xy+y^2 - 6x)} + (x + 2y)^2 e^{(x^2+xy+y^2 - 6x)}$

$f_{xy}(x,y) = e^{(x^2+xy+y^2 - 6x)} + (2x + y - 6)(x + 2y)e^{(x^2+xy+y^2 - 6x)}$

point	A	B	C	D	conclusion
$(4, -2)$	$2e^{24}$	$2e^{24}$	e^{24}	+	local minimum

17. $G_x(x,y) = 12x^2 - 26xy + 4y^2$
$G_y(x,y) = -13x^2 + 8xy - 16y + 52$
$G_x(2,1) = 48 - 52 + 4 = 0$
$G_y(2,1) = -52 + 16 - 16 + 52 = 0$
$G_x(2,12) = 48 - 624 + 576 = 0$

$$G_y(2,12) = -52 + 192 - 192 + 52 = 0$$

$$G_x(\tfrac{26}{35}, \tfrac{156}{35}) = (8112 - 105{,}456 + 97344)/(35)^2 = 0$$

$$G_y(\tfrac{26}{35}, \tfrac{156}{35}) = (-8788 + 32448 - 87360 + 63700)/(35)^2 = 0$$

$$G_x(\tfrac{-26}{9}, \tfrac{-13}{9}) = (8112 + 8788 + 676)/81 = 0$$

$$G_y(\tfrac{-26}{9}, \tfrac{-13}{9}) = (-8788 + 2704 + 1872 + 4212)/81 = 0$$

$$G_{xx}(x,y) = 24x - 26y$$
$$G_{yy}(x,y) = 8x - 16$$
$$G_{xy}(x,y) = -26x + 8y$$

point	A	B	C	D	conclusion
(2,1)	22	0	44	$-$	saddle point
(2,12)	-264	0	44	$-$	saddle point
$(\tfrac{26}{35}, \tfrac{156}{35})$	$-\tfrac{3432}{35}$	$-\tfrac{352}{35}$	$\tfrac{572}{35}$	$+$	local maximum
$(\tfrac{-26}{9}, \tfrac{-13}{9})$	$-\tfrac{286}{9}$	$-\tfrac{352}{9}$	$\tfrac{572}{9}$	$-$	saddle point

19. $P(q_D, q_S) = 50q_D + 86q_S - 3.5q_D^2 + 5q_D q_S - 2.8q_S^2 - 2000$

$$\frac{\partial P}{\partial q_D} = -7q_D + 5q_S + 50$$

$$\frac{\partial P}{\partial q_S} = 5q_D - 5.6q_S + 86$$

The critical point occurs when $q_D = 50$ and $q_S = 60$
$A = -7,\ B = -5.6,\ C = 5,\ D > 0.$

The maximum daily profit occurs when $q_D = 50$ and $q_S = 60$. The maximum daily profit is $1,830.

21.

$$lwh = 5120$$

$$C(l,w,h,\lambda) = 12lw + 2(wh)(4.8) + 2(lh)(4.8) - \lambda(lwh - 5120)$$

(1) $\dfrac{\partial C}{\partial l} = 12w + 9.6h - \lambda wh = 0 \Rightarrow \lambda = \dfrac{12w + 9.6h}{wh} = \dfrac{12}{h} + \dfrac{9.6}{w}$ $\left.\begin{array}{c} \\ \\ \end{array}\right\} \Rightarrow l = w$

$\dfrac{\partial C}{\partial w} = 12l + 9.6h - \lambda lh = 0 \Rightarrow \lambda = \dfrac{12l + 9.6h}{lh} = \dfrac{12}{h} + \dfrac{9.6}{l}$

$\dfrac{\partial C}{\partial h} = 9.6w + 9.6l - \lambda lw = 0 \Rightarrow \lambda = \dfrac{19.2}{w}$

$\dfrac{\partial C}{\partial \lambda} = -lwh + 5120 = 0 \Rightarrow h = \dfrac{5120}{w^2}$

From (1) $12w + 9.6\left(\dfrac{5120}{w^2}\right) - \dfrac{19.2(5120)}{w^2} = 0$

$$12w^3 = 49152$$
$$w^3 = 4096$$

$$w = 16,\ l = 16,\ h = 20$$

The dimensions which minimize the cost are $w = 16m,\ l = 16m,\ h = 20m$.

23. $d = \sqrt{(x-3)^2 + (y-6) + (z-2)^2}$ subject to $2x + 4y + 5z + 5 = 0$

Let $F(x,y,z,\lambda) = (x-3)^2 + (y-6)^2 + (z-2)^2 - \lambda(2x + 4y + 5z + 5) = 0$

$$\frac{\partial F}{\partial x} = 2(x-3) - 2\lambda = 0 \Rightarrow \lambda = (x-3) \tag{1}$$

$$\frac{\partial F}{\partial y} = 2(y-6) - 4\lambda = 0 \Rightarrow \lambda = \frac{y-6}{2} \tag{2}$$

$$\frac{\partial F}{\partial z} = 2(z-2) - 5\lambda = 2z - 5\lambda - 4 = 0 \Rightarrow \frac{2z-4}{5} = \lambda \tag{3}$$

$$\frac{\partial F}{\partial \lambda} = -2x - 4y - 5z - 5 = 0 \tag{4}$$

From (1) and (2) $\quad y - 6 = 2x - 6$
$\qquad\qquad\qquad\qquad y = 2x$

From (1) and (3) $\quad 5(x-3) = 2z - 4$
$\qquad\qquad\qquad\qquad 5x - 15 = 2z - 4$
$$z = \frac{5x-11}{2}$$

From (4) $\qquad\qquad -2x - 8x - 5\left(\dfrac{5x-11}{2}\right) = 5$
$\qquad\qquad\qquad\quad -20x - 25x + 55 = 10$
$\qquad\qquad\qquad\qquad\quad -45x = -45$
$\qquad\qquad\quad x = 1,\ y = 2,\ z = -3,\ \lambda = -2$

The minimum distance is

$$\sqrt{(1-3)^2 + (2-6)^2 + (-3-2)^2} = \sqrt{4 + 16 + 25} = 3\sqrt{5}$$

25. $375 L^{2/3} K^{1/3} = 112{,}500$

$C(L,K,\lambda) = 55L + 220K - \lambda(375 L^{2/3} K^{1/3} - 112{,}500)$

$$\frac{\partial C}{\partial L} = 55 - 250\lambda\frac{K^{1/3}}{L^{1/3}} = 0 \Rightarrow \lambda = \frac{11 L^{1/3}}{50 K^{1/3}}$$

$$\frac{\partial C}{\partial K} = 220 - \frac{125\lambda L^{2/3}}{K^{2/3}} = 0 \Rightarrow \lambda = \frac{44 K^{2/3}}{25 L^{2/3}}$$

$$\frac{\partial C}{\partial \lambda} = -375 L^{2/3} K^{1/3} + 112{,}500 = 0$$

$\dfrac{11L}{50} = \dfrac{44K}{25},\ L = 8K$

$\qquad -375(8K)^{2/3} K^{1/3} = 112{,}500$
$\qquad\qquad\qquad\qquad\quad K = 75$
$\qquad\qquad\qquad\qquad\quad L = 600$

The minimum cost of total labor is

$$55(600) + 220(75) = \$49{,}500$$

27. (a)

i	x_i	y_i	x_iy_i	x_i^2
1	2	5	10	4
2	4	4	16	16
3	6	2	12	36
4	8	1	8	64

$n=4$ $X=20$ $Y=12$ $P=46$ $S=120$

$$m = \frac{4(46) - 20(12)}{4(120) - 400} = \frac{-56}{80} = \frac{-7}{10} = -.7$$

$$b = \frac{120(12) - 46(20)}{4(120) - 400} = \frac{520}{80} = 6.5$$

The equation of the regression line is $y = -.7x + 6.5$

(b)

29. $z = 3e^{u^3v^2}$, $u = x^2 + 4y^3$, $v = \sqrt{x^3} + y^2$

$$\frac{\partial z}{\partial u} = 9u^2v^2e^{u^3v^2} \qquad\qquad \frac{\partial z}{\partial v} = 6u^3ve^{u^3v^2}$$

$$\frac{\partial u}{\partial x} = 2x \qquad\qquad \frac{\partial u}{\partial y} = 12y^2$$

$$\frac{\partial v}{\partial x} = \frac{3}{2}x^{1/2} \qquad\qquad \frac{\partial v}{\partial y} = 2y$$

$$\frac{\partial z}{\partial x} = \frac{\partial z}{\partial u}\cdot\frac{\partial u}{\partial x} + \frac{\partial z}{\partial v}\cdot\frac{\partial v}{\partial x}$$

$$= 9u^2v^2e^{u^2v^2}(2x) + 6u^3ve^{u^3v^2}(\tfrac{3}{2}x^{1/2})$$

$$= 18(x^2 + 4y^3)^2x(x^{3/2} + y^2)^2e^{(x^2+4y^3)^3(\sqrt{x^3}+y^2)^2}$$

$$+ 9(x^2 + 4y^3)^3(\sqrt{x^3} + y^2)\sqrt{x}\;e^{(x^2+4y^3)^3(\sqrt{x^3}+y)^2}$$

$$\frac{\partial z}{\partial y} = \frac{\partial z}{\partial u}\cdot\frac{\partial u}{\partial y} + \frac{\partial z}{\partial v}\cdot\frac{\partial v}{\partial y}$$

$$= 9u^2v^2e^{u^3v^2}(12y^2) + 6u^3ve^{u^3v^2}(2y)$$

$$= e^{u^3v^2}(108u^2v^2y^2 + 12u^3vy)$$

$$= e^{(x^2+4y^3)^3(\sqrt{x^3}+y^2)^2}(108(x^2 + 4y^3)^2(\sqrt{x^3} + y^2)^2y^2 + 12(x^2 + 4y^3)^3(\sqrt{x^3} + y^2)y)$$

31. (a) $\displaystyle\int (12x^3y^2 + 13xy^3)\,dy = 4x^3y^3 + \frac{13}{4}xy^4 + C(x)$

(b) $\displaystyle\int 3x^3 \cdot \sqrt[5]{y}\; dx = \frac{3}{4}x^4 \sqrt[5]{y} + K(y)$

(c) $\int_{1}^{2}(4x^3y^2+10xy^3)\,dy=\frac{4}{3}x^3y^3+\frac{10}{4}xy^4\big|_{1}^{2}$

$=\frac{32}{3}x^3+\frac{5}{2}x(16)-\frac{4}{3}x^3-\frac{5}{2}x$

$=\frac{28}{3}x^3+\frac{75x}{2}$

(d) $\int_{1}^{2}\big[\int_{3y}^{y^2}15x^3y^4\,dx\big]\,dy$

$=\int_{1}^{2}\Big(\frac{15}{4}x^4y^4\big|_{3y}^{y^2}\Big)\,dy=\int_{1}^{2}\frac{15}{4}(y^{12}-81y^8)\,dy$

$=\frac{15}{4}\Big(\frac{1}{13}y^{13}-\frac{81}{9}y^9\big|_{1}^{2}\Big)=\frac{15}{52}(2^{13})-\frac{135(2^9)}{4}-\frac{15}{52}+\frac{135}{4}$

$=\frac{15}{52}(8191)-\frac{135}{4}(511)\doteq-14{,}883$

33. $\displaystyle\iint_{R}f(x,y)\,dA=\int_{-1}^{3}\int_{1}^{2}(2x^2+5xy)\,dy\,dx$

$=\int_{-1}^{3}(2x^2y+\frac{5}{2}xy^2\big|_{1}^{2})\,dx=\int_{-1}^{3}(4x^2+10x-2x^2-\frac{5}{2}x)\,dx$

$=\int_{-1}^{2}(2x^2+\frac{15}{2}x)\,dx=\frac{2}{3}x^3+\frac{15}{4}x^2\big|_{-1}^{3}$

$=(18+\frac{135}{4})-(-\frac{2}{3}+\frac{15}{4})=\frac{56}{3}+30=\frac{146}{3}$

35.

$\displaystyle\int_{0}^{10}\int_{.3x+3}^{6}(3x^2+xy)\,dy\,dx+\int_{0}^{10}\int_{0}^{-.3x+3}(3x^2+xy)\,dy\,dx$

$=\int_{0}^{10}3x^2y+\frac{x}{2}y^2\big|_{.3x+3}^{6}\ dx+\int_{0}^{10}3x^2y+\frac{x}{2}y^2\big|_{0}^{-.3x+3}dx$

$=\int_{0}^{10}18x^2+18x-3x^2(.3x+3)-\frac{x}{2}(.3x+3)^2+3x^2(-.3x+3)+\frac{x}{2}(-.3x+3)^2dx$

$=\int_{0}^{10}18x^2+18x+3x^2(-.3x+3-.3x-3)+\frac{x}{2}(.09x^2-1.8x+9-.09x^2-1.8x-9)\,dx$

$=\int_{0}^{10}18x^2+18x-1.8x^3-1.8x^2\,dx=5.4x^3+9x^2-.45x^4\big|_{0}^{10}$

$=5400+900-4500=1800$

37.

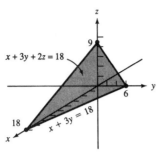

$$V = \int_{y=0}^{6} \int_{x=0}^{18-3y} \frac{18-x-3y}{2} dx\, dy$$

$$= \int_{0}^{6} \left(\int_{0}^{18-3y} 9 - \frac{x}{2} - \frac{3}{2}y\, dx \right) dy$$

$$= \int_{0}^{6} 9x - \frac{x^2}{4} - \frac{3}{2}xy \Big|_{0}^{18-3y} dy$$

$$= \int_{0}^{6} 9(18-3y) - \frac{1}{4}(18-3y)^2 - \frac{3}{2}y(18-3y)\, dy$$

$$= \int_{0}^{6} 162 - 27y^2 - \frac{9}{4}(6-y)^2 - 27y + \frac{9}{2}y^2\, dy$$

$$= 162y - 27y^2 + \frac{3}{2}y^3 + \frac{3}{4}(6-y)^3 \Big|_{0}^{6}$$

$$= 162(6) - 27(36) + \frac{3}{2}(216) + 0 - \frac{3}{4}(216)$$

$$= 162$$

39. $x + y \le 300,\ x \ge 0,\ y \ge 0$

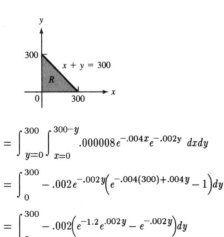

$$P(x+y \le 300) = \iint_{R} f(x,y)\, dA\, dx$$

$$= \int_{y=0}^{300} \int_{x=0}^{300-y} .000008 e^{-.004x} e^{-.002y}\ dx\, dy$$

$$= \int_{0}^{300} -.002 e^{-.002y} \left(e^{-.004(300)+.004y} - 1 \right) dy$$

$$= \int_{0}^{300} -.002 \left(e^{-1.2} e^{.002y} - e^{-.002y} \right) dy$$

$$= -\left(e^{-1.2} e^{.002y} + e^{-.002y} \right) \Big|_{0}^{300} = .2035$$

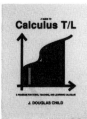